Modern Methods of Construction

emerald PUBLISHING

ice Publishing

Modern Methods of Construction:

Foundations, Processes and Applications

Emmanuel I. Daniel and Fidelis A. Emuze

Published by Emerald Publishing Limited,
Floor 5, Northspring, 21-23 Wellington Street, Leeds LS1 4DL.

ICE Publishing is an imprint of Emerald Publishing Limited

Other ICE Publishing titles:
Structural Design of Buildings: Fundamentals in Design, Management and Sustainability
Edited by Feng Fu and David Richardson. ISBN 978-1-83549-577-3
Automation and Manufacturing for the Built Environment
Steve Thompson. ISBN: 978-1-83608-599-7
Temporary Works Part Two: Further Principles of Design and Construction
Edited by Ray Filip. ISBN: 978-1-83608-525-6

A catalogue record for this book is available from the British Library

ISBN 978-1-83708-437-1

Cover photo: iStock/Darya Ostrenko
Commissioning Editor: Michael Fenton
Content Development Editor: Ryan Molyneux
Books Production Lead: Emma Sudderick
Project Manager: Rajachitra S

Typeset by KnowledgeWorks Global Limited

Printed and bound by CPI Group (UK) Ltd, Croydon, CR0 4YY

Contents

Part III – Global case studies of MMC in action 239

Foreword

As someone who has spent more than ten years advocating for reform of construction industry practices, I have come to recognise that the path to positive change is complex and littered with multiple obstacles and blockers. My perspective on this has been informed both as an advisor to government in relation to policy making and public sector procurement, as well as being an industry practitioner advising a variety of mostly private-sector clients on routes to securing better outcomes by way of suitable modernisation strategies.

My association with industry change is perhaps most strongly forged by my 2016 government review of the UK construction industry's labour model, provocatively titled 'Modernise or Die'. The increased adoption of modern methods of construction (MMC) and the at-scale increasing of project level pre-manufactured value (PMV) were positioned as central themes of that review, founded in the growing concern that industry's reliance on site labour intensity was increasingly becoming a threat to long-term resiliency as structural declines in available and competent workforce threaten the historic ability of industry to both shed and grow labour resources in line with economic cycles.

Today, in 2026, all of the forecasts made regarding labour market declines and persistent low productivity have been borne out and mean the burning platform for change remains as relevant as ever. The immediacy of the problem is masked temporarily in some economies by output declines but the underlying issue remains. Despite that, in large part, the construction industry of 2016 is not dissimilar from that of 2026. We must therefore understand why that is and what are the initiators of true change rather than aspirational hopes for a future state. We must find the routes to business model and production model reform that avoid us doing the same thing again and again.

At the heart of ongoing business as usual in construction is reliance on a flawed model of bespoke design, lowest price-led procurement and labour-intensive production methods. Wherever I go in the world, I see a similar prognosis for declining resiliency in labour markets, amplified by demographics and societal change, and a general consensus that construction needs to do things differently. I equally see frustration that those markets are unable to pivot in the long run to a different approach despite multiple attempts through government interventions or industry campaigns.

It is worth remembering that construction's core operating model has not emerged by accident and the highly volatile project-based pipeline and fragmented client base that shapes construction sector demand is inherently predisposed to industrialised production. There is a clear mismatch between the fixed capital commitment required to invest in industrialised MMC production and what is often variable, short-term workload.

The need for better demand planning and smoothing lies at the heart of how new delivery models can flourish in the future and has been part of my dialogue with government to create a demand-led transformation. This has largely been derailed in the UK by a post-Covid economic correction which has seriously impacted the housing and construction sectors.

It appears increasingly evident that any move towards a manufacturing mindset has to respect the commercial models, behavioural drivers and many institutional constraints which stand in the way of implementing the models that we see operate in other manufacturing-led sectors such as automotive or aerospace. I would suggest this is a clear health warning for those advocating that the way forward is a pivot from 'MMC' language to 'industrialisation', implying a wider root and branch ecosystem change. My view is that this is just not possible for large parts of the construction operating system, especially in the private sector, and any attempt at a system-level overhaul is doomed to failure due to the industry's deep-rooted commercial and behavioural realities that cannot be changed. MMC adoption should therefore seek to respect and embrace rather than ignore or override broader contextual constraints.

It is against this imperfect market context that Dr Daniel and Professor Emuze are publishing this important piece of work, which is a valuable addition to the ongoing discourse that needs to take place between industry, policymakers and academia. The last decade has seen extensive failure in certain parts of the global MMC supply chain. It is therefore timely that we learn lessons and understand which sustainable models can square the circle of increased productisation with the economic need to be fleet of foot, agile and price competitive.

Setting utopian visions of the future or engaging in prolonged navel gazing do not help in this debate, and the goal must be to understand what works and what doesn't in the real world and to remain current on how ongoing societal and technological changes may further shape the landscape ahead. This feels particularly relevant for the rapidly accelerating world of artificial intelligence and robotics.

This book covers the ground in a logical and thorough way, exploring the foundational principles of MMC, including definition and context and then understanding, with practical examples, how MMC can be applied in practice, as a combination of product and process and how this can link to better outcomes when done well.

I am particularly pleased to see international case studies being included in this book. It has been a real privilege, since I authored 'Modernise or Die', to have developed a global network of collaborators across governments, academic institutions and practitioners in North America, Europe and Australasia. Shared learning is key to avoiding repeating mistakes and, more importantly, being able to shine a light on and propagate where success has been achieved and how this can be rolled out on a wider basis.

I hope this book serves as a suitable provocation for those interested in seeing industry improvement take hold in a sustainable, deliverable way.

Mark Farmer
Founder and Board Executive, Cast Consultancy, United Kingdom

About the Editors

Dr Emmanuel I. Daniel (PhD), University of Wolverhampton, UK
Dr Emmanuel I. Daniel is an Associate Professor (Reader) in sustainable construction management and the Deputy Research Group Leader for digital construction in the Faculty of Science and Engineering at the University of Wolverhampton, UK. He is a Chartered Construction Manager with over ten years of experience in the construction industry and a Senior Fellow of the Higher Education Academy, UK. His research covers modern methods of construction, off-site and digital construction, Lean construction, digitalisation, maritime port infrastructure, social value and brownfield regeneration, health and safety management, and workforce capability development. He has contributed to various funded projects for organisations including the British Council, Lloyd's Register Foundation and National Highways, UK. He is an active member of the International Group for Lean Construction, a guest editor for the Management, Procurement and Law journal and the Journal of Marine Science and Engineering, and a sector editor for the Building Engineering Journal. He has over 100 academic publications and has examined PhD work in India, South Africa, Italy, the Republic of Ireland and the UK. He currently supervises more than ten PhD students and is an adjunct professor in construction technology and management at the University of Gondar, Ethiopia.

Professor Fidelis A. Emuze (PhD), Texas State University, USA
Professor Fidelis Emuze is a professor of instruction in construction science and management at Texas State University, San Marcos, Texas, USA, specialising in Lean construction, sustainability, health and safety, and advanced research methods. He holds honorary visiting professor roles at the Central University of Technology, Free State, and Nelson Mandela University, both in South Africa. He serves as the coordinator of CIB W123 – People in Construction. Fidelis is the Editor-in-Chief of the Proceedings of the Institution of Civil Engineers – Management, Procurement and Law. He also serves on the editorial boards of several international journals and has evaluated more than 80 postgraduate research reports worldwide. Having taught and supervised undergraduate and postgraduate research, including 39 graduate students (24 doctoral), in construction management, Lean construction, and health and safety, he also led the development of ten new construction qualifications in South Africa, demonstrating his versatility and dedication to advancing construction education.

Introduction to Modern Methods of Construction: Foundations, Processes and Applications

The rapid advancement of technologies, digital systems and industrialised production methods has rekindled global interest in modern methods of construction (MMC). Although the concept has been discussed across policy, industry and academic circles for decades, its definition, scope and implications remain subjects of ongoing debate.

In addition, the global construction industry is undergoing a major transformation driven by the need for increased efficiency, sustainability, safety and productivity. Traditional construction, characterised by fragmentation, inefficiency and high waste, faces challenges from innovative approaches collectively known as MMC. Generally understood as industrialised, digitally enabled and process-integrated methods of project delivery, MMC combines advanced manufacturing techniques, digital technologies, off-site production and Lean construction to enhance how buildings and infrastructure are designed, built and delivered. As governments and industry players work to tackle housing and infrastructure shortages, environmental issues and labour challenges, MMC has emerged as a key pathway to a more resilient and future-ready built environment.

This book, *Modern Methods of Construction: Foundations, Processes and Applications*, explores the developing field of MMC and its impact on construction practice, policy and research. It consolidates multidisciplinary perspectives on the theoretical foundations, technological advances, operational procedures and practical applications of MMC across diverse global contexts. By bringing together theoretical insights, technological developments and empirical case studies, the collection highlights how emerging tools and approaches, such as digitalisation, generative computational design and building information modelling (BIM), alongside off-site manufacturing and Lean construction principles, are transforming the construction value chain.

The book is organised into three thematic parts that collectively present a comprehensive understanding of MMC.

Part I: Digitalisation and practical implementation of modern methods of construction examine the technological and managerial frameworks that enable MMC adoption. Chapters in this section explore topics such as generative computational design for modular construction, the integration of Lean and simulation approaches in housing delivery, and the organisational factors that facilitate or hinder the adoption of digital technologies such as BIM within MMC environments.

Part II: Performance impact and benefits of modern methods of construction focuses on the outcomes and performance improvements associated with these construction methods. Contributions in this section investigate quality management in MMC systems, cost optimisation in off-site construction, collaborative costing approaches and the role of timber-based MMC in sustainable construction. Additionally, the section highlights the often-overlooked social dimension by examining the potential benefits of off-site construction for workers' mental health and well-being.

Part III: Global case studies of modern methods of construction in action presents empirical insights from diverse geographical contexts. Through case studies from Europe, China, Australia, Africa and developing economies, this section demonstrates how MMC principles are being implemented across different institutional, economic and cultural settings. These chapters reveal both the opportunities and challenges associated with scaling MMC, including issues related to policy frameworks, institutional barriers and industry readiness.

This book combines theoretical perspectives, technological advancements and global case studies, offering a comprehensive view of the evolving MMC system. It aims to support academics, students, practitioners, policymakers and industry stakeholders in understanding how modern construction methods can address urgent industry challenges and enable more efficient, sustainable delivery of buildings and infrastructure projects. Collectively, the contributions in this volume demonstrate how MMC is transforming construction practices and offer insights into the systemic changes needed to enable a more productive, sustainable and resilient construction industry.

Emmanuel I Daniel and Fidelis A Emuze
ISBN 978-1-83708-437-1
https://doi.org/10.1108/978-1-83708-436-420261001

Chapter 1
Modern methods of construction: evolution, meaning and underpinning theories

Emmanuel I. Daniel[1], Christiana Ada Adah[2] and Fidelis A. Emuze[3]

[1]*The University of Wolverhampton, Wolverhampton, United Kingdom* 0000-0002-5675-1845
[2]*Baze University, Abuja, Nigeria* 0000-0003-1562-7749
[3]*Texas State University, Texas, United States of America* 0000-0001-7714-4457

Abstract

Modern methods of construction (MMC) have gained increasing attention as a response to persistent challenges in the construction industry, including low productivity, skills shortages, housing deficits and sustainability pressures. Despite growing interest from industry and policymakers, the concept of MMC remains ambiguously defined and weakly theorised, with much of the existing literature focusing on practices rather than underlying explanations. This study addresses this gap by critically examining the definitions, classifications, historical evolution and theoretical foundations of MMC. A qualitative critical literature review was conducted, drawing on peer-reviewed academic studies, government publications and industry reports. The review synthesises diverse and often conflicting interpretations of MMC. It traces its evolution from early twentieth-century non-traditional construction systems to contemporary frameworks shaped by industrialisation, digitalisation and net-zero ambitions. The study identifies seven core principles of MMC. It demonstrates that these principles are underpinned by four complementary theories: lean production theory, design for manufacture and assembly (DfMA) theory, systems theory and industrialisation theory. Based on this synthesis, a theoretically grounded framework for MMC is developed to explain how and why MMC operates as an integrated socio-technical system rather than a collection of isolated techniques. The findings provide conceptual clarity and theoretical rigour, highlighting the risks of partial or technology-led adoption. The study contributes to construction management research by strengthening the theoretical foundation of MMC and offering practical insights for policymakers and practitioners seeking more effective and sustainable implementation.

Keywords: digitalisation; lean construction; modern methods of construction; off-site construction; theory

1.1. Introduction

Traditional construction methods have been used in the construction industry for centuries across the globe (Helpgeek, 2023; Mba *et al.*, 2025). These methods rely heavily on labour-intensive processes performed by skilled craftsmen, resulting in longer completion times and higher costs. Additionally, implementing modifications and changes after completion can be challenging. Traditional construction methods often rely on locally available materials, which may not be renewable or sustainable, posing environmental challenges (Helpgeek, 2023). Another feature of traditional construction practices is that production occurs on building sites, leaving little room for flexibility in construction (Sardén and Engström, 2010). Thus, traditional construction practices are confronted with challenges such as cost overruns, construction delays and construction waste (Layon *et al.*, 2024). Despite technological innovations and calls for change over the years, the construction industry remains entrenched in traditional practices (Hughes, 2024). The industry has been plagued by underperformance and resistance to change and innovation, resulting in poor product quality and poor health and safety records compared with other industries (Nadim, 2012; Saad *et al.*, 2023a).

The resistance to change in the industry is chiefly due to the nature of construction projects and incentivisation for innovation (Hughes, 2024). While these limitations have long prompted calls for innovation, they reveal a deeper challenge: the construction industry struggles not only to adopt new methods but also to define and systematise what 'modernisation' in construction means.

The construction industry has been evolving, with emerging construction techniques improving building efficiency and sustainability. In recent years, traditional methods have gradually been replaced by modern approaches that prioritise speed, cost-effectiveness and environmental impact (Helpgeek, 2023). Innovations such as reinforced concrete, prefabrication techniques and new structural systems have emerged, revolutionizing the way buildings are designed and constructed (Vagtholm *et al.*, 2023). These innovations have emerged to overcome the challenges of traditional construction methods and improve project efficiency. These emergent techniques are called modern methods of construction (MMC) (Layon *et al.*, 2024; Saad *et al.*, 2023). MMC is reputed to deliver faster and more efficient construction (Kyjaková *et al.*, 2014). MMC utilises technological advancements and automation to streamline processes and optimise resource use. However, despite these reported benefits, MMC is often treated as a set of techniques rather than as a coherent production logic or socio-technical system, which contributes to inconsistent implementation and mixed performance outcomes.

Daniel *et al.* (2024) describe MMC as methods supporting efficient construction delivery and process improvement. Duncheva and Bradley (2019) add that it includes strategies like off-site construction, lean, simulation, BIM and digital twins, all aiding production improvements. Daniel and Oshodi (2022) found that studies on MMC have focused on approaches and techniques such as lean, off-site, simulation, building information modelling, digital twins, virtual reality and augmented reality. This shows that MMC is not limited to off-site construction. These approaches highlight the lack of a boundary for MMC, emphasizing the need to differentiate technologies, processes and foundational principles.

Kozlovská *et al.* (2014) and Sardén and Engström (2010) noted that MMC tackles the uncertainties inherent in traditional construction in the areas of time, defects, safety, environmental impact, costs, profits and lifecycle performance. MMC was introduced on a commercial scale in the aftermath of the Second World War, driven by the need to rapidly expand housing output (NHBC Foundation, 2016; Williamson *et al.*, 2019). Green (2022) further noted that MMC emerged due to the high demand for housing arising from population growth and war. Since the end of the Second World War, MMCs have been utilised in countries facing housing crises. With the construction industry also facing various housing shortages, productivity issues and sustainability challenges over the years, MMC emerged as an effective alternative to meet housing demands, enhance construction sustainability through productivity and quality improvement, mitigate environmental impact, increase client satisfaction and shorten construction time (Duncheva *et al.*, 2020; Hernández *et al.*, 2023; O'Hegarty *et al.*, 2025; Švajlenka and Kozlovská, 2018).

When governments of countries struggled to meet high housing demand, they often turned to the private sector to manage these high-risk projects. These countries include the United Kingdom, the Netherlands, Japan, Sweden and Malaysia (Layon *et al.*, 2024; Williamson *et al.*, 2019). Maslova and Burgess (2023) and Rafa and Khalid (2024) reported that problems such as housing shortages and construction delays in the UK led to MMC being seen as a solution for quicker, safer and greener delivery of new homes. Data show that MMC has been effective in addressing key construction challenges in the UK (Saad *et al.*, 2023). Understanding this history is crucial, as cycles of rebranding and reinvention have shaped modern views of MMC, often obscuring its production and organisational foundations.

Over the past decade, interest in MMC has grown across industry and academia. In academia, publications on MMC in construction management research have increased (Saad *et al.*, 2023a). For the construction industry, the appeal of MMC has been amplified by its association with modernisation and a transformative approach to construction processes (Green, 2022; Li and mok Paik, 2025). According to Layon *et al.* (2024), MMC enables buildings to be constructed more quickly, with better quality and at a lower cost. The numerous benefits of MMC are increasingly recognised in many developed nations, yet widespread adoption still has a long way to go. Many other developed countries, such as Germany, Scandinavia, Japan, the USA and the UK, are gradually adopting MMC on a broader scale (Layon *et al.*, 2024). However, the meaning of MMC and the theories and principles underpinning it remain unclear. As a result, MMC research has largely prioritised implementation narratives over explanatory frameworks, thereby limiting cumulative knowledge development and weakening the translation of MMC into policy and practice.

This chapter aims to explore the evolution, conceptual meaning and theoretical underpinnings of MMC by addressing the following research questions.

(a) What are the current definitions and classifications of MMC?
(b) How have MMC evolved over the years?
(c) What are the principles and theories that underpin the MMC?
(d) How can a theoretical framework be used to explain the working and practice of the MMC?

Addressing the questions sequentially facilitates the progression of the chapter from definitional clarity, through historical understanding, to a principled and theoretical explanation of MMC practice.

1.2. Perspectives on modern methods of construction and research gap

This section critically examines how MMC has been defined and interpreted in the literature, revealing deep conceptual fragmentation that limits both research coherence and effective policy implementation.

Despite the advantages of MMC and the growing research interest, there are conflicting opinions about its definition and a general lack of understanding of what it entails. Many people are unaware of MMC's true meaning, which may stem from the fact that various terms are used to describe MMC worldwide. According to a UK House of Commons report, confusion and inconsistencies were found in the terminology used to describe MMC (House of Commons, 2019). Narrow, off-site-centred definitions risk reducing MMC to a construction location decision, whereas broader interpretations shift attention toward production logic, integration and system performance, which are central to understanding MMC outcomes. Lack of clear definition and understanding of what MMC has implications for both industry and policy makers, such as difficulty in progress monitoring and policy formation that will support its uptake.

Some of the terminologies used to depict MMC include prefabrication, off-site construction and industrialised construction (O'Hegarty *et al.*, 2025). Although MMC is the most common name used in many countries, it has been given different names in different countries and is often used interchangeably (Goulding *et al.*, 2014; Hernández *et al.*, 2023; Layon *et al.*, 2024) – for instance, MMC is known as off-site construction in the UK, China and Australia, as industrial building systems in Malaysia, Japan and Sweden, as prefabricated construction in Hong Kong and Singapore, and as modern methods of construction (MMC) in the USA (Layon *et al.*, 2024; Švajlenka and Kozlovská, 2018). Thus, diverse definitions of MMC are found in existing literature – for example, most definitions in existing literature narrow MMC down to simply off-site construction (see Dowsett *et al.*, 2022; Layon *et al.*, 2024; Maqbool *et al.*, 2023; Sweeney, 2024). In contrast, other authors (see Rushton, 2021; Sánchez-Garrido *et al.*, 2024; Williamson *et al.*, 2019) offer a broader interpretation that includes activities beyond off-site processes.

It is crucial to provide a clear explanation of what MMC is. While these terminological variations reflect local policy and industry contexts, their interchangeable use has blurred conceptual boundaries, making it challenging to compare evidence, evaluate performance or develop consistent MMC policy frameworks.

Existing MMC research predominantly concentrates on implementation-oriented themes, such as benefits, barriers, housing delivery and sustainability outcomes – for instance, studies such as Rushton (2021), Government of Jersey (2022), Saad *et al.* (2023a) all reported increasing implementation of MMC in the construction industry. There are studies (Ofori-Kuragu *et al.*, 2022; Shibani *et al.*, 2021) that have explored MMC as a solution to the housing deficit experienced in various countries. Existing studies (Bernold, 2016; Layon *et al.*, 2024 Sweeney, 2024; Williamson *et al.*, 2019) have explored barriers to MMC implementation in the construction industries of various countries. Several studies (see Hu and Chong, 2021; O'Hegarty *et al.*, 2025; Rafa and Khalid, 2024) have investigated the environmental sustainability of MMC (e.g. embodied carbon and net zero carbon) implementation. Although this body of work offers valuable insights into MMC adoption, it provides a limited explanation of the underlying mechanisms by which MMC achieves or fails to achieve these outcomes.

Furthermore, few studies have attempted to relate MMC to theories. Some studies have attributed MMC to the diffusion of innovation (DOI) theory (Saad *et al.*, 2023). While diffusion-based perspectives provide understanding of adoption dynamics, they do not clarify the internal production logic, system integration, or performance features that set MMC apart from traditional method construction. However, their study focuses only on low MMC uptake.

There are many implications for MMC practice and research if it is not built on solid theory – for instance, Gibb and Isack (2003) note that much MMC research lacks a conceptual model that integrates technical, managerial and policy aspects, which limits cumulative knowledge. It could be argued that, without a theoretical base, MMC studies may be seen as focusing on describing technologies rather than explaining why or how these methods succeed or fail in different contexts. According to Pan and Sidwell (2011), MMC research often lacks theoretical alignment, resulting in fragmented insights that fail to build cumulative understanding.

Additionally, without theoretical rigour, research findings can be too narrow or situational to inform strategic decision making. As Harty (2005) rightly points out, construction innovation must be studied through socio-technical and organisational change theories to be effectively applied in policy. This is further supported by Ivory (2005), who argues that without theoretical integration, construction management research risks being technically rich but conceptually poor. The absence of theoretical grounding has led MMC to be treated primarily as a set of technologies or techniques rather than as a coherent production and organisational system. Consequently, MMC research has struggled to generate cumulative knowledge or transferable lessons across projects and national contexts.

From the foregoing, there is a need for a comprehensive synthesis of the evolution of MMC and to establish the theoretical foundations underpinning MMC. This chapter aims to explore the evolution, conceptual meaning and theoretical underpinnings of MMC by addressing the above-mentioned research questions. Accordingly, it undertakes a structured synthesis of MMC's definitions and classifications, traces its historical evolution, identifies core operational principles and maps these principles to complementary theories.

1.3. Methodological approach

This study employed a qualitative critical literature review to examine the definitions, classifications, evolution and theoretical foundations of MMC. Given the fragmentation and limited integration in existing MMC research (Section 1.2), a thorough qualitative review was essential to challenge assumptions, clarify differing interpretations and advance theory development rather than just providing a description. A critical review was considered appropriate because the study aimed to move beyond descriptive summarisation and interrogate conceptual ambiguity, theoretical gaps and dominant narratives within MMC research. Such an approach is well-suited to construction management research, where knowledge is often fragmented and practice-led, and where theory-building is required to support cumulative understanding (Grant and Booth, 2009; Daniel and Pasquire, 2019).

The review drew on a wide range of sources, including peer-reviewed journal articles, academic books, government reports, policy documents and industry publications. Grey literature was intentionally included because government and industry publications play a central role in shaping MMC terminology, classifications and policy-driven adoption (MHCLG, 2019; NAO, 2005). Rather than treating grey literature as supplementary, this study positions policy and industry documents as constitutive sources that actively shape MMC definitions, classifications and adoption narratives.

Sources were included if they explicitly addressed MMC concepts, definitions, classifications, historical development, principles or theoretical explanations. No strict temporal boundary was applied, allowing the integration of early historical documents with contemporary studies to enable a longitudinal analysis of MMC's evolution (Green, 2022; Ministry of Health, 1926).

Data analysis followed an iterative and interpretive process. Firstly, the selected literature was reviewed to identify recurring definitions, classifications and areas of conceptual disagreement. Secondly, a historical synthesis was undertaken to trace key phases in the development of MMC, highlighting shifts in terminology, policy emphasis and industry practice. Thirdly, theoretical mapping was conducted to align identified MMC principles with established theories in construction and production management, including lean production theory, design for manufacture and assembly (DfMA) theory, systems theory and industrialisation theory (Boothroyd *et al.*, 2011; Gibb and Isack, 2003; Koskela, 2000). Together, these stages enabled the transition from fragmented descriptions of MMC toward an integrated conceptual and theoretical explanation of its operation as a production system.

The study ensured transparency in source selection, triangulation across academic and grey literature, and reflexive interpretation of findings. While the study does not involve primary empirical data, this was an intentional methodological choice aligned with the chapter's aim of conceptual clarification and theory development rather than hypothesis testing.

1.4. Results and discussion

This section synthesises the findings of the literature review in five stages. Firstly, it clarifies the current definitions and classifications of MMC. Secondly, it traces the historical evolution of MMC to explain how contemporary interpretations emerged. Thirdly, it distils seven core principles underpinning MMC practice. It then maps these principles into four complementary theories. Finally, it integrates these insights into a theoretical framework that explains MMC as a socio-technical production system.

1.4.1 Current definitions and classifications of modern methods of construction

In terms of definition, several authors (including Layon *et al.*, 2024; Saad, 2023; Williamson *et al.*, 2019) noted that there is no universally accepted definition of MMC, with various definitions propounded over the years. Currently, researchers are divided over the definition of MMC, with two distinct schools of thought. Some authors limit MMC to off-site construction, while others view it as encompassing broader construction activities beyond just off-site methods. The various definitions of the concept of MMC by research publications are presented in Table 1.1. The table presents the definitions of MMC and the emerging themes found in the several definitions. The definition in Table 1.1 categorises into two categories: government publications (non-research publications) and research publications.

Table 1.1 Meaning of MMC (continued on next page)

Sources	Authors	Definitions of MMC	Emerging Themes
Government publication	Government of Jersey (2022), p. 2	MMC is 'a generic term used to describe a building process which uses prefabrication and factory assembly as a fast way of delivering new buildings. It maximises the efficient use of building materials, construction workforce and resources'	Prefabrication and factory assembly
Government publication	MHCLG (2019)	The MMC definition framework 'is a new seven category definition framework that enables a full and future-proofed range of 'Modern Methods of Construction' used in homebuilding to be better understood with regularised terminology. The definition framework spans all types of pre-manufacturing, site-based materials and process innovation'	Comprises seven category framework definitions ▪ Pre-manufacturing (3D primary structural systems) ▪ Pre-manufacturing (2D primary structural systems) ▪ Pre-manufacturing components (non-systematised primary structure) ▪ Additive manufacturing (structural and non-structural) ▪ Pre-manufacturing (non-structural assemblies and sub-assemblies) ▪ Traditional building product led site labour reduction / productivity improvements ▪ Site process led site labour reduction / productivity / assurance improvements
Government publication	NHBC Foundation (2016), p. 2	MMC 'embraces a number of approaches involving off-site manufacture or assembly'	Off-site manufacture or assembly
Research publication	Dowsett *et al.* (2022)	MMC is 'a term largely synonymous with the prefabrication of product components in factories'	Prefabrication
Research publication	Layon *et al.* (2024)	MMC is 'prefabrication and off-site construction products or processes'	Prefabrication and off-site construction

Table 1.1 Continued

Sources	Authors	Definitions of MMC	Emerging Themes
Research publication	Sweeney (2024)	MMC as 'a broad term that captures a diverse range of building processes and products'	Diverse building processes and products
Research publication	Sánchez-Garrido *et al*. (2024)	MMC 'presents an innovative paradigm for designing and constructing infrastructure and buildings more efficiently, using conventional materials with unconventional techniques'	Combination of conventional materials with unconventional techniques
Research publication	Rushton (2021)	MMC is 'an umbrella term describing non-traditional building methods and materials'	Non-traditional building methods and materials
Research publication	Duncheva *et al*. (2020)	MMC are 'all the approaches that aim to optimise the construction process to obtain better products in less time'	Optimisation of construction process
Research publication	Williamson *et al*. (2019)	MMC is a 'collective term, used to describe a number of unconventional and unfamiliar innovative construction methods'	Unconventional and unfamiliar innovative construction methods
Research publication	Lovell (2012)	MMC 'involves the manufacture of house parts off-site in a specially designed factory'	Off-site

Source: Synthesis from Literature review

1.4.1.1 A review of MMC definitions: critical evaluation of the definition and meaning of MMC

Clearly, MMC includes, but is not limited to, off-site construction (Duncheva *et al.*, 2020). In the same vein, MMC encompasses a range of on- and off-site techniques, including non-traditional on- and off-site construction methods that serve as alternatives to traditional construction practices (Government Commercial Function, 2022; Suliman and Rankin, 2021). Due to discrepancies in the definition of MMC and its misrepresentation, there have been proposals for a standard definition (Saad, 2023). Taken together, these definitions reveal that MMC has been framed variously as a construction location, a set of technologies, or a process improvement strategy, with limited agreement on its underlying production logic.

To address this ambiguity, this chapter adopts a definition that foregrounds integration, production logic and system performance rather than individual technologies. MMC is thus defined in this chapter as a non-traditional approach to construction project delivery, which could be off-site and on-site, including process, methods and material innovation, that supports efficient design and delivery of the construction project leading to an improved process efficiency and sustainable construction with enhanced value for the customer.

1.4.1.2 Classifications and types of MMC

MMC has different classifications, ranging from the product type produced to the location of production (Nadim, 2012). Some authors, such as Švajlenka and Kozlovská (2018) and Layon *et al.* (2024), classified MMC into two categories: off-site construction (OSC) and non-off-site construction. Off-site construction is the prefabrication of elements or parts of structures, which are constructed in a factory and then transported and assembled on-site (Kyjaková *et al.*, 2014). Duncheva *et al.* (2020) described off-site construction as a method that significantly enhances a product by manufacturing and pre-assembling components, elements, or modules in a factory before their installation at the final location.

Duncheva *et al.* (2020) pointed out that while off-site construction is often regarded as a valuable strategy for achieving MMC, simply adopting off-site techniques does not inherently ensure the realisation of the process efficiencies associated with MMC. Off-site construction is a method of construction executed outside project sites (Layon *et al.*, 2024). Off-site construction methods include panel building systems, volumetric or modular construction, hybrid or semi-volumetric and sub-assemblies and components (Duncheva *et al.*, 2020; Layon *et al.*, 2024). The non-off-site construction involves building methods and techniques that take place on-site (Kyjaková *et al.*, 2014; Layon *et al.*, 2024). MMC classifications can be broadly distinguished between binary schemes (e.g. off-site versus on-site) and multi-category frameworks that capture both technological and process-based innovation. The UK provides a comprehensive framework for defining MMC (MHCLG, 2019) and categorises them into seven, as shown in Table 1.2.

Some authors outlined various MMC types, as highlighted in Table 1.3.

The UK MMC framework categorisation and the types of MMC presented in Tables 1.2 and 1.3 show that MMC is not limited to off-site construction activities but also includes other innovative on-site processes, including digitalisation and process improvement activities such

Table 1.2 UK MMC framework categorisation

Category	Definition	Description
Category 1	Pre-manufacturing 3D primary structural systems	The production of volumetric units in controlled factory conditions, before final installation. The modules have structural performance.
Category 2	Pre-manufacturing 2D primary structural systems	The manufacture of panelised systems with structural performance in a factory environment.
Category 3	Pre-manufacturing components (non-systemised primary structure)	The use of structural members such as framed or engineered timber, cold or hot rolled steel and pre-cast concrete.
Category 4	Additive manufacturing (structural and non-structural)	The printing of building parts through various materials based on digital design and manufacturing techniques.
Category 5	Pre-manufacturing (non-structural assemblies and sub-assemblies)	The manufacture of non-loadbearing 2D and 3D elements and components.
Category 6	Traditional building product-led site labour reduction/ productivity improvements	Include traditional building methods that have been optimised to reduce site labour (e.g. pre-cut configurations, easy jointing features).
Category 7	Site process led labour productivity/assurance improvements	Include approaches utilizing innovative site-based construction techniques such as lean construction techniques.

Source: Synthesis from Literature Review

Table 1.3 Types of MMC

S/N	MMC Types	Source
1	Modular construction	Li and mok Paik, 2025; Layon *et al.*, 2024; Dewagoda *et al.*, 2024; Musa *et al.*, 2016; Kyjaková *et al.*, 2014
2	Additive manufacturing	Josa and de la Fuente, 2024
3	Industrialised building system (IBS)	Al-Aidrous *et al.*, 2021; Rashidi and Ibrahim, 2017; Ramli *et al.*, 2014; Kamar *et al.*, 2009
4	Fibre reinforced polymers (FRPs)	Evernden and Toby Mottram, 2012
5	Prefabrication	Li and mok Paik, 2025; Liu *et al.*, 2018; Musa *et al.*, 2016; Kozlovská and Spišáková, 2013
6	On-site manufacturing	Li and mok Paik, 2025
7	Design for manufacture and assembly (DfMA)	Li and mok Paik, 2025

Source: Synthesis from Literature Review

as lean (MHCLG, 2019). The UK framework is particularly significant because it formally recognises that MMC extends beyond factory manufacture to include site-based process improvement and digital integration. According to Duncheva *et al.* (2020), MMC includes all the approaches that aim to optimise the construction process.

1.4.2 Evolution of modern methods of construction

1.4.2.1 Timeline of the evolution of MMC

The academic and professional discourse surrounding MMC remains divided. Some scholars contend that the term is redundant or conceptually uncomfortable, suggesting that the underlying techniques are far from 'new' (Green, 2022). Conversely, an opposing school of thought views MMC as a vital umbrella term for describing diverse, innovative approaches that optimise the construction process (Daniel *et al.*, 2024; Duncheva *et al.*, 2020). Despite this, the current debate on MMC is very superficial and lacks discourse that supports its conceptualisation.

Although industrialised construction has developed globally, the UK provides the most continuous and well-documented policy and terminological record of MMC, making it a useful anchor chronology for analysis. Historically, the term MMC first emerged in technical literature during the 1920s as a response to the post-World War I housing crisis in the United Kingdom. The conceptual groundwork was laid by the Tudor Walters Report (1918), which sought to establish standards for British public housing (Walters, 1918). While the report itself primarily used the descriptor 'non-traditional', referring to the departure from wet trades like brick-and-mortar, it provided the catalyst for industrialisation. Indeed, Francis Crittall, a pioneer of metal window frames, explicitly used the phrase 'modern methods of construction' in his 1918 proposals to the Tudor Walters Committee regarding the industrialisation of the sector.

The term was formally codified in the Ministry of Health Report (1926), which is arguably the first official public document to coin 'modern methods of construction' as a collective noun. In this context, it described a suite of innovative techniques, specifically pre-cast concrete and steel framing, designed to accelerate housing delivery (Ministry of Health, 1926). The timeline below (Figure 1.1) illustrates this evolution, tracing the term from its 1918 conceptualisation to its current 2026 framework.

1918–1926: EMERGENCE OF THE TERM 'MODERN METHODS OF CONSTRUCTION'

Early uses of the term MMC emerged in response to post-WWI housing shortages in the UK, where precast concrete and steel framing were promoted to improve productivity in labour-intensive wet trades (Bowley, 1926; Ministry of Health, 1926). This early MMC discourse focused on incremental efficiency rather than systemic transformation of construction processes. Early literature focused on steel and concrete frames as replacements for traditional masonry. However, these were often critiqued for poor thermal performance and higher costs relative to established 'wet trades' (Crittall, 1918).

1940S–1950S: PREFABRICATION AND OFF-SITE CONSTRUCTION

Following WWII, prefabrication and off-site construction became the dominant expression of industrialised construction, with government policy prioritizing speed and scale to address

Figure 1.1 Timeline of MMC evolution

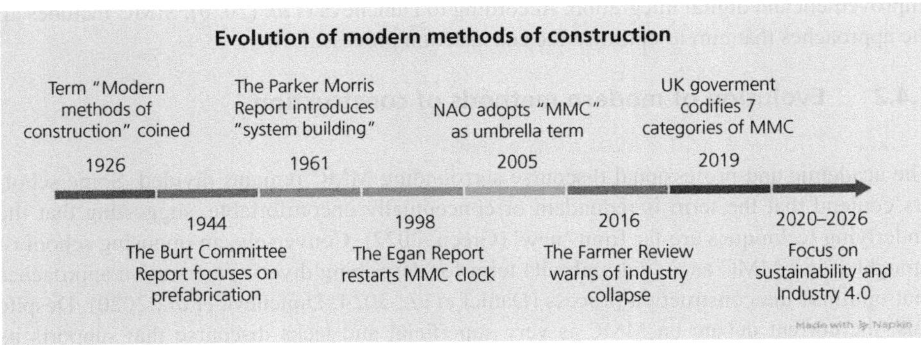

Source: Authors' creation

acute housing shortages (Burt Committee, 1944). Although technically successful, prefabrication later became culturally associated with impermanence and inferior quality, generating enduring scepticism toward non-traditional construction methods (Burnett, 1986). This era left a damaging legacy, with prefabrication becoming synonymous with temporary and low-quality social housing. It is worth mentioning that the use of prefabrication to describe a non-traditional approach to housing delivery was also reported in the United States of America (Kelly, 1951). Much later, in the 2010s, off-site construction was used by the National Institute of Building Sciences (NIBS), USA, to describe the traditional approach.

This means different names have been used to describe non-traditional approaches to housing delivery in various parts of the world. This period reinforced the association between MMC and speed, while underplaying long-term system integration.

1960S–1970S: SYSTEM BUILDING

System building expanded industrialisation ambitions through large-panel concrete systems embedded in state-led housing programmes (Parker Morris, 1961). However, structural defects and socially problematic high-rise environments undermined confidence, reinforcing perceptions of technical and social risk and contributing to a long-term retreat from industrialised housing solutions in the UK (Dunleavy, 1981).

1998: THE EGAN REPORT–REBRANDING OF MODERN METHODS OF CONSTRUCTION

The Egan Report rebranded MMC around lean manufacturing principles, emphasizing integration, efficiency and quality rather than specific technologies (Egan, 1998). This constituted a deliberate discursive reset that distanced MMC from post-war failures. However, critics argue the report privileged process efficiency and technological optimism while underplaying institutional fragmentation and historical lessons (Green, 2023).

2005: NATIONAL AUDIT OFFICE ADOPTION

The formal adoption of MMC by the National Audit Office (NAO) institutionalised the term within government policy and procurement (NAO, 2005). MMC became an umbrella concept encompassing diverse techniques, enhancing flexibility but reducing analytical precision. Their report, 'Using Modern Methods of Construction…', was the first to institutionalise the abbreviation 'MMC' as a formal policy category. The literature shifted from discussing 'materials' to discussing 'process' and 'predictability'. While this legitimised innovation, it blurred distinctions between incremental improvement and radical system change (NAO, 2005; Pan *et al.*, 2012).

2016: THE FARMER REVIEW

The Farmer Review reframed MMC as an existential necessity, linking its adoption to labour shortages and productivity decline (Farmer, 2016). The 2019 MMC categorisation prioritised standardisation to reassure lenders and insurers. The evolution culminated in the official seven-category MMC framework, which provides a standardised technical language that enables better insurance, valuation and digital integration through Industry 4.0 (MHCLG, 2019). The categories one to five of the MMC framework tend to focus on off-site or pre-manufactured processes, while categories six and seven focus on on-site improvement processes such as the lean approach and digitalisation. This shows that off-site construction is part of the MMC. Pan *et al.* (2007) confirm this when they explore house builders' views on the use of off-site MMC in the UK. However, the MMC framework prioritised market confidence and governance clarity over critical evaluation of performance outcomes.

2020–2026: THE INTERSECTION OF NET ZERO AND MARKET VOLATILITY

Contemporary MMC discourse aligns strongly with sustainability and digitalisation agendas, positioning MMC as a central mechanism for achieving net zero housing (CIOB, 2024; UKERC, 2023). However, the period following Covid has been defined by a 'market correction' that runs counter to this optimism. The high-profile collapses of Urban Splash (House) in 2022, Ilke Homes in 2023 and the cessation of production at Legal and General's modular factory in 2023 have exposed deep-seated financial vulnerabilities.

Critics argue that these failures stem from a broken business model characterised by high fixed factory overheads and a lack of a consistent, predictable pipeline of demand (Saad *et al.*, 2024). Despite the technical legitimacy of the seven-category framework (MHCLG, 2019), the volumetric sector (Category 1) remains highly sensitive to interest rate volatility and inflation. This suggests that the policy ambition of the Farmer Review (2016) may have underestimated the capital-intensive nature of modular manufacturing, where firms must maintain 100% capacity to remain solvent, a feat made impossible by the UK's fragmented planning system (House of Lords, 2024).

1.4.2.2 Evaluation of the evolution of MMC

A critical evaluation of the historical evolution of MMC reveals that the term predates contemporary descriptors such as 'prefabrication' and 'off-site construction'. The transition toward 'prefabrication' and 'industrialised building' in the mid-twentieth century can be interpreted as a strategic attempt to distance the sector from the early technical limitations of 1920s systems. However, this terminological shift did not prevent further systemic failures, most notably the structural and social shortcomings observed during the 1960s high-rise boom (Griffiths *et al.*, 1968). Also, the use of prefabrication to describe the non-traditional approach to housing production was not limited to the UK; it was also used in the USA, while in Hong Kong, modular integrated construction is used, and in Canada, off-site construction is used. This again shows that different terminology is used in the traditional approach to housing production.

Furthermore, this evolution demonstrates that MMC is not limited to off-site manufacturing. Instead, it serves as an expansive framework encompassing all non-traditional processes aimed at systemic improvement. This includes on-site advancements such as digitalisation and lean construction, as codified in categories six and seven of the current frameworks (MHCLG, 2019). The current framework (MHCLG, 2019) clarifies that MMC is not synonymous with off-site construction alone. By incorporating categories six and seven, the literature now recognises that digitalisation, lean improvements and site-based productivity are as vital to factory-built modules as they are to other manufacturing sectors. However, a critical tension remains while policy literature (CIOB, 2024; Farmer, 2016) pushes for rapid adoption to meet net zero targets, empirical evidence regarding long-term durability and cost-efficiency remains inconsistent (Sánchez-Garrido *et al.*, 2023).

The transition from 1920s 'non-traditional' systems to the contemporary MMC Definition Framework signifies a fundamental shift from material-based terminology to process-oriented classification. While the Ministry of Health (1926) initially utilised the term to address housing shortages in 'material terms', the modern resurgence, catalysed by the Egan (1998) and Farmer (2016) reviews, reflects an industrial and digital evolution. Today, the term has moved beyond the historical 'brick versus frame' debate to encompass a sophisticated spectrum of manufacturing and site-based improvements, where digital precision and lean methodology are as critical as the physical components themselves.

Thus, this chapter argues that MMC should be conceptualised as a holistic integration of the lean off-site construction approach and digitalisation, among other methods, to support process improvements that transcend the binary of on-site versus off-site delivery. Figure 1.2 shows the conceptualisation of MMC that includes off-site production, lean management and digitalisation.

1.4.3 Exploration of the principles and theory underpinning modern methods of construction

1.4.3.1 Principles of modern methods of construction

Unlike MMC types or classifications, which describe what is built and where, MMC principles describe the underlying logics that govern how construction is designed, produced and delivered. The Oxford English Dictionary defines principles as 'a fundamental truth or proposition that serves as the foundation for a system of belief or behaviour or for a chain of reasoning' (Oxford English Dictionary, 2025). In the context of MMC, this means the principle is the fundamental proposition that underpins the reasoning and practice of MMC. According to Frank and McLarnon (2008), principles constitute the value system that guides and influences people's actions. Principles provide a framework that supports consistency (Benachio *et al.*, 2020). However, Labbé and Daudin (2015) argued that when principles are isolated from practice and context, they risk becoming dogmatic. This underscores the need to link principles to practice. The MMC principles identified in this chapter are drawn from a

Figure 1.2 Conceptualisation of MMC

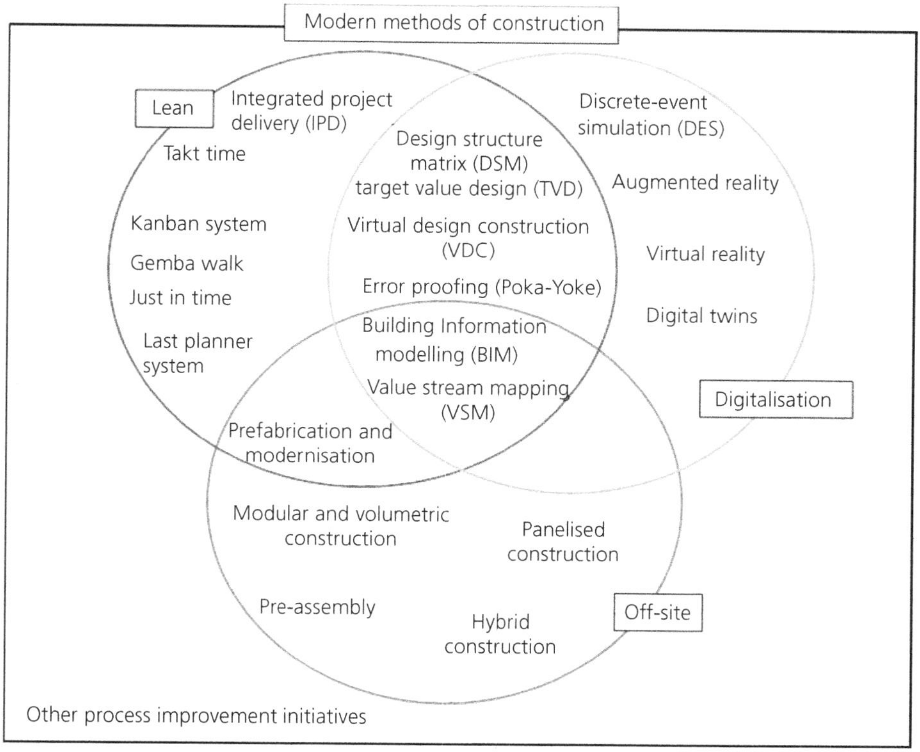

Source: Authors' creation

synthesis and evaluation of their meaning and definitions, as well as from their evolution as reported in peer-reviewed academic publications, reports, books, policy documents and other grey literature. The seven identified MMC principles are as follows.

- Principle one (P1) – off-site construction and DfMA.
- Principle two (P2) – lean production and waste minisation.
- Principle three (P3) – digitalisation.
- Principle four (P4) – controlled environment production.
- Principle five (P5) – standardisation and platform interoperability
- Principle six (P6) – just-in-time logistics and assembly.
- Principle seven (P7) – holistic sustainability and performance.

Table 1.4 shows the seven principles of MMC and their relation to theory. The table reveals that MMC represents a departure from the traditional approach to construction project delivery. According to Koskela (2000), the conventional approach to construction project management focuses on inputs and outputs, with less emphasis on workflow between activities.

Table 1.4 MMC principles and associated theories

MMC Theory	Key Citations	Associated MMC Principles
Design for manufacture and assembly (DfMA) theory	Boothroyd, 1994	P1, P5 and P6
Lean production theory	Ohno (1988); Womack and Jones (1996); Koskela (1992)	P2, P4 and P6
Systems theory (socio-technical systems)	Von Bertalanffy (1968)	P3, P5 and P7
Industrialisation theory of construction	Gibb and Isack (2003); Lessing (2006)	P1, P4 and P5

Source: Synthesis from literature review

PRINCIPLE 1. OFF-SITE CONSTRUCTION AND DESIGN FOR MANUFACTURE AND ASSEMBLY

Off-site construction and design for manufacture and assembly are among the principles that guide the MMC approach. MMC front-loads design through off-site construction (OSC) and design for manufacture and assembly (DfMA), ensuring components are optimised for precision before reaching the site (Daniel *et al.*, 2024; Gao *et al.*, 2020). According to Gao *et al.* (2020), DfMA reduces part count and assembly errors. However, it requires a higher upfront investment in design. Off-site construction and DfMA are grouped because large-scale factory production requires design rationalisation and assembly-focused decisions.

PRINCIPLE 2. LEAN PRODUCTION AND WASTE MINIMISATION

Conventional construction focuses on inputs and outputs, leading to fragmentation and an increase in non-value-adding activities. However, the lean production approach introduced the flow and value perspectives, as advocated in the transformation–flow–value (TFV) model developed by Koskela (2000). The application of lean production principles ensures that customer and end-user value is considered in both the design and delivery of the product, while also maintaining workflow continuity during execution.

PRINCIPLE 3. DIGITALISATION

Digitalisation is central to the MMC approach. It ensures that all components are digitally connected, supporting communication and information exchange among stakeholders. Building information modelling (BIM) supports the digitalisation of construction elements and processes (Onyenokporo *et al.*, 2024). This eliminates the manual information-transfer errors prevalent in traditional construction. However, it requires a digitally literate workforce. Linking BIM to computer-aided manufacturing (CAM) and automated fabrication, MMC significantly reduces errors associated with manual data re-entry, a known risk in traditional construction delivery (Eastman *et al.*, 2011). Digitalisation also enables real-time information exchange between off-site and on-site activities through a digital twin and a material passport.

PRINCIPLE 4. CONTROLLED ENVIRONMENT PRODUCTION

Shifting construction operations to a controlled factory environment is a core feature of the MMC. According to Goodier and Gibb (2007) and Liu *et al.* (2024), this approach ensures consistent quality, tighter tolerances and improved health and safety for workers. In contrast to traditional outdoor sites, where weather, variable workmanship and sequencing conflicts compromise quality, factory production delivers repeatable outputs and measurable performance improvements (Goodier and Gibb, 2007; NHBC Foundation, 2016). However, the factory infrastructure requires maintenance, which could be costly.

PRINCIPLE 5. STANDARDISATION AND PLATFORM INTEROPERABILITY

MMC emphasises standardisation through product platforms rather than bespoke, project-specific solutions (Farmer, 2016; Mujahed *et al.*, 2025). Buildings are configured using interoperable systems and components, enabling scalability, adaptability and ease of maintenance. Platform-based approaches reduce design duplication and enable long-term asset flexibility, addressing a key limitation of traditional construction's one-off design culture (Bryden Wood, 2018; Lessing *et al.*, 2015). They also support mass customisation and production.

PRINCIPLE 6. JUST-IN-TIME (JIT) LOGISTICS AND ASSEMBLY

The just-in-time (JIT) approach ensures components arrive exactly when needed, transforming the site into an assembly point rather than a storage yard (Bidhendi *et al.*, 2025). According to Vrijhoef and Koskela (2000), this logistics strategy reduces environmental disturbance, improves safety and increases installation efficiency compared with traditional sites characterised by overlapping trades and uncoordinated deliveries. It also reduces local congestion and noise. However, it could leave the project vulnerable to supply chain disruptions.

PRINCIPLE 7. SUSTAINABILITY AND PERFORMANCE

Sustainability within MMC is embedded at the system level rather than treated as a post-design add-on. Digital modelling enables early-stage evaluation of embodied carbon, thermal performance and lifecycle energy use (Onyenokporo *et al.*, 2024). Additionally, factory precision improves airtightness and thermal consistency, while optimised material selection reduces both operational and embodied carbon (Raqqad, 2024). Studies consistently show that MMC can deliver superior environmental and energy performance compared with traditional construction methods (HM Government, 2022; Pan *et al.*, 2012).

1.4.3.2 Evaluation of the seven principles of modern methods of construction

The seven principles of MMC collectively signify a fundamental reconfiguration of construction delivery, shifting the industry from fragmented, site-based practices towards an integrated, industrialised production system. Their core contribution lies in relocating decision-making upstream, where design integration, digital coordination and process standardisation exert the most significant influence on cost, quality and performance outcomes (Koskela, 2000). Rather than operating independently, the principles function as an interdependent continuum, meaning that partial or selective adoption risks undermining their intended benefits (Pan and Sidwell, 2011).

The principles also expose an inherent tension between efficiency and flexibility. Off-site manufacture, controlled environments and standardisation enhance predictability and reduce waste, yet they also increase reliance on early design certainty, supply chain reliability and organisational maturity (Farmer, 2016; Vrijhoef and Koskela, 2000). In this sense, MMC redistributes risk rather than eliminating it, transferring vulnerability from the construction site to the design and production phases. Sustainability and performance emerge not as isolated objectives but as system-level outcomes of digitalisation, precision manufacturing and lifecycle thinking, reinforcing evidence that MMC's environmental advantages depend on holistic integration rather than technological substitution alone (Pan *et al.*, 2012; HM Government, 2022).

This implies that the seven principles should be understood as a socio-technical transformation that requires alignment between technology, processes and people. Without such alignment, MMC risks replicating traditional inefficiencies in an industrialised form rather than delivering meaningful and lasting change.

1.4.4 Theory of modern methods of construction

Building on the identified MMC principles, this section describes how well-established production and organisational theories explain MMC's operational logic and performance characteristics. This section presents the MMC theories that underpin the seven MMC principles discussed in the previous section. From the literature synthesis, four relevant theories have been identified to support MMC. Table 1.4 presents the MMC theories and briefly introduces the associated principles and theories.

1.4.4.1 Organisational theories underpinning the MMC principles

DESIGN FOR MANUFACTURE AND ASSEMBLY (DFMA) THEORY

The DfMA theory posits that the cost, quality and performance of a product are largely determined at the design stage (Boothroyd and Dewhurst, 1983; Boothroyd *et al.*, 2010). This ensures that the manufacturing and assembly of the components are factored into the decision making at the design stage. This theory directly explains Principle 1 (off-site construction and design for manufacture and assembly) – for example, MMC prioritises front-loaded design to optimise components for factory manufacture and rapid on-site assembly. It also supports Principle 5 (standardisation and platform interoperability) by encouraging reduced part variation and the use of repeatable interfaces.

Additionally, DfMA underpins P6 (just-in-time logistics and assembly) by ensuring that components are designed for predictable sequencing, efficient transportation and rapid installation.

LEAN PRODUCTION THEORY

Lean production theory originated in manufacturing, particularly in the Toyota production system, and emphasises the elimination of waste, continuous flow and value generation from the client's perspective (Ohno, 1988; Womack and Jones, 1996). In construction, Koskela (1992) adapted lean thinking to highlight the limitations of traditional input-to-output production models.

Lean theory strongly underpins P2 (lean production and waste minimisation) by viewing MMC as a flow-based system that systematically reduces material waste, rework and inefficiencies common in traditional construction. It also explains P6 (just-in-time logistics and assembly), in which materials and components are delivered precisely when required, minimising inventory and site congestion. Furthermore, P4 (controlled environment production) aligns with lean principles by enabling predictable workflows, takt planning and continuous improvement through factory-based production.

SYSTEMS THEORY

Systems theory holds that organisational and production processes are linked within a system comprising technical, social and organisational components (Von Bertalanffy, 1968). For performance and growth to occur, there must be interaction among the elements of the system. This theoretical view underpins MMC P3, P5 and P7. For MMC (P3 – digitalisation), building information modelling can be seen as the point of interaction and integration, linking the facility's design, fabrication, assembly and operation. MMC P5 (standardisation and platform interoperability) also demonstrates systems thinking by enabling different elements to fit together to form a structure, such as a building, that functions as a unit, as in modular construction. Systems thinking supports P7 (sustainability and performance) by enabling consideration of the building's whole life cycle, including material selection, embodied carbon and performance in use.

INDUSTRIALISATION THEORY OF CONSTRUCTION

Industrialisation theory views construction as the transition from the traditional craft-based approach to a production-based approach centred on component standardisation that is repeatable and can be automated (Gibb and Isack, 2003; Lessing, 2006). The theory supports component prefabrication and underpins P1, P4 and P5. P1 (off-site construction and DfMA) treats off-site manufacture as a core industrialisation strategy. It also supports P4 (controlled environment production), in which factory settings replace weather-dependent site activities to ensure consistency and quality control. Furthermore, P5 (standardisation and platform Interoperability) is based on industrialisation theory, as it supports the development of standardised, repeatable and mass-produced modules.

The four theories of MMC have helped to explain the seven principles of MMC. No single theory fully explains MMC; instead, it arises at the intersection of these complementary perspectives.

1.4.4.2 Evaluation and discussion of MMC theories with associated principles

The four theoretical perspectives underpinning MMC, lean production theory, design for manufacture and assembly (DfMA) theory, systems theory and industrialisation theory, provide a coherent explanation for the seven MMC principles and their divergence from traditional construction practice, and are summarised as follows.

- *Lean production theory* explains MMC's emphasis on minimising waste, maintaining workflow continuity and just-in-time logistics. However, its success depends on organisational culture and managerial capability rather than technical tools alone (Koskela, 2000).
- *DfMA* theory accounts for the front-loading of design to support off-site manufacture and standardisation, while also exposing tensions between early design certainty and flexibility (Boothroyd and Dewhurst, 2011).
- *Systems theory* underpins digitalisation, interoperability and lifecycle sustainability, emphasising that MMC performance emerges from integrated socio-technical interactions rather than isolated interventions (Von Bertalanffy, 1968).
- *Industrialisation theory* situates MMC within a broader shift from craft-based construction to repeatable, factory-based production, while highlighting risks associated with capital intensity and scale dependency (Gibb and Isack, 2003).

These theories demonstrate that MMC principles operate as an interdependent system, requiring holistic, theory-informed implementation to avoid reproducing traditional inefficiencies in an industrialised form.

1.4.5 Theoretical framework for modern methods of construction

The framework explains MMC outcomes as emergent properties of alignment between design integration, production control, digital coordination and organisational capability, mediated by market and policy conditions. Figure 1.3 shows the theoretical framework of the MMC.

The theoretical framework reveals that the seven MMC principles are neither ad hoc nor purely technological but are grounded in well-established production and systems theories. Lean production theory, DfMA theory, systems theory and industrialisation theory together explain how MMC restructures construction into an integrated, design-led and performance-oriented production system (Boothroyd, 1994; Koskela, 1992; Von Bertalanffy, 1968; Womack and Jones, 1996). It can be argued that these theories address long-standing deficiencies in traditional construction, particularly fragmentation, inefficiency and poor knowledge integration (Koskela, 2000).

The framework reveals that MMC success depends on the alignment of theory and practice. Lean production theory emphasises waste minimisation, workflow continuity and just-in-time logistics, but its effectiveness depends on cultural and organisational maturity rather than on technical adoption alone (Womack and Jones, 1996). Also, DfMA theory highlights the strategic shift of risk and value creation to the design stage, supporting off-site manufacture and standardisation. Yet, it also exposes vulnerabilities associated with early design fixation and reduced flexibility (Boothroyd and Dewhurst, 2011). Systems theory provides

Figure 1.3 Theoretical framework of MMC

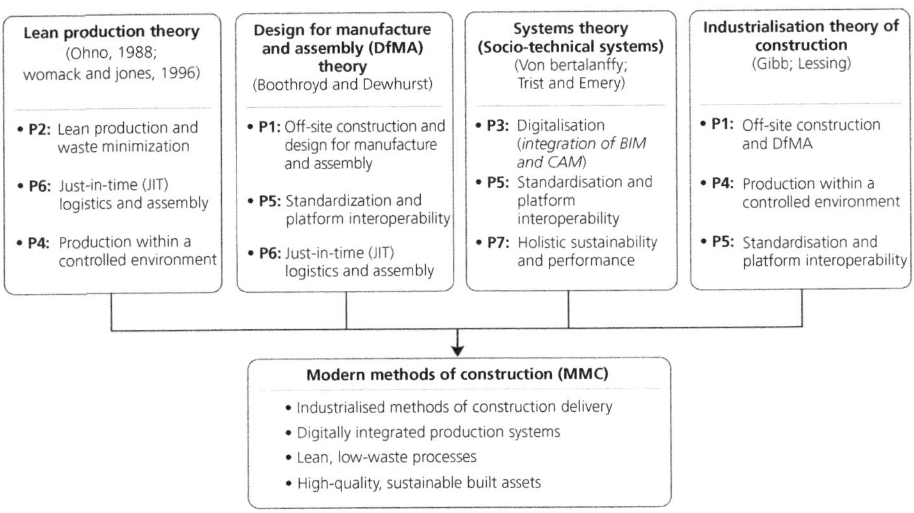

Source: Authors' creation

the intellectual foundation for digitalisation, interoperability and lifecycle sustainability, reinforcing that MMC performance emerges from system-wide integration rather than isolated interventions (Von Bertalanffy, 1968). Industrialisation theory contextualises MMC as a structural transformation from craft-based construction to repeatable, factory-based production, while also underscoring the economic risks of high capital investment and dependency on scale (Gibb and Isack, 2003).

The theoretical framework confirms that MMC principles operate as a continuum of interdependent mechanisms. Selective or superficial application risks reproducing traditional inefficiencies in an industrialised form. Meaningful transformation, therefore, requires not only technological capability but sustained alignment between theoretical intent, organisational capacity and industry practice. This is important, given that some MMC companies have gone into administration in the UK in recent times (Saad *et al.*, 2024). These failures illustrate misalignment between industrialisation theory and fragmented demand pipelines, rather than deficiencies in MMC technologies themselves.

1.5. Implications of the study for research and practice

The findings of this study have significant implications for both construction practice and future research. These implications are drawn from the chapter's theoretical framing of MMC as an integrated socio-technical production system rather than a discrete set of construction technologies.

1.5.1 Implications for practice

From a systems and industrialisation perspective, MMC implementation requires alignment across design, production, logistics, digital infrastructure and organisational capability. For practice, the study demonstrates that MMC should not be approached as a single technological solution, such as off-site or modular construction, but as an integrated socio-technical system. Practitioners who adopt MMC selectively, for example, by focusing solely on volumetric construction without aligning design, procurement, digital workflows and organisational capability, risk reproducing traditional inefficiencies in an industrialised form. Selective adoption thus disrupts the interdependence between MMC principles, undermining the performance benefits predicted by lean, DfMA and systems-based approaches.

The evidence shows that successful MMC implementation depends on early design integration, standardisation and digital coordination underpinned by lean and DfMA principles. This implies a need for industry-wide capability development, including upskilling in digital literacy, design-led decision-making and systems thinking, as well as procurement models that support early collaboration and long-term pipelines rather than project-by-project delivery. This implies a shift away from transactional procurement toward models that support early contractor involvement, design integration and long-term demand certainty.

1.5.2 Implications for research

The dominance of descriptive, practice-oriented MMC studies has limited theoretical accumulation and constrained the comparability of findings across contexts. For research, the study highlights a persistent gap between MMC practice-focused studies and theoretically grounded explanations of how and why MMC succeeds or fails. The identification of four underpinning theories provides a foundation for more cumulative and explanatory research, moving beyond descriptive accounts of benefits and barriers.

Future research should operationalise the framework by examining how combinations of design maturity, digital integration, standardisation and pipeline stability influence cost certainty, quality outcomes, carbon performance and business viability. Comparative, longitudinal and mixed-methods studies are particularly suited to capturing the dynamic interactions between organisational capability, market conditions and MMC performance.

Together, these implications reinforce the need to reposition MMC as a theory-informed, system-level transformation rather than a technology-led solution, a point consolidated in the concluding section.

1.6. Conclusions

This chapter addressed four interrelated research questions concerning the definition, classification, evolution and theoretical foundations of MMC, as well as how theory explains MMC practice. In doing so, the chapter responds directly to persistent ambiguity and theoretical fragmentation that have characterised MMC discourse across both academic and policy literature. Through a critical qualitative review of academic literature, policy documents and industry reports, the chapter demonstrates that MMC is not a recent or narrowly defined concept, but an evolving, process-oriented approach that has been repeatedly reframed in response to economic, social and technological challenges.

The study confirms the absence of a universally accepted definition of MMC. The literature reveals a clear divide between narrow interpretations that equate MMC solely with off-site construction and broader perspectives that view it as an umbrella concept encompassing both off-site and on-site innovations, including lean construction, digitalisation and process optimisation. In response, this study proposes a definition of MMC as a non-traditional, manufacturing-led approach that integrates process, design and material innovation to enhance efficiency, sustainability and customer value, thereby addressing persistent conceptual ambiguity. Importantly, this definition does not equate MMC with innovation in general, but anchors it in industrialised production logic, design-led integration and system-level performance outcomes.

The historical analysis found that MMC has evolved through distinct phases, from early twentieth-century non-traditional systems and post-war prefabrication to contemporary frameworks shaped by lean production, digital technologies and sustainability agendas. This recurring pattern of reinvention shows that MMC is less a discrete technological breakthrough than an evolving response to enduring structural challenges within construction.

The study identifies seven core principles of MMC. It establishes that these principles are underpinned by four complementary theories: lean production theory, design for manufacture and assembly (DfMA) theory, systems theory and industrialisation theory. Together, the principles and theories explain MMC not as a linear improvement pathway, but as an interdependent configuration in which performance gains depend on alignment across design, production, digital systems and organisational capability.

This chapter proposed a theoretical framework that demonstrates that MMC should be understood as a socio-technical system rather than a collection of isolated techniques. The theoretical framework explains why MMC outcomes vary widely in practice, demonstrating that success depends less on the adoption of specific technologies and more on the degree of systemic and organisational alignment. While this chapter is intentionally based on a synthesis of existing literature rather than new empirical data, it provides conceptual clarity, historical insight and theoretical rigour that is currently underrepresented in MMC research. By repositioning MMC as a theory-informed socio-technical system, the chapter offers a stable foundation for more coherent research, policy formulation and industry implementation.

REFERENCES

Al-Aidrous AHMH, Rahmawati Y, Wan Yusof K, Omar Baarimah A and Alawag A M (2021) Review of industrialized buildings experience in Malaysia: an example of a developing country. *IOP Conference Series: Earth and Environmental Science* **682(012003)**, 10.1088/1755-1315/682/1/012003.

Benachio GLF, Freitas M do CD and Tavares SF (2020) Circular economy in the construction industry: a systematic literature review. *Journal of Cleaner Production* **260**: 121046.

Bernold L (2016) Discussion of 'Barriers of Implementing Modern Methods of Construction' by M. Motiar Rahman. *Journal of Management in Engineering* **32(2)**: 07015002.

Bidhendi A, Herrera RF and Poshdar M (2025) Simulating JIT implementation in prefabricated construction: a digital framework for supply chain optimization. In *Proceedings of the 33rd annual conference of the international group for lean construction (IGLC 33)* (Seppänen O, Koskela L and Murata K (eds.)), pp. 657–668, 10.24928/2025/0158.

Boothroyd G and Dewhurst P (1983) *Design for Assembly: A Designer's Handbook.* University of Massachusetts, Amherst, MA, USA.

Boothroyd G (1994) Product design for manufacture and assembly. *Computer-Aided Design* **26(7)**: 505–520, 10.1016/0010-4485(94)90082-5.

Boothroyd G, Dewhurst P and Knight W (2010) *Product Design for Manufacture and Assembly* (3rd ed.). CRC Press, Boca Raton, FL, USA.

Bowley M (1926) *The British Building Industry.* Cambridge University Press, Cambridge, UK.

Bryden Wood (2018) *Bridging the Gap Between Construction + Manufacturing, Cambridge.* Bryden Wood Technology Limited, London, UK.

Burnett J (1986) *A Social History of Housing 1815–1985.* Methuen, London, UK.

Burt Committee (1944) *Report of the Committee on Housing.* HMSO, London, UK.

CIOB (2024) Modern Methods of Construction: Barriers and Benefits for Irish Housing. https://www.ciob.org/media/2341/download (accessed 02/01/2026).

Crittall F (1918) *Evidence Submitted to the Tudor Walters Committee on the Industrialisation of Housing Construction.* HMSO, London, UK.

Daniel EI and Oshodi O (2022) Lean-offsite-simulation nexus for housing construction: a state-of-the-art review of the existing knowledge. *Construction Innovation: Information, Process, Management* **23(5)**: 994–1017, 10.1108/CI-03-2022-0051.

Daniel EI and Pasquire C (2019) Creating social value within the delivery of construction projects: the role of lean approach. *Engineering, Construction and Architectural Management*, 10.1108/ECAM-06-2017-0096

Daniel EI, Konara AN and Oshodi O (2024) Exploring the use of lean, offsite and simulation in housing delivery: evidence from the UK. *Buildings* **14(6)**: 1642, 10.3390/buildings14061642.

Dewagoda KG, Ng, ST, Kumaraswamy MM and Chen J (2024) Design for Circular Manufacturing and Assembly (DfCMA): synergising circularity and modularity in the building construction industry. *Sustainability (Switzerland)* **16(21)**, 10.3390/su16219192

Dowsett RM, Green MS and Harty CF (2022) Speculation beyond technology: building scenarios through storytelling. *Buildings and Cities* **3(1)**: 534–553, 10.5334/bc.213.

Duncheva M, Calcagno C and Henquel V (2020) Offsite fundamentals. In *Offsite Ready Booklets* (pp. 1–36). Construction Industry Training Board.

Duncheva T and Bradley FF (2019) A multi-faceted productivity comparison of offsite timber manufacturing strategies in mainland Europe and the UK. *Journal of Construction Engineering and Management* **145(8)**: 04019043, 10.1061/(ASCE)CO.1943-7862.0001641.

Dunleavy P (1981) *The Politics of Mass Housing in Britain 1945–1975*. Clarendon Press, Oxford, UK.

Eastman CM, Teicholz P, Sacks R and Liston K (2011) *BIM Handbook: A Guide to Building Information Modeling for Owner, Manager, Designer, Engineers and Contractors*. John Wiley and Sons, Hoboken, NJ, USA.

Egan J (1998) *Rethinking Construction: The Report of the Construction Task Force*. Department of the Environment, Transport and the Regions, London, UK.

Evernden MC and Toby Mottram J (2012) A case for houses to be constructed of fibre reinforced polymer components. *Construction Materials* **165(1)**: 3–13. https://doi.org/10.1680/coma.2012.165.1.3

Farmer M (2016) *The Farmer Review of the UK Construction Labour Model*. Construction Leadership Council, London, UK.

Frank J and McLarnon J (2008). *Young Carers, Parents and Their Families: Key Principles of Practice*. Include Project, The Children's Society www.youngcarer.com/pdfs/Key%20Principles%20of%20Practice%20with%20cover.pdf (accessed 31/12/2025).

Gao S, Jin R and Lu W (2020) Design for manufacture and assembly in construction: a review. *Building Research and Information* **48(5)**: 538–550, 10.1080/09613218.2019.1660608

Gibb A and Isack F (2003) Re-engineering through pre-assembly: client expectations and drivers. *Building Research and Information* **31(2)**: 146–160, 10.1080/0961321

Goodier C and Gibb A (2007) Future opportunities for offsite in the UK. *Construction Management and Economics* **25(6)**: 585–595.

Goulding JS, Pour Rahimian F, Arif M and Sharp MD (2014) New offsite production and business models in construction: priorities for the future research agenda. *Architectural Engineering and Design Management* **11(3)**: 163–184, 10.1080/17452007.2014.891501

Government Commercial Function (2022) *Modern Methods of Construction: Guidance Note*. https://assets.publishing.service.gov.uk/government/uploads/system/uploads/attachment_data/file/1102387/20220901-MMC-Guidance-Note.pdf

Government of Jersey (2022) *Modern Methods of Construction: Housing Delivery Innovation*.

Grant MJ and Booth A (2009) A typology of reviews: an analysis of 14 review types and associated methodologies. *Health Information and Libraries Journal* **26(2)**: 91–108.

Green SD (2022) Modern methods of construction: reflections on the current research agenda. *Buildings and Cities* **3(1)**: 653–662, 10.5334/bc.265.

Green SD (2023) *Making Sense of Construction Improvement*. Routledge, London, UK.

Griffiths H, Pugsley A and Saunders O (1968) *Structural Safety and Failures in Large Panel System Buildings*. HMSO, London, UK.

Harty C (2005) Innovation in construction: a sociology of technology approach. *Building Research and Information* **33(6)**: 512–522, 10.1080/09613210500288605

Hernández H, Ossio F and Silva M (2023) Assessment of sustainability and efficiency metrics in modern methods of construction: a case study using a life cycle assessment approach. *Sustainability* **15(6267)**: 1–25, 10.3390/su15076267.

HM Government (2022) The Construction Playbook. https://assets.publishing.service.gov.uk/media/6312222de90e075880923330/14.116_CO_Construction_Playbook_Web.pdf (accessed 31/12/2025).

House of Commons (2019) Housing Communities and Local Government Committee Oral Evidence: Modern Methods of Construction. HC 1831. http://data.parliament.uk/writtenevidence/committeeevidence.svc/evidencedocument/housing-communities-and-local-government-committee/modern-methods-of-construction/oral/98436.pdf (accessed 31/12/2025).

House of Lords (2024) Modern Methods of Construction in the Housing Industry. https://lordslibrary.parliament.uk/modern-methods-of-construction-in-the-housing-industry/ (accessed on 31/12/2025).

Hu X and Chong HY (2021) Environmental sustainability of off-site manufacturing: a literature review. *Engineering, Construction and Architectural Management* **28(1)**: 332–350, 10.1108/ECAM-06-2019-0288.

Hughes D (2024) The Immutable Nature of the Construction Industry: Understanding the Resistance to Change. https://medium.com/@hangadave/the-immutable-nature-of-the-construction-industry-understanding-the-resistance-to-change-efd65ea84fea (accessed 23/04/26)

Ivory C (2005) The cult of customer responsiveness: Is design innovation the price of a client-focused construction industry? *Construction Management and Economics* **23(8)**: 861–870.

Josa I and de la Fuente A (2024) Traditional and modern methods of construction: Comparative study of the sustainability of single-family homes. *Structural Concrete, October* 1–19, 10.1002/suco.202400802.

Kamar KAM, Alshawi M and Hamid Z (2009) Barriers to industrialized building system (IBS): the Case of Malaysia. *Built and Human Environment 9th International Postgraduate Research Conference*, 1–16.

Kelly B (1951) *The Prefabrication of Houses*. McGraw-Hill, New York, USA.

Koskela L (1992) Application of the new production theory to construction (*CIFE Technical Report #72*). Stanford University, USA.

Koskela L (2000) *An Exploration Towards a Production Theory and Its Application to Construction*. VTT Technical Research Centre of Finland, Espoo, Finland.

Kozlovská M and Spišáková M (2013) Construction waste generation across construction project life-cycle. *Organisation, Technology and Management in Construction: An International Journal* **5(1)**: 687–695, 10.5592/otmcj.2013.1.5.

Kozlovská M, Kyjaková L and Mačková D (2014) Comparison of low-energy houses selected parameters made by traditional and modern methods of construction. *Advanced Materials Research* **1041**: 51–54, 10.4028/www.scientific.net/AMR.1041.51.

Kyjaková L, Mandičák T and Mesároš P (2014) Modern methods of constructions and their components. *Journal of Engineering and Architecture* **2(1)**: 27–35,https://jea.thebrpi.org/journals/jea/Vol_2_No_1_March_2014/3.pdf.

Labbé J and Daudin P (2015) Applying the humanitarian principles: reflecting on the experience of the International Committee of the Red Cross. *International Review of the Red Cross* **97(897-898)**:183–210, 10.1017/S1816383115000715.

Layon LB, Said M, Al Siyabi A, Garcia RM and Umar AA (2024) Barriers affecting the implementation of modern methods of construction (MMC) in Oman. *Journal of Civil Engineering Research and Technology* **6(6)**: 1–6, 10.47363/JCERT/2024(6)165.

Lessing J (2006) *Industrialised House-Building: Conceptual Orientation and Strategic Perspectives*. Lund University, Lund, Sweden.

Lessing J, Stehn L and Ekholm A (2015) Industrialised house-building – development and conceptual orientation of the field. *Construction Innovation* **15**: 378–399.

Li, Z. and mok Paik S (2025) Resource cost optimization strategies in off-site manufacturing for modern construction methods: a lean theory perspective case study in Ireland. *International Journal of Construction Management 1–9*, 10.1080/15623599.2025.2490169.

Liu H, Li S and Wen H (2024) Effect of smart construction technology characteristics on the safety performance of construction projects: an empirical analysis based on structural equation modeling. *Buildings* **14**: 1894.

Liu K, Su, Y. and Zhang S (2018) Evaluating supplier management maturity in prefabricated construction project-survey analysis in China. *Sustainability* **10(3046)**: 1–21, 10.3390/su10093046.

Lovell H (2012) Modern methods of construction. In and *International Encyclopedia of Housing and Home* (Smith SJ, Elsinga M, O'Mahony LF, Eng OS and Wachter S (eds)). Elsevier Ltd, Vol. **4**, pp. 312–316, 10.1016/B978-0-08-047163-1.00588-9.

Maqbool R, Namaghi JR, Rashid Y and Altuwaim A (2023) How modern methods of construction would support to meet the sustainable construction 2025 targets, the answer is still unclear. *Ain Shams Engineering Journal* **14(4)**: 101943, 10.1016/j.asej.2022.101943.

Maslova S and Burgess G (2023) Delivering human-centred housing: understanding the role of post-occupancy evaluation and customer feedback in traditional and innovative social housebuilding in England. *Construction Management and Economics* **41(4)**: 277–292, 10.1080/01446193.2022.2111694.

Mba EJ, Ozigbo IW, Oforji PI *et al.* (2025) Impact of traditional building techniques on modern construction in Southeastern Nigeria. *International Journal of Research and Innovation in Applied Science (IJRIAS):* **10(5)**: 60–66, 10.51584/IJRIAS.

MHCLG (Ministry of Housing, Communities and Local Government) (2019) Modern methods of construction: introducing the MMC definition framework. In *Ministry of Housing, Communities and Local Government*, https://www.cast-consultancy.com/wp-content/uploads/2019/03/MMC-I-Pad-base_GOVUK-FINAL_SECURE.pdf.

Ministry of Health (1926) *Housing (Financial Provisions) Act 1924: Circular 38a*. HMSO, London, UK.

Mujahed L, Feng G and Wang J (2025) A dual-method analysis of P-DfMA adoption in the AEC industry through the TOE framework: insights from interviews and policy analysis. *Buildings* **15(22)**: 4063.

Musa MF, Yusof MR, Mohammad MF and Samsudin NS (2016) Towards the adoption of modular construction and prefabrication in the construction environment: a case study in Malaysia. *ARPN Journal of Engineering and Applied Sciences* **11(13)**: 8122–8131.

Nadim W (2012) Modern methods of construction. In and *Construction Innovation and Process Improvement*, 1st ed. (Akintoye A, Goulding JS and Zawdie G (Eds.)). Blackwell Publishing Ltd, Oxford, UK.

NAO (2005) *Using Modern Methods of Construction to Build Homes More Quickly and Efficiently*. National Audit Office, London, UK.

NHBC Foundation (2016) *Modern Methods of Construction: Views from the Industry*. IHS BRE Press, https://www.nhbc.co.uk/insights-and-media/foundation/publications/modern-methods-of-construction-views-from-the-industry (accessed 23/4/26).

O'Hegarty R, McCarthy A, O'hagan J *et al.* (2025) Understanding the embodied carbon credentials of modern methods of construction. *Buildings and Cities* **6(1)**: 70–89, 10.5334/bc.515.

Ofori-Kuragu J K, Osei-Kyei R and Wanigarathna N (2022) Offsite construction methods—what we learned from the UK Housing Sector. *Infrastructures* **7(12)**: 1–22, 10.3390/infrastructures7120164.

Ohno T (1988) *Toyota Production System: Beyond Large-Scale Production*. Productivity Press, Cambridge, MA, USA.

Onyenokporo NC, Tzortzopoulos P, Biscaya S, Collington N and O'Connor S (2024) Net-zero and digitalisation in off-site construction. In and *Proceedings of the 32nd Annual Conference of the International Group for Lean Construction (IGLC32)* (Costa DB, Drevland F and Florez-Perez L (eds)), pp. 1–12, 10.24928/2024/021.

Oxford English Dictionary (2025) https://doi.org/10.1093/OED/6961664122 (accessed 31/12/2025).

Pan W and R. Sidwell (2011) Demystifying the cost barriers to offsite construction in the UK. *Construction Management and Economics* **29**(11): 1081–1099, 10.1080/01446193.2011.637938.

Pan W, Gibb AG and Dainty AR (2007) Perspectives of UK housebuilders on the use of offsite modern methods of construction. *Construction Management and Economics* **25(2)**: 183–194.

Pan W, Gibb AG and Dainty AR (2012) Strategies for integrating the use of off-site production into housebuilding. *Engineering, Construction and Architectural Management* **19(1)**: 5–31.

Parker Morris (1961) *Homes for Today and Tomorrow*. HMSO, London, UK.

Rafa N and Khalid R (2024) *Modern Methods of Construction (MMC) for Net Zero Housing: Implications from the Social Sciences and Humanities*, 10.13140/RG.2.2.20749.96487.

Ramli N A, Abdullah C S and Nawi M N M (2014) Definition and new directions of IBS Load bearing masonry (LBM) system in construction industry. *Advances in Environmental Biology* **8(5)**: 1864–1867.

Raqqad ZK (2024) Environmentally sustainable construction by mitigating embodied carbon emissions of large-scale projects: a study on off-site practices on high-rise buildings. *Architecture, Structures and Construction* **4**: 1–14, 10.1007/s44150-023-00099-4.

Rashidi A and Ibrahim R (2017) Industrialized construction chronology: the disputes and success factors for a resilient construction industry in Malaysia. *The Open Construction and Building Technology Journal* **11(1)**: 286–300, 10.2174/1874836801711010286.

Rushton T (2021) Modern methods of construction. *Journal of Building Survey, Appraisal and Valuation* **10(4)**: 312–316, 10.1016/B978-0-08-047163-1.00588-9.

Saad A M (2023) Exploring contractual arrangements associated with adoption of modern methods of construction (MMC) in the UK. *Journal of Legal Affairs and Dispute Resolution (ASCE)*, **15(4)**, 10.1061/JLADAH.LADR-986.

Saad AM, Dulaimi M and Zulu S L (2023). Examining the influence of UK public clients ' characteristics on their own innovation-decision towards the modern methods of construction (MMC). *Sustainability Article* **15(4159)**, 10.3390/su15054159.

Saad AM, Dulaimi M, Arogundade S, Zulu SL and Gorse C (2024) Modern methods of construction (MMC) and innovation negativism in the UK public sector. *Built Environment Project and Asset Management* **14(2)**: 147–163, 10.1108/BEPAM-06-2023-0108.

Sánchez-Garrido AJ, Navarro IJ and Yepes V (2024) Sustainable preventive maintenance of MMC-based concrete building structures in a harsh environment. *Journal of Building Engineering* **95(2)**:110155, 10.1016/j.jobe.2024.110155.

Sánchez-Garrido AJ, Navarro IJ, García J and Yepes V (2023) A systematic literature review on modern methods of construction in building: An integrated approach using machine learning. *Journal of Building Engineering* **73**: 106725, 10.1016/j.jobe.2023.106725.

Sardén Y and Engström S (2010) Modern methods of construction: A solution for an industry characterised by uncertainty? *Association of Researchers in Construction Management, ARCOM 2010 - Proceedings of the 26th Annual Conference*, 1101–1110.

Shibani A, Agha A, Saidani M *et al.* (2021) The effectiveness of using modern construction methods as a solution to assist the social housing shortage in the United Kingdom Engineering and Knowledge Management. *Proceedings of the International Conference on Industrial Engineering and Operations Management, July*, 2868–2883, 10.46254/an11.20210516.

Suliman A and Rankin J (2021) Maturity-based mapping of technology and method innovation in off-site construction: conceptual frameworks. *Journal of Information Technology in Construction* **26**: 381–408, 10.36680/j.itcon.2021.021.

Švajlenka J and Kozlovská M (2018) Perception of user criteria in the context of sustainability of modern methods of construction based on wood. *Sustainability (Switzerland)*: **10(2)**, 10.3390/su10020116.

Sweeney R (2024) *Modern Methods of Construction: Barriers and Benefits for Irish housing*. Dublin: TASC (Think-tank for Action on Social Change), in collaboration with the Chartered Institute of Building (CIOB). Dublin, Eire.

UKERC (2023) *Decarbonising the Built Environment: The Role of MMC*. UK Energy Research Centre, London, UK.

Vagtholm R, Matteo A, Vand B and Tupenaite L (2023) Evolution and current state of building materials, construction methods and building regulations in the U.K.: implications for sustainable building practices. *Buildings* **13(1480)**, 10.3390/buildings13061480.

Von Bertalanffy L (1968) *General System Theory: Foundations, Development*. George Braziller, New York, NY, USA.

Vrijhoef R and Koskela L (2000) The four roles of supply chain management in construction. *European Journal of Purchasing and Supply Management* **6**: 169–178, 10.1016/S0969-7012(00)00013-7.

Walters JT (1918) *The Tudor Walters Report*. HMSO, London, UK.

Williamson M, Ganah A and John G (2019) Barriers to adopting modern methods of construction in the UK. *Journal of Construction Engineering, Management and Innovation* **2(1)**: 30–39, 10.31462/jcemi.2019.01030039.

Womack JP and Jones DT (1996) *Lean Thinking: Banish Waste and Create Wealth in Your Corporation*. Simon and Schuster, New York, NY, USA.

Part I

Digitalisation and practical implementation of modern methods of construction

emerald PUBLISHING ice

Emmanuel I Daniel and Fidelis A Emuze
ISBN 978-1-83708-437-1
https://doi.org/10.1108/978-1-83708-436-420261002

Chapter 2

Multi-objective optimisation in modular construction: a generative computational design approach for panelised floors

Fatemeh Arabnia[1], Abdul-Majeed Mahamadu[2], Yiming Xiang[3] and Faris Elghaish[4]

[1]*Shahid Beheshti University, Iran* 0009-0001-9655-642X
[2]*University College London, United Kingdom* 0000-0001-7757-8562
[3]*University College London, WC1E 7HB, United Kingdom* 0000-0002-1080-6118
[4]*Queen's University Belfast, United Kingdom* 0000-0002-7558-6291

Abstract

Design for manufacturing and assembly (DfMA) has been widely recognised as an approach in the application of modern methods of construction (MMC) for its ability to enhance productivity. Particularly, it is applied in modular construction to reduce waste and improve efficiency. Yet its systematic application to prefabricated floor slabs remains underdeveloped, with limited tools to guide early-stage design decisions. This study proposes a generative computational design approach that embeds DfMA principles into multi-objective optimisation to support informed decision-making from the outset. Adopting a design science research methodology, a parametric optimisation model was developed in Rhino and Grasshopper to explore variations in floor dimensions, panel sizes and modularity levels. To reduce construction cost and time while improving sustainability, the evolutionary solvers Octopus and Wallacei were employed to identify optimal solutions. Results show Wallacei achieved superior time efficiency, whereas Octopus delivered better cost optimisation and balanced performance across objectives. Expert evaluation (n=6) confirmed the model's applicability, usability and potential for industry uptake. This study offers a practical design optimisation tool that embeds DfMA principles in early-stage decision making. Designers and scholars can employ this model for cost-effective and efficient modular construction solutions.

Keywords: modular construction; design for manufacturing and assembly; multi-objective optimisation; generative design; computational design; panelised floor

2.1. Introduction

Modular construction (MC) falls within the broad umbrella of modern methods of construction (MMC). Homes England (2024) described MMC as a collective term for building methods that are designed to be more efficient and effective than the traditional methods. This method includes prefabrication, off-site manufacturing and modular construction. There has been a rise in popularity of MMC within the construction field owing to its capacity to boost efficiency and productivity (Thurairajah *et al.*, 2023). By manufacturing building elements in specialised facilities (Li *et al.*, 2014), MC increases delivery speed, minimises construction hazards and material waste and improves material recyclability and reusability (Bertram *et al.*, 2019).

The design phase is recognised as having a critical influence on modular construction, affecting project costs, element fabrication and onsite assembly, as design decisions establish the key constraints for subsequent stages (Liu *et al.*, 2025). Conventionally, designers rely on the hands-on experience and manual checking to ensure modular element design fulfils the requirements of production, transportation and onsite assembly. However, insufficient detail in the pre-made parts or their junctions, unrealistic division of these elements, or an excessively intricate design that obstructs ongoing enhancements during the production stage frequently occur, challenging the MC application (Yuan *et al.*, 2018).

To address this issue, the approach of DfMA was introduced to enhance the buildability of building products at early stages (Widanage and Kim, 2024). Accompanied by innovative tools like computational and parametric design, generative design and building information modelling (BIM), DfMA is expected to integrate design and manufacturing considerations, address complex requirements in a systematic manner and deliver efficient, cost-effective and sustainable MC design solutions (Montazeri *et al.*, 2024).

This approach has been applied in the optimisation of module layout (Wei *et al.*, 2022), vertical structure design (Dong *et al.*, 2023), façade panel division (Deng *et al.*, 2021) and tiling plans (Wu *et al.*, 2022) for better design proficiency (Ghannad and Lee, 2022), buildability (Yuan *et al.*, 2018), affordability and sustainability (Pibal *et al.*, 2022). However, existing studies focused on geometric-based optimisation (e.g. module layout or architectural/structural forms), lacking the consideration of specific real-word constraints such as constructability, site logistics and material limitations, as well as broader goals including cost, schedule and sustainability targets. This bias has led to inefficient designs, higher costs and project delays in real-world applications.

Additionally, the application of DfMA-oriented design optimisation is limited for large-span horizontal structural elements, particularly prefabricated floor panels. Therefore, a comprehensive, automated optimisation approach for prefabricated panels warrants further investigation. This research aims to develop a novel generative computational design methodology to optimise prefabricated panelised construction by answering the question, 'How can a novel computational design methodology optimise the design of prefabricated panelised floors in modular construction while incorporating DfMA principles?'

2.2. Perspectives and/concepts

2.2.1 Modular construction

Modular construction (MC) also known as off-site construction or prefabrication, involves producing building components in controlled environments to reduce onsite work and improve quality (Ghannad *et al.*, 2019). Tatum *et al.*, (1986) identified four modularisation levels, from individual elements to full volumetric structures – for example, prefabricated prefinished volumetric construction, which uses off-site-built modules for onsite assembly (Gao *et al.*, 2018). Although modularity often refers to volumetric or off-site-producible units, its definition remains inconsistent in construction literature (Cao *et al.*, 2022), yet it is key to improving construction efficiency.

The MMC industry has adopted mass customisation strategies from manufacturing to deliver tailored products while maintaining the efficiency, cost-effectiveness and quality associated with mass production (Da Rocha *et al.*, 2015). This approach balances product diversity with large-scale output through flexible processes and structures (Pine, 1993), requiring informed decisions in product design, processes and supply chains, especially during the design phase (Da Rocha *et al.*, 2015).

Product modularisation is key to achieving mass customisation in MC, as highlighted by Nahmens (2007) and Erixon (1998). It involves using a limited set of modules to create diverse product variations (Gershenson *et al.*, 2003), enabling high production volumes and customisation simultaneously. Together, mass customisation and modularisation align customer needs with industry efficiency and quality.

The platform approach in MC balances standardisation with mass customisation. According to the Product Platform Rulebook (Construction Innovation Hub, 2023), a product platform integrates key components, processes and relationships to deliver varied products through standardised elements and interfaces. This method boosts productivity, efficiency and predictability, while reducing costs, rework and labour needs. It also enhances quality, safety and environmental performance by minimising waste and carbon footprint.

2.2.2 Design for manufacture and assembly

Design is a comprehensive, cross-disciplinary process requiring meticulous research, fine-tuning and evaluation to build the intended structure. It explores various alternatives in the solution and problem spaces that significantly impact the project's economic viability (Li, H. and Lachmayer, 2018). The design phase, therefore, involves rigorous scrutiny of multiple variables for feasibility and the development and testing of numerous variants to find the optimal solution (Pibal *et al.*, 2022).

Particularly in MC, the design phase is crucial due to its implications on operational costs and the necessity for a robust design system for the seamless production and installation of prefab structures (Yuan *et al.*, 2018). All these facets underscore the intricate and impactful nature of design in MC, making it an endeavour of profound criticality.

DfMA is a recognised approach that began in the manufacturing sector to enhance the ease of product fabrication and assembly (Gao *et al.*, 2019; Lu *et al.*, 2021). This approach is divided into two fundamental parts: design for manufacturing (DFM) and design for assembly (DFA). DFM emphasises cost-effective and efficient production, whereas DFA seeks to optimise assembly processes by way of design simplification, component reduction and standardisation (Tan *et al.*, 2020).

The value of design for manufacture and assembly (DfMA) is particularly pronounced in MC design, where early integration of manufacturing and assembly knowledge can significantly influence project outcomes (Yuan *et al.*, 2018). A central challenge in this field is bridging the gap between the design stages and practical manufacturing constraints, including constructability, material availability and assembly processes (Gao, 2019). Incorporating these considerations at the initial design stage is crucial to enhance efficiency, cost-effectiveness and overall project quality. DfMA thus extends beyond structural requirements to encompass economic and constructability factors, supported by evaluation systems that guide decision-making (Rankohi, 2022).

DfMA principles promote simplifying designs with fewer precast elements, standardised and pre-manufactured components and minimal connectors. They also advocate for homogeneous, eco-friendly materials and modular, error-resistant designs suitable for automation. These principles aim to achieve DfMA goals, including time and cost savings, environmental preservation and quality improvement (Kuzmanovska, 2019; Tan, 2022).

Despite its demonstrated benefits, existing studies reveal gaps in the practical application of DfMA, particularly in automated optimisation of prefabricated panels and the integration of real-world constraints such as site logistics and large-scale structural spans. These gaps highlight the need for a comprehensive, automated DfMA-based framework for prefabricated panels, justifying the proposed study.

2.3. Methodological approach

This study adopts a parametric, multi-objective optimisation approach to determine the optimal modularity of prefabricated concrete floor systems. The methodology covers three main steps: *(a)* problem formulation, *(b)* parametric model development and optimisation, and *(c)* model evaluation.

2.3.1 Problem formulation

The optimisation problem is designed around the key principles of design for manufacture and assembly (DfMA), with specific attention to manufacturing and assembly efficiency. It is defined based on the following assumptions.

(a) Panel sizes follow standardisation principles, allowing up to four unique sizes, each being an integer multiple of the smallest panel.
(b) The number of manufacturing machines equals the number of panel sizes, with each machine producing one type simultaneously.
(c) Structural analysis, column positioning and reinforcement are excluded to simplify the problem.

In addition to these assumptions, the optimisation problem incorporates the following parameters, which act as decision variables and directly influence the modularity of the system

c

- the number of distinct panel sizes
- the modularity level, expressed through the number of divisions along the floor's length and width
- the distribution ratio of each panel size within the given floor area.

The optimisation is subject to a set of constraints arising from practical considerations such as factory production, transportation and on-site assembly capacity. These include

- maximum and minimum panel dimensions imposed by factory machinery
- transportation restrictions on maximum panel sizes
- crane lifting capacities
- minimum and maximum distances between connectors.

The performance of solutions is estimated utilising Equations 2.1 to 2.8.

$$T_a = N_c \times T_c \quad (2.1)$$

$$C_a = N_c \times C_c \quad (2.2)$$

$$A_i = U_i \times V_i \quad (2.3)$$

$$T_i = 15 + (A_i - 0.25) \times 10 / 0.25 \quad (2.4)$$

$$Tm = \sum T_i \times Np_i \quad (2.5)$$

$$C_p = C_{cast} + C_m \quad (2.6)$$

$$C_{cast} = (Af - \sum A_i \times Np_i) \times C_{C\_sqr} \quad (2.7)$$

$$Cm = \sum Tm_i \times Cm_i \quad (2.8)$$

The primary optimisation objectives are to minimise assembly time, assembly cost, manufacturing time and panel manufacturing costs. These objectives are formulated as Equation 2.9 and the variables are defined in Table 2.1.

$$Minimise\{T_a, C_a, T_m, Cm\} \quad (2.9)$$

2.3.2 Parametric model development and optimisation

The optimisation model was developed using Rhinoceros (Rhino) and Grasshopper, which together provide a flexible environment for parametric and algorithmic design. Rhino is a 3D modelling platform that allows precise geometric representation, while Grasshopper is its visual programming plugin, enabling the definition of relationships between geometric parameters and performance criteria through node-based scripting. These tools allow dynamic updates of design geometry in response to changing parameters, facilitating efficient iteration during optimisation.

As illustrated in Figure 2.1, the workflow of the proposed model involves nine main steps.

Table 2.1 Variable definitions

Symbol	Description
N_C	Number of connections
T_C	Time required for the assembly of each connection
C_C	Cost per connection assembly
C_a	Total assembly cost
A_i	Area of panel *i*
U_i, V_i	Dimensions of panel *i*
T_i	Manufacturing time of panel *i*
N_{p_i}	Number of panels of type *i*
T_m	Total manufacturing time
C_m	Manufacturing cost of panels
C_{cast}	On-site casting cost
C_p	Total material cost
A_f	Total floor area
C_{C_sqr}	Unit casting cost per m^2
T_{m_i}	Operating time of machine *i*
C_{m_i}	Unit cost of machine *i*

Source: Authors' creation

(a) *Defining the base geometry*. A rectangular floor surface is created with adjustable length and width, subdivided in two directions to establish the modularity level.
(b) *Panel generation*. Up to four panel sizes are generated as integer multiples of the smallest unit, with assigned percentage distributions ensuring full floor coverage.
(c) *Mapping configurations*. Panel sizes and distributions are mapped onto the subdivided grid and the configuration is refined to match the total panel count.
(d) *Panel identification*. Each panel's dimensions and area are extracted; unique sizes are identified and visualised in Rhino using distinct colours.
(e) *Labelling system*. Panels are automatically labelled according to relative size and labels are displayed at their centres for clarity.
(f) *Connector definition*. Panel edges are converted into connectors; subdivision density is adjusted and duplicate connectors are removed to compute the total number.
(g) *Cost estimation*. Manufacturing and casting costs are calculated based on the casting area and the operation time of manufacturing machines.
(h) *Time estimation*. The total manufacturing duration is determined by calculating panel production times per machine and selecting the maximum value.
(i) *Optimisation*. The Wallacei and Octopus solvers are applied to explore optimal design solutions, balancing modularity, cost and manufacturing efficiency.

Figure 2.1 Flow chart of optimisation model (continued on next page)

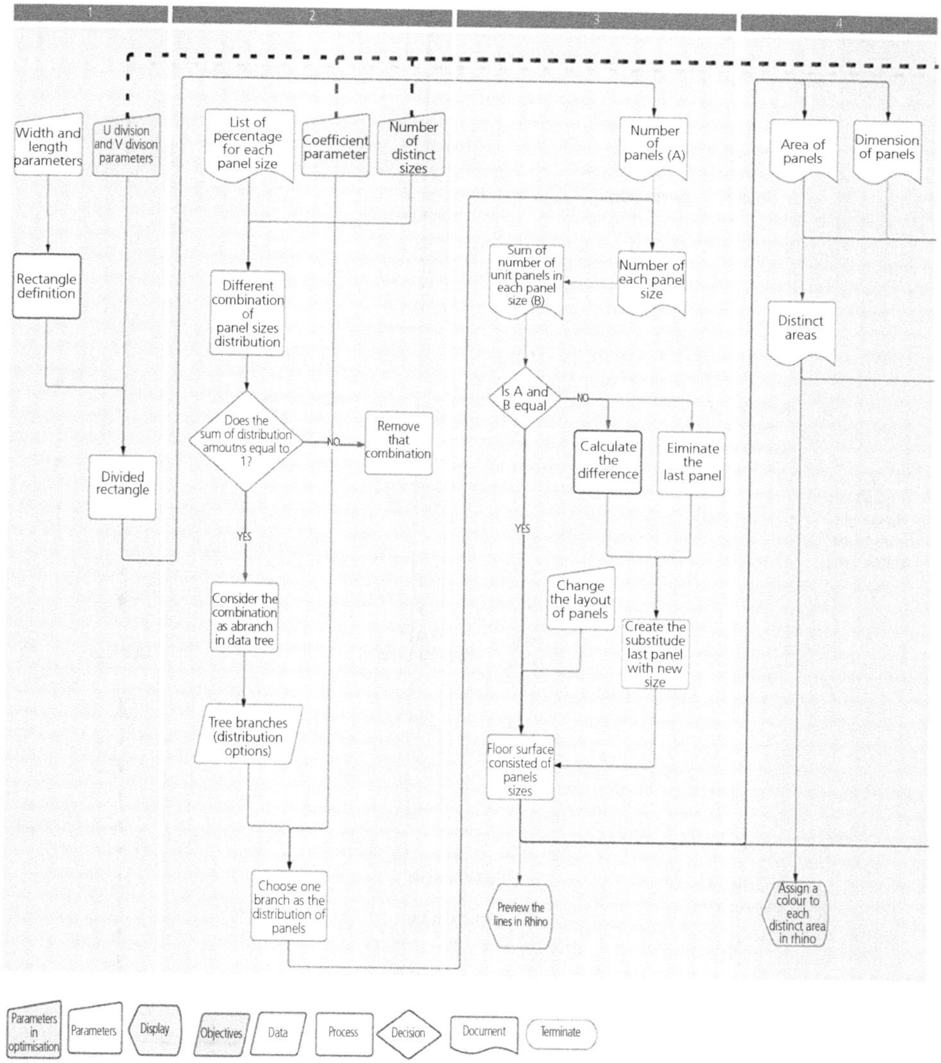

Figure 2.1 Continued

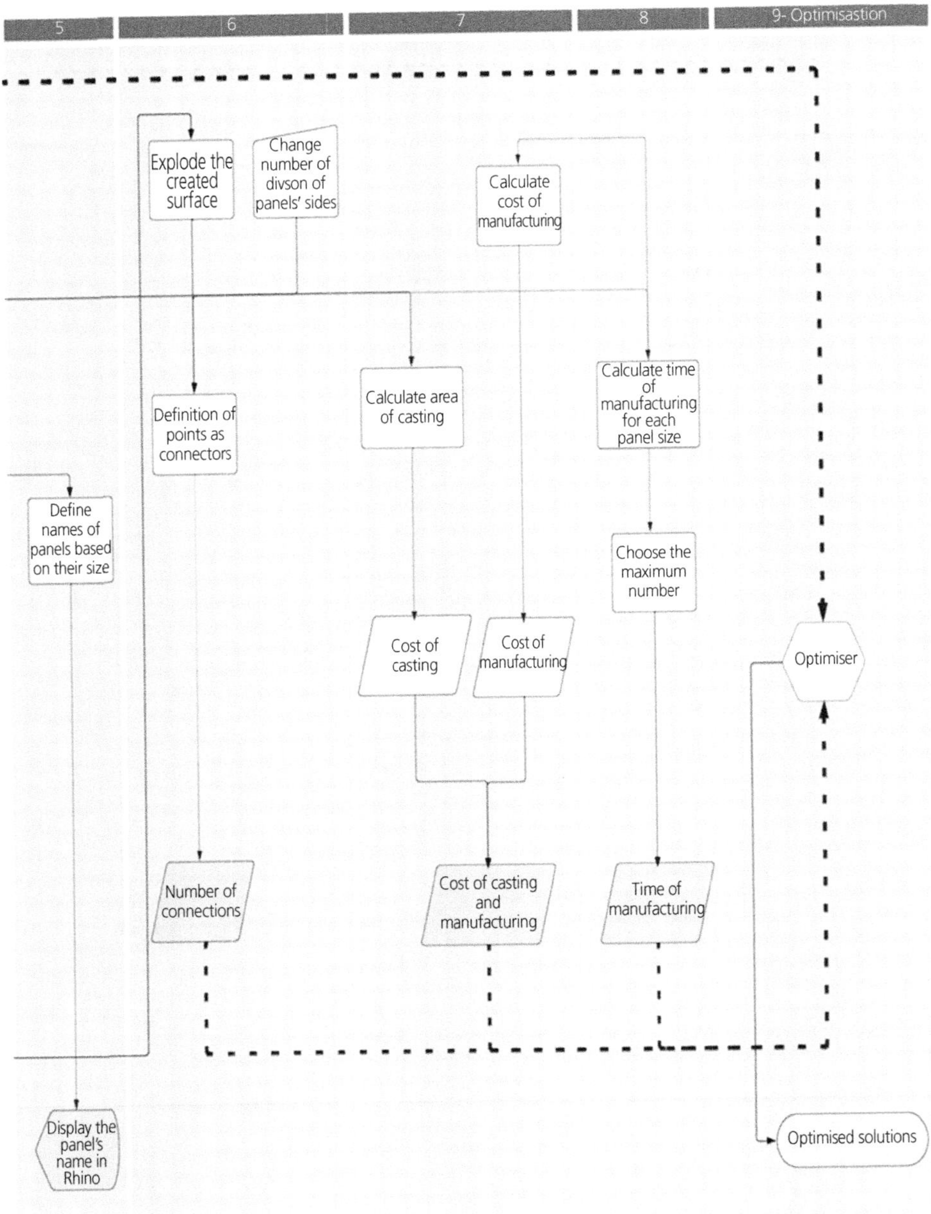

Source: Authors' creation

This process integrates parametric design flexibility with multi-objective optimisation, ensuring that design variations are automatically evaluated against DfMA-oriented performance metrics.

2.3.3 Model evaluation

To evaluate the developed generative computational design model, a structured questionnaire was employed to gather expert feedback on its functionality, usability and capability. The use of a questionnaire was justified by the need to obtain both quantitative and qualitative assessments from specialists who could provide informed judgments on the model's performance, interpretability and practical relevance within multi-objective optimisation for modular construction.

A five-point Likert scale (ranging from 1 – strongly disagree to 5 – strongly agree) was used alongside multiple-choice and open-ended questions. This combination allowed for a comprehensive evaluation by quantifying expert perceptions while also capturing detailed qualitative insights and improvement suggestions. The questions covered key aspects such as model accuracy, computational efficiency, design flexibility, user-friendliness and potential applicability in real-world modular construction scenarios.

The sample comprised six experts ($n = 6$), selected through purposive sampling based on their expertise in relevant domains. Four participants were academics – lecturers or PhD candidates – specialising in parametric design, digital construction and optimisation methods, while two participants were professionals from architectural and construction companies involved in off-site and prefabrication projects. Their ages ranged between 27 and 33 years and all held postgraduate or doctoral qualifications. This diverse mix of academic and industry perspectives ensured balanced and critical evaluation of the model.

For the quantitative analysis, mean scores were computed to summarise the central tendency of responses for each evaluation criterion. Each Likert-scale response was assigned a numerical value from one to five, and the average mean for each question was derived by summing all responses and dividing by the number of participants ($n = 6$).

2.4. Results and discussion

2.4.1 Optimisation performance

A population size of 100 and a maximum of ten generations were defined in both Wallacei and Octopus optimisers to explore the design solution space effectively. In this context, population refers to the total number of design alternatives (individuals) generated and evaluated in each iteration (generation) of the optimisation process. Each individual represents a unique combination of the model's decision variables – such as the number of panel divisions, panel sizes and distribution ratios – evaluated against the defined objectives of cost, time and assembly efficiency. Through iterative evolution across generations, the optimisers identify improved solutions based on the performance of previous populations.

From the final generation, four optimised solutions were selected as representative outcomes: those yielding the fewest connectors, lowest cost, shortest construction time and best overall balance among competing objectives as shown in Table 2.2.

Table 2.2 Performance of optimised results

Optimisation direction	Construction time (min)	Construction cost (£)	Number of connections	Figure
Construction time	1032	9126	119	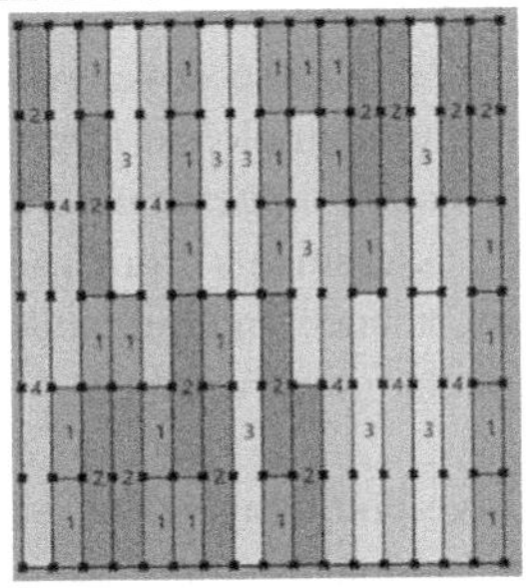
Construction cost	1240	9171	126	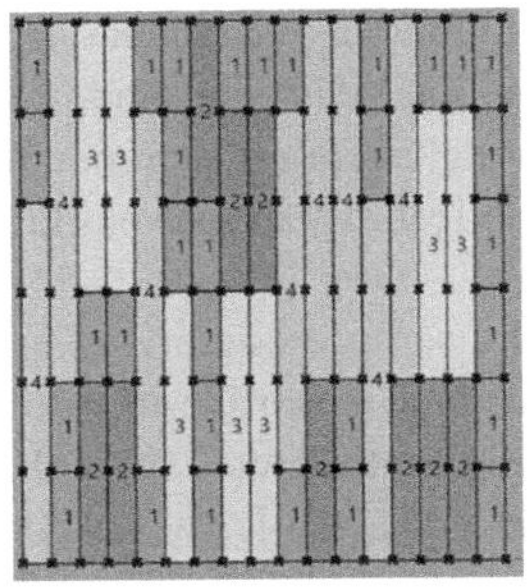
Number of connections	2190	9181	42	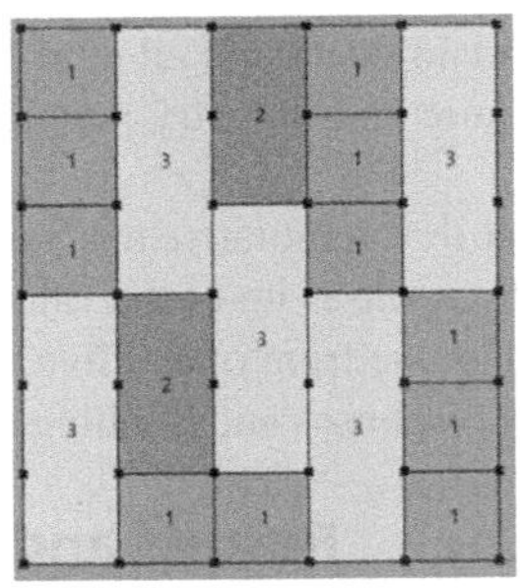
Balanced	1540	9160	56	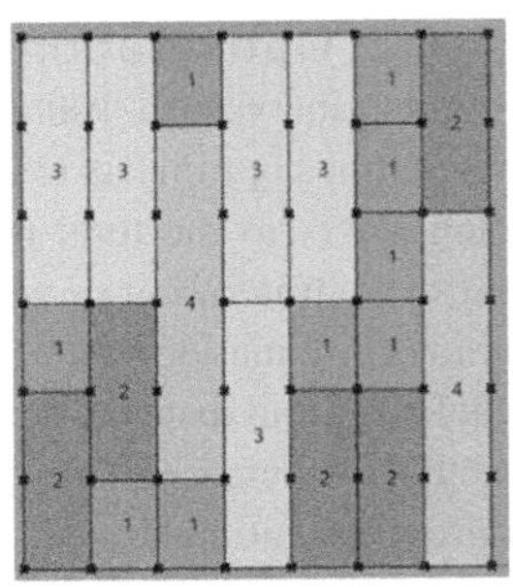

Source: Authors' creation

2.4.2 Questionnaire analysis

Table 2.3 presents the options related to functionality, usability and capability alongside the scores received on a Likert scale from the experts.

2.4.3 Discussion

This study responds to the growing demand for a structured and automated approach to optimising panelised floor design in MC, a key subset of MMC known for its efficiency and sustainability benefits (Homes England, 2024; Thurairajah *et al.*, 2023). While earlier studies primarily focused on geometric or layout-based optimisation (Deng *et al.*, 2021; Dong *et al.*, 2023; Wei *et al.*, 2022; Wu *et al.*, 2022), this research extends the scope by integrating DfMA principles into the generative computational process to address real-world production and assembly considerations. The developed framework thus bridges the gap identified in the literature, where automated optimisation rarely incorporated practical DfMA constraints such as constructability, transport limitations and onsite assembly efficiency (Widanage and Kim, 2024; Yuan *et al.*, 2018).

The findings corroborate and expand upon prior works that highlighted the critical role of design decisions in modular construction performance (Liu *et al.*, 2025; Pibal *et al.*, 2022). By adopting a two-step methodology – parametric modelling followed by multi-objective optimisation – the study aligns with computational approaches proposed by Deng *et al.* (2021) for façade panel optimisation and Wu *et al.* (2022) for tiling layouts but advances these by incorporating *degrees of modularisation* and *panel configuration diversity* as decision variables. This approach operationalises the theoretical DfMA principles discussed by Gao *et al.* (2019) and Tan *et al.* (2020), particularly those advocating design simplification, component reduction and efficient assembly.

The integration of standardisation, mass customisation and platform thinking in the model reflects key modular construction strategies discussed by Nahmens (2007), Da Rocha *et al.* (2015) and the Construction Innovation Hub (2023). By using two, three, or four panel sizes as repeatable multiples, the model supports platform-based modularisation that balances manufacturing efficiency with design flexibility – addressing the 'mass customisation' challenge noted in earlier studies (Cao *et al.*, 2022; Pine, 1993). Unlike traditional manual design approaches (Yuan *et al.*, 2018), the Grasshopper-based model enables systematic evaluation of modular configurations while ensuring compliance with assumed DfMA parameters such as machine limits, crane capacities and transportation constraints (Li and Lachmayer, 2018).

The comparative optimisation results between Octopus and Wallacei illustrate nuanced trade-offs between time, cost and connection efficiency – an advancement over previous studies that focused on single-objective optimisation (Wei *et al.*, 2022). Specifically, Octopus provided more balanced multi-objective solutions, while Wallacei prioritised time reduction. This aligns with Gerber and Lin (2013), who emphasised that evolutionary solvers could support integrated design decision-making in complex systems.

Expert evaluation outcomes further validated the model's functionality and usability, with mean Likert scores above 4 for most parameters, indicating strong agreement on its practical potential. Similar to observations by Ghannad and Lee (2022) and Pibal *et al.* (2022), experts

Table 2.3 Expert evaluation results

Item	Mean Score	Rating
Functionality		
Applicable to the modular construction Industry	4.33	Agree
Provides a sufficient level of detail for a modular construction project	4.00	Agree
Capable of integrating with BIM-based tools	4.50	Agree
Adjusts well to changes in project parameters or requirements	4.50	Agree
Encompasses all the necessary constraints related to DfMA	3.00	Neutral
Model's graphical representations and visuals are clear and comprehensible	4.67	Agree
Useability		
Model's interface is easy to use	4.00	Agree
Responds quickly to user inputs or changes in parameters	4.50	Agree
Adjusting or modifying parameters within the model is easy	4.83	Agree
Reliable in delivering accurate and consistent results	4.40	Agree
Capability		
Helps to reduce the cost and time of manufacturing of modular construction	4.50	Agree
Helps sustainability by decreasing the amount of casting	4.50	Agree
Streamlines the design process and reduces unnecessary efforts	4.67	Agree
Facilitates innovation in design of modular construction	4.33	Agree
Enhances the accuracy of the design of modular construction	3.66	Neutral

Scoring scale:
$1 \leq$ Mean Score $< 2 \rightarrow$ Strongly Disagree
$2 \leq$ Mean Score $< 3 \rightarrow$ Disagree
$3 \leq$ Mean Score $< 4 \rightarrow$ Neutral
$4 \leq$ Mean Score $< 5 \rightarrow$ Agree
Mean Score $= 5 \rightarrow$ Strongly Agree

Source: Authors' creation

recognised the value of algorithm-aided design frameworks in improving modular construction efficiency, while also noting limitations related to model scalability and standards compliance. The identification of improvement areas – including performance validation, user experience and integration with BIM environments – supports previous calls for digital standardisation and interoperability within MC (Gao *et al.*, 2018; Montazeri *et al.*, 2024).

In addressing the research question '*How can a novel computational design methodology optimise prefabricated panelised floors in modular construction while incorporating DfMA principles?*', this study demonstrates a practical and replicable workflow that directly tackles the literature gap surrounding automated DfMA-based optimisation of prefabricated panels. By focusing on panel-level modularisation, rather than overall building form, it complements earlier geometric optimisation works (Deng *et al.*, 2021; Wei *et al.*, 2022) and introduces a new layer of performance evaluation integrating cost, time and connector efficiency.

Nevertheless, consistent with findings by Kamali and Hewage (2016) and Kovacic and Zoller (2015), the research acknowledges limitations such as simplified structural assumptions, hypothetical data use and exclusion of environmental parameters. Future studies should therefore expand on these findings by validating models with real project data, integrating carbon assessment metrics and aligning with international construction standards. Furthermore, cross-platform integration with tools such as Dynamo and Revit could enhance industry applicability, advancing the vision of fully automated, sustainable and DfMA-compliant modular construction systems.

In summary, this study builds on the theoretical foundations of DfMA and generative design and provides an applied, data-driven methodology for modular floor optimisation. It contributes to closing the gap between computational research and construction practice by embedding manufacturability and assembly knowledge into the early design stage, thereby promoting efficiency, cost-effectiveness and design innovation in modular construction.

2.5. Implications for practice and research

This research significantly enriches the prevailing understanding of MC design by presenting a novel computational generative design methodology that bridges the gap between conceptual modelling and practical manufacturability. Unlike earlier studies that predominantly addressed architectural form, layout optimisation, or structural configuration (e.g. Deng *et al.*, 2021; Dong *et al.*, 2023; Wei *et al.*, 2022), this study extends optimisation to the realms of manufacturing and assembly, aligning closely with DfMA principles (Gao *et al.*, 2019; Lu *et al.*, 2021). By embedding evolutionary solvers such as Octopus and Wallacei within a DfMA-oriented framework, it represents one of the first systematic efforts to integrate multi-objective optimisation with modular design decision-making, thus contributing a new methodological and theoretical dimension to the field.

From a pragmatic standpoint, this research offers substantial implications for the modular construction sector, particularly within prefabricated floor panel design and production. The proposed methodology enables data-driven decision-making during the early design stages, allowing designers and engineers to identify optimal modularity levels, minimise connectors and balance cost, time and constructability. Such outcomes directly enhance manufacturing precision, reduce rework and material waste and improve coordination between design and production teams. These advancements translate into tangible industrial benefits, including shortened project timelines, lower production costs and improved quality control, in line with the efficiency objectives of MMC (Homes England, 2024; Thurairajah *et al.*, 2023). Furthermore, although sustainability was not the central focus of this study, the optimisation approach inherently supports environmental performance by minimising concrete usage and reducing the number of joints, contributing to lower embodied carbon and resource consumption (Pibal *et al.*, 2022; Wu *et al.*, 2022).

From a theoretical perspective, the study advances the discourse on computational design integration in MC by operationalising DfMA principles within generative design environments. It demonstrates how design automation can move beyond geometry-based optimisation toward multi-dimensional objectives that include manufacturability, assembly and performance efficiency. This positions the research at the intersection of design computation theory and construction management practice, offering a replicable model for future investigations into AI- and algorithm-driven design optimisation. Additionally, it sets the foundation for further theoretical development in performance-based design frameworks, where sustainability, resilience and flexibility can be incorporated as core optimisation objectives.

Future studies could extend this theoretical foundation by validating the model with real-world project data, embedding additional parameters such as logistics, life cycle impacts and labour productivity and integrating the workflow into BIM-based environments like Revit or Dynamo to strengthen industry applicability and academic contribution.

2.6. Conclusions

This research has addressed a significant gap in the literature by developing a novel generative computational design methodology for optimising prefabricated panelised floor systems in MC. Grounded in DfMA principles, the study proposed a parametric and multi-objective optimisation framework capable of determining the optimal level of modularisation while minimising cost, time and connector use.

The developed parametric model, built in Grasshopper, enables high flexibility in adjusting design parameters such as floor size, panel configuration and modularity levels. Through the application of two evolutionary optimisers – Octopus and Wallacei – the study generated and compared multiple design solutions. The results revealed that both optimisers effectively reduced connectors and manufacturing time, while Octopus provided more balanced trade-offs across objectives, making it more suitable for holistic design optimisation.

Expert evaluation through a structured questionnaire confirmed the model's strong performance in functionality (mean = 4.3), usability (mean = 4.2) and capability (mean = 4.1), with participants highlighting its potential to enhance design efficiency, decision-making and innovation in modular design workflows. Key findings demonstrated that the model

(a) enables automated assessment of modularity levels to balance standardisation and customisation

(*b*) reduces potential manufacturing and assembly time through optimised panel configurations
(*c*) minimises the number of connectors, supporting ease of assembly and lower material consumption
(*d*) provides a replicable framework for integrating parametric modelling with DfMA-based optimisation.

Nevertheless, limitations remain, including the simplified structural assumptions, lack of compliance with existing standards and reliance on hypothetical datasets. Future research should validate the framework using real-world project data, integrate sustainability parameters (e.g. embodied carbon) and expand optimisation criteria to include logistics and structural constraints.

Overall, this study makes a meaningful contribution to both theory and practice by advancing computational design methods for MMC, offering a replicable and scalable approach that bridges design innovation with manufacturability and assembly efficiency.

REFERENCES

Bertram N, Fuchs S, Mischke J *et al.* (2019) *Modular Construction: From Projects to Products*. McKinsey & Company, NY, USA.

Cao J, Bucher D F, Hall DM and Eggers M (2022) A graph-based approach for module library development in industrialized construction. *Computers in Industry* **139**, 10.1016/j.compind.2022.103659.

Construction Innovation Hub (2023) *Product Platform Rulebook*.

Da Rocha CG, Formoso CT and Tzortzopoulos P (2015) Adopting product modularity in house building to support mass customisation. *Sustainability (Switzerland)* **7(5)**, 10.3390/su7054919.

Deng M, Tan Y, Singh J, Joneja A and Cheng JCP (2021) A BIM-based framework for automated generation of fabrication drawings for façade panels. *Computers in Industry*, **126**, 10.1016/j.compind.2021.103395

Dong D, Zou Y, Pan H *et al.* (2023) DFMA-oriented modular and parametric design and secondary splitting of vertical PC components. *Scientific Reports* **13(1)**, 10.1038/s41598-023-30192-z

Erixon G (1998) *Modular Function Deployment*. KTH Royal Institute of Technology.

Gao S (2019) Design for manufacture and assembly in construction. *Proceedings of the Institution of Civil Engineers - Management, Procurement and Law* **172(3)**: 113–122.

Gao S, Jin R and Lu W (2019) Design for manufacture and assembly in construction: a review. *Building Research and Information* **0(0)**: 1–13, 10.1080/09613218.2019.1660608

Gao S, Low SP and Nair K (2018) Design for manufacturing and assembly (DfMA): a preliminary study of factors influencing its adoption in Singapore. *Architectural Engineering and Design Management* **14(6)**, 10.1080/17452007.2018.1502653

Gerber DJ and Lin SH (2013) Design optioneering: multi-disciplinary design optimization through parameterization, domain integration and automation of a genetic algorithm. *Advanced Engineering Informatics* **27(4)**: 610–624.

Gershenson JK, Prasad GJ and Zhang Y (2003) Product modularity: definitions and benefits. *Journal of Engineering Design* **14(3)**, 10.1080/0954482031000091068

Ghannad P and Lee YC (2022) Automated modular housing design using a module configuration algorithm and a coupled generative adversarial network (CoGAN). *Automation in Construction* **139**, 10.1016/j.autcon.2022.104234.

Ghannad P, Lee Y-C and Choi JO (2019) Investigating stakeholders' perceptions of feasibility and implications of modular construction-based post-disaster reconstruction. *Modular and Offsite Construction (MOC) Summit Proceedings*, 10.29173/mocs132.

Kamali M and Hewage K (2016) Life cycle performance of modular buildings: a critical review. *Renewable and Sustainable Energy Review* **62**: 1171–1183.

Kovacic I and Zoller V (2015) Building life cycle optimization tools for early design phases. *Energy* **92**: 409–419.

Kuzmanovska I (2019) *DfMA in Building Design and Construction: Uses and Abuses.* DFMA Forum.

Li H and Lachmayer R (2018) Generative design approach for modeling creative designs. *IOP Conference Series: Materials Science and Engineering* **408(1)**, 10.1088/1757-899X/408/1/012035.

Li Z, Shen GQ and Xue X (2014) Critical review of the research on the management of prefabricated construction. *Habitat International* **43**: 240–249, 10.1016/j.habitatint.2014.04.001.

Liu L, Tai H-W, Wang T, Qiao L and Cheng K-T (2025) Analyzing cost impacts across the entire process of prefabricated building construction. *Scientific Reports* **15(1)**: 12345.

Lu W, Tan T, Xu J *et al.* (2021) Design for manufacture and assembly (DfMA) in construction: the old and the new. *Architectural Engineering and Design Management* **17(1–2)**: 77–91, 10.1080/17452007.2020.1768505

Montazeri S, Lei Z and Odo N (2024) Design for Manufacturing and Assembly (DfMA) in construction: a holistic review of current trends and future directions. *Buildings* 14(1). Multidisciplinary Digital Publishing Institute (MDPI), 10.3390/buildings14010285

Nahmens I (2007) *Mass Customization Strategies and Their Relationship to Lean Production in the Homebuilding Industry* [University of Central Florida], http://library.ucf.edu https://doi.org/10.1061/JCCEAZ.0000639

Pibal SS, Khoss K and Kovacic I (2022) Framework of an algorithm-aided BIM approach for modular residential building information models. *International Journal of Architectural Computing* **20(4)**, 10.1177/14780771221138320

Pine BJ (1993) Mass customizing products and services. *Planning Review* **21(4)**, 10.1108/eb054420

Rankohi S, Carbone C, Iordanova I and Bourgault M (2022) Design-for-manufacturing and assembly (DfMA) for the construction industry: a review, in Al-Hussein, M. (ed.) *Proceedings of the 2022 Modular and Offsite Construction (MOC) Summit*, Edmonton, Alberta, Canada, 27–29 July, pp. 1–8

Tan T (2022) Digital-enabled Design for Manufacture and Assembly (DfMA) in offsite construction: a modularity perspective for the product and process integration. *Construction Innovation* **22(1)**: 1–20.

Tan T, Lu W, Tan G and *et al.* (2020) Construction-oriented design for manufacture and assembly guidelines. *Journal of Construction Engineering and Management* **146(8)**, 10.1061/(asce)co.1943-7862.0001877.

Tatum CB., Vanegas JA and Williams JM (1986) *Constructability Improvement Using Prefabrication, Preassembly and Modularization.*

Thurairajah N, Rathnasinghe A, Ali M and Shashwat S (2023) Unexpected challenges in the modular construction implementation: are UK contractors ready? *Sustainability (Switzerland)* **15(10)**, 10.3390/su15108105.

Wei Y, Choi H and Lei Z (2022) A generative design approach for modular construction in congested urban areas. *Smart and Sustainable Built Environment* **11(4)**, 10.1108/SASBE-04-2021-0068

Widanage C and Kim K P (2024) Integrating Design for Manufacture and Assembly (DfMA) with BIM for infrastructure. In *Automation in Construction*. Elsevier B.V, Vol. 167, 10.1016/j.autcon.2024.105705

Wu S, Zhang N, Luo X and Lu WZ (2022) Multi-objective optimization in floor tile planning: coupling BIM and parametric design. *Automation in Construction* **140**, 10.1016/j.autcon.2022.104384

Yuan Z, Sun C and Wang Y (2018) Design for manufacture and assembly-oriented parametric design of prefabricated buildings. *Automation in Construction* **88**: 13–22, 10.1016/j.autcon.2017.12.021.

Emmanuel I Daniel and Fidelis A Emuze
ISBN 978-1-83708-437-1
https://doi.org/10.1108/978-1-83708-436-420261003

Chapter 3
Modern method of construction: the impact of lean, off-site and simulation in housing delivery in the UK

Emmanuel I. Daniel[1], Ashendra Nikeshala Konara Mudiyanselage[1], Olalekan Shamsideen Oshodi[2], Esther Fasan[3] and Evette Prout[4]

[1]*University of Wolverhampton* 0000-0002-5675-1845; 0009-0003-1339-5312
[2]*Anglia Ruskin University* 0000-0001-6106-7717
[3]*Solihull College* 0000-0003-3834-7045
[4]*Kind & CO Ltd, London, United Kingdom* *0009-0005-0760-3167*

Abstract

The UK housing sector faces significant challenges, including high costs, delays, quality issues and inefficiencies in sustainability. This study examines the impact of combining lean principles, off-site construction methods and simulation (LOS) for housing delivery in the UK. Using a qualitative approach, eighteen construction stakeholders from all four nations of the United Kingdom participated in the study. Unlike prior studies that primarily focused on the theoretical integration of LOS approaches, this research delivers an empirical foundation demonstrating their combined impact on productivity, quality improvements, stakeholder involvement, shorter construction timelines, improved cost efficiency and carbon reductions supporting net-zero carbon targets in UK housing construction. Additionally, the study finds that implementing LOS encourages innovation, collaboration and more efficient processes, leading to fewer design errors, optimised supply chain logistics and higher stakeholder engagement. Despite these benefits, the study emphasises the need for cultural and operational change within the construction industry to enable broader adoption of LOS, including investing in skills development and advanced technologies. This research provides a foundation for policymakers, industry leaders and stakeholders to adopt LOS as a key strategy for delivering a standardised, collaboratively digitalised housing sector in the UK.

Keywords: construction industry; housing; lean; offsite; simulation; United Kingdom

3.1. Introduction

The availability of housing influences the well-being and quality of life of people. Studies have shown that a shortage of housing contributes to income inequality (Foster and Kleit, 2015). Rapid urbanisation has exacerbated the global housing shortage problem (Zhang, 2016). To address this problem, the government needs to develop and implement policies that improve housing supply and promote sustainable urbanisation. The housing construction industry has long faced criticism for its lack of innovative construction techniques, its limited usage of such methods and its continuous efforts to increase productivity (Hayward, 2012). In the UK, the previous government advised that for it to meet its target of delivering 300 000 houses per year, an innovative approach such as modern methods of construction (MMC) must be adopted.

MMC are innovative approaches that aim to improve the efficiency, quality and sustainability of building delivery through the integration of advanced technologies, prefabrication and process optimisation (Hwang *et al.*, 2023). MMC encompasses a range of techniques, including volumetric modular construction, panelised systems and hybrid solutions that enable significant reductions in on-site labour, construction time and waste generation (Abanda *et al.*, 2022). Furthermore, MMC contributes to addressing housing shortages and achieving net-zero carbon targets by promoting sustainable materials, circular economy principles and reduced life-cycle emissions in construction (Durdyev *et al.*, 2022).

Recently, the construction industry, though slow to change, has been adopting innovative methods to address some of these limitations and improve efficiency in the production process. Studies have explored approaches such as LOS in the industry (Daniel and Oshodi, 2022; Goh and Goh, 2019; Mostafa *et al.*, 2016). However, most of these studies are based on literature reviews – for example, Mostafa *et al.* (2016) examined lean, off-site and agile methodologies through a review. Similarly, Daniel and Oshodi (2022) reviewed the application of LOS in housing production, while Goh and Goh (2019) explored lean production-based theory simulation.

However, the majority of existing literature has remained primarily theoretical or focused on isolated elements of LOS (e.g. lean tools in off-site manufacturing or BIM-driven simulations) rather than examining their integrated deployment in UK housing. This indicates that there are limited empirical studies on the combined application of LOS to understand their impact on housing delivery in the UK context. Daniel and Oshodi (2022) also pointed out that limited attention has been given to integrating lean construction, off-site construction and simulation, highlighting the need for further research.

Given this, the current study has been designed to fill this gap by exploring the impact of the combined application of LOS in housing delivery in the UK. Studies have shown that gathering contextualised evidence would support the reception of a new approach (O'Connor *et al.* 2015). This means the evidence from this study would help the future use of LOS in the UK construction industry. The main research question is 'What is the impact of the combined application of lean, off-site and simulation in housing delivery in the UK?'

3.2. Perspectives on LOS

A substantial body of literature has been published on the outcomes of construction projects. These studies have led to calls for identifying strategies to improve project results – for example, it has been shown that implementing lean and off-site methods positively impacts cost, time and safety performance in construction projects (Mao *et al.*, 2013; Demirkesen, 2020; Bajjou and Chafi, 2021). Additionally, research has acknowledged that using simulation offers a better understanding of the construction process and highlights potential areas for improvement (Li *et al.*, 2018).

Regarding the individual impact of off-site construction on housing development, advancements in design, productivity, efficiency, building performance, cost reductions, risk mitigation and minimising negative environmental impacts (Gusmao Brissi, 2021; Sutrisnaa *et al.*, 2022; Tapia, 2023), lean construction plays a vital role in the industry economically, environmentally and socially. It achieves this by eliminating waste and delays, reducing carbon emissions, defects, additional costs, overproduction and labour involvement, while maximising overall product value and customer satisfaction (Awad *et al.*, 2021; Heravi *et al.*, 2020; Shaqour, 2022). Additionally, a key advantage of simulation, as highlighted by Jankovic *et al.* (2021), is that newly built housing can more easily achieve net-zero carbon targets by following net-zero standards from the early design stages. The literature suggests that applying LOS could enhance the outcomes of construction projects as follows.

3.2.1 Productivity

According to Sacks *et al.* (2017), lean construction transforms projects into more predictable production systems, improving both labour productivity and project delivery performance. Tools such as the Last Planner System (LPS) create more reliable short-term planning and reduce waiting/hand-off delays (Ballard and Tommelein, 2021). Value stream mapping, 5S and pull-based logistics reduce non-value activities and on-site material waste, improving output per input on tasks or work areas (Agrawal *et al.*, 2024).

When it comes to off-site construction, it is directly leading the construction projects to high productivity levels through different dimensions such as higher labour productivity with the controlled factory environments (Ehwi *et al.*, 2022), shorter onsite schedules and faster handover by reducing onsite operations and speeding up overall programme durations (Broadhead *et al.*, 2023), less rework and better quality control with standardised factory quality assurance (Broadhead *et al.*, 2023), and, finally, improved supply chain and logistics (Nesarnobari *et al.*, 2024). Apart from that, simulation is contributing to the productivity by optimising the workflow and reducing idling time, enhancing schedule reliability and reducing variability, testing resource/technology deployment, pre-project planning and decision support (Biruk and Rzepecki, 2021; Taiwo *et al.*, 2022; Wu *et al.*, 2022).

3.2.2 Sustainability

Lean construction contributes positively to sustainability in construction projects by reducing material consumption and waste, improving quality, lowering rework which directly contributes carbon emission reductions, shortening programmes with reductions in energy use in execution and by enabling strategies that support circularity by designing for reuse with off-site manufacture (Krajewska and Siewczyńska, 2025; Moradi and Sormunen, 2024; Taie *et al.*, 2024). Similarly, when considering off-site construction, it contributes to sustainability through environmental aspects such as lower carbon emissions and effective material consumption (Akinade *et al.* 2020; Li *et al.* 2023), economic aspects such as cost reductions and fewer construction defects (Hwang *et al.* 2020; Yuan *et al.* 2022), improved safety as social aspects (Hosseini *et al.*, 2020), and circularity aspects such as designed for reuse (Pan and Hon, 2021). On the other hand, simulation contributes to sustainability in construction projects by optimising resource use, minimising energy consumption and reducing environmental impacts through virtual testing of design, logistics and construction scenarios before implementation (Darko *et al.*, 2021; Lu *et al.*, 2020; Marzouk and Abdelkader, 2022).

3.2.3 Stakeholder engagement

Daniel *et al.* (2024) highlighted that the simulation is steering lean concepts in the housing construction sector in a positive direction by motivating relevant stakeholders and emphasising significant benefits such as waste minimisation through maximising material efficiency, proper project planning and speeding up construction, among others. Moreover, over the past five years, the use of a combination of off-site construction, lean construction and BIM as a simulation method has been promoted and adopted in the construction sector, offering advantages such as improvements in the project management triangle (time, cost and quality), increased involvement of stakeholders and client satisfaction, and workflow management benefits, among others (Marte Gómez *et al.*, 2021).

Studies on the application of LOS in construction projects are an emerging area of research within the field of construction management. Some researchers have investigated the application of one technique. For instance, it was shown that the application of lean (Demirkesen, 2020) or off-site (Mao *et al.*, 2013) improved project outcomes. Also, Moghadam and Al-Hussein (2013) provide evidence that the use of LOS is beneficial for construction projects. Despite the advances in knowledge, the impact of LOS on the outcomes of construction projects is not well-understood. This could be the underlying reason for its limited application in the real world. An understanding of the potential impact of LOS could encourage its adoption.

3.3. Methodological approach

This study adopted an interpretive view based on a qualitative research strategy to explore the impact of LOS applications on housing delivery in the UK. This approach was adopted as it enabled the study to gain deep insight into the phenomenon by engaging with the actual actors and the construction stakeholders in the UK. According to Maxwell (2023) and Bryman (2016), the interpretive view, which adopts a qualitative strategy, is appropriate when a study seeks to unearth an issue from the opinions of those involved in the process. The respondent demography is stated in the Table 3.1 below.

Data for the study were collected from England, Scotland, Wales and Northern Ireland, as shown in Table 3.1. Purposive sampling was used to select participants. This method was adopted to ensure that only those with relevant knowledge and experience in using LOS for

Table 3.1 Respondent demography

Participant	Position	Years of experience in housing sector	Company	Nation of operation in the UK
P01	Technical director	20	Architectural	England, Wales
P02	Contract manager	21	Contracting	Scotland
P03	Consultant	10	Consulting	England
P04	Consultant	8	Consulting	Northern Ireland, England
P05	Bid coordinator and best practice manager	6	Contracting	England
P06	Director (AMC)	28	Developer	England, Wales
P07	Consultant	25	Consulting	England
P08	Managing director	25	Contracting	Scotland
P09	Delivery project manager	19	Client	England
P10	Process improvement manager	25	Consulting	Wales
P11	Innovation programme manager	10	Client (Council)	England
P12	Project manager (OSM)	11	Design consultancy	England, Northern Ireland
P13	Manufacturing manager	12	Manufacturing and contracting	England
P14	Productivity and performance manager	30	Contracting	England, Wales, Scotland
P15	Site manager	6	Contracting	Northern Ireland,
P16	Site manager housing	8	Contracting	England, Wales, England
P17	Planned project officer	8	Client	England
P18	Quantity surveyor in housing	7	Contracting	England, Wales, Scotland, Northern Ireland

Source: Authors' creation

delivering construction projects participated in the study (Campbell *et al.*, 2020; Bryman, 2016). Participants were screened through their LinkedIn profiles and later invited to participate. They came from client organisations, property developers, consulting firms, contracting organisations and other relevant entities. The respondents held various job titles, including construction directors, innovation managers and construction managers. All participants had between five and 20 years of experience using these approaches in the delivery of construction projects in the UK.

Semi-structured interviews were used to gather evidence on the impact of LOS on housing delivery in the UK. The questions were open-ended, allowing study participants to express their views fully. According to Bryman (2016), semi-structured interviews enable two-way communication between the researcher and respondents, thereby enriching the research findings. The research instrument includes two sections: background information about the respondents and a question on how LOS is used in their projects. Ethical approval was obtained from the university and a consent form was issued to participants before the interviews.

All the interviews were conducted online by way of Microsoft Teams and were recorded, transcribed and analysed. The average interview time was 50 min and the interview continued until a saturation point was reached (Urquhart, 2013). This occurred when no new theme was observed, specifically with some respondents in this study, and the interview was stopped. The data were manually analysed using Braun and Clark's (2016) recommended six procedures for analysing qualitative data. Following the transcription and familiarisation, the data were subsequently coded based on the emerging themes. After the initial analysis, the initial themes were reviewed with the research teams to agree on the final themes.

3.4. Results and discussion

3.4.1 The impact of using LOS in housing delivery in the UK

Figure 3.1 highlights the significant impacts of using LOS in housing delivery in the UK identified.

3.4.2 Impact on time, cost and quality

The study reveals that LOS can dominate housing delivery in the UK with its positive impacts. Furthermore, when combined, these three approaches will deliver greater efficiency. One of the respondents stated that

> 'So, if you are using off-site manufacture but you are not using digital to enhance that, you are not going to achieve the same level of efficiencies as if you use digital tools to help you with that'.

While analysing the use of LOS in housing delivery in the UK, it highlighted the need to consider the effects on cost, time and quality by applying these three approaches. The study found that the final product or construction process directly influences several factors related to cost, time and quality in housing delivery with the implementation of LOS. Most importantly, one

Figure 3.1 The impacts of using LOS in UK housing delivery

Source: Authors' creation

respondent emphasised that, in some projects, customer satisfaction depends on time, cost and quality, also known as the three project management constraints. Additionally, another respondent stated that

> 'Quality is a factor. Cost, that's a debatable one now. For a net-zero product, you may pay a premium upfront, but the long-term benefits will pay for themselves. While utilising those three approaches, lean, offsite and simulation, safety should be better within the controlled environment of a factory as an essential consideration'.

Further to the discussion on lean approaches, a respondent stated that

> 'To identify how to achieve the time, cost and quality required for a selected housing project, a desk study needs to be conducted and through that, the established project goals for a particular project can be accomplished'.

Apart from that, one of the respondents highlighted an essential factor related to the key performance indicators in the construction, noting that '[t]o achieve monthly KPIs, the … safety, quality, cost, people and delivery need to be monitored; it's just a measurement tool used in the factory'.

3.4.2.1 Time

Time management has been identified as a crucial factor in ensuring the successful completion of a construction project (Shehu, 2021). Regarding the impacts of LOS on the timeline of construction projects, the respondents held different perspectives.

One respondent highlighted that time is a significant factor, as it directly impacts not only the cost factor but also on-time, successful project delivery, customer satisfaction, proper project flow and ease for all stakeholders involved. Furthermore, the respondent stated that implementing LOS, especially off-site construction, leads to faster project delivery in housing construction by streamlining construction procedures compared to traditional construction methodologies.

Furthermore, some respondents noted that when housing construction projects facilitate MMC, such as LOS, the construction time can be reduced, leading to early project completions. Similarly, Irfanullah *et al.* (2023) highlighted time as a critical factor in building construction projects and it must be appropriately managed to ensure the timely project delivery.

In addition to that, a respondent highlighted the positive impacts of implementing LOS for prototype housing, noting that

> 'especially in a case where you are dealing with, say, prototype housing, so you're delivering, say, 200 houses and all of them look the same, with lean construction you're also able to save significant amount of time on the project because all the components of the building will be manufactured similarly. You will have gone through a mass-production phase'.

3.4.2.2 Cost

According to Ajayi *et al.* (2024), cost management is vital for a building project's success and sustainability. Similarly, Okereke *et al.* (2022) stated that both the strategic performance of construction companies and their project execution are improved through efficient cost management. Therefore, when considering the cost factor for LOS implementation, it stands out as a key influencing element (Xue *et al.*, 2018). This is because the cost will also affect the final project budget. In particular, the budget is a crucial aspect of housing construction projects, as it helps keep the project on track to meet specific goals.

The study also clearly highlighted that the implementation of LOS is waving off the difficulties in traditional housing project pricing processes. As an example, consider that constructing a house without LOS usage can lead to defects due to incorrect techniques, mistakes made by inexperienced labourers and a lack of knowledge and understanding about the modern methods of construction, and it probably may be costlier to rectify those defects during or after the completion of the project (Daniel, *et al.*, 2024). One respondent observed

> 'but then it's even costlier to rectify the problems by not introducing these systems in the first place, so the likes of L&G, who supply modular housing, you know, it's a very well-known, publicised fact that they've lost tens of millions of pounds because they got it wrong. They didn't set the factory up correctly. They were using the wrong techniques and they probably weren't doing lean or simulation or anything very similar to that, and they had the wrong skill set in there doing the wrong thing'.

3.4.2.3 Quality

Quality has been identified as a critical success factor in the building construction sector at every stage of the life cycle to maintain final product quality (Wawak *et al.*, 2020; Ashokkumar, 2014). The application of LOS construction has been reported to improve the quality and efficiency of construction (Bajjou and Chafi 2018; Marte Gómez *et al.*, 2021). By adopting construction in a factory-controlled environment, the likelihood of achieving a more consistent standard and systematic quality assurance process can be increased. Regarding the impact of the quality of the projects using LOS in housing delivery, a respondent stated that

> 'if you're manufacturing in a factory and you're putting everything into a unit in a factory and you've got quality assurance in place in a factory, then it should make things a lot better on the ground. You should be able to achieve higher standards more quickly and deliver projects more efficiently'.

Furthermore, another respondent observed that 'it has been discussed that being constructed under a controlled, supervised quality cannot be taken with the single constructed building due to the complications with the human resources, such as labourers and site managers, etc'. At the same time, the study highlighted that the National House Building Council (NHBC) is only offered the option to go on-site and conduct random checks. From the foregoing, it could be argued that the quality can be monitored and controlled to a certain extent.

Nevertheless, one of the respondents highlighted a vital fact connecting building construction with car manufacturing as an excellent example stating that

> 'What you can't be doing, like you can do with a car, you get a product recall on a car or on a toy or whatever, yes, it's expensive to do that because then they've got to come back to the garage, or they've got to go back to the manufacturer and it's got to be fixed. How do you do a product recall on a house? You know, you can't just send the house back, yes?'

As a conclusion from that statement, when it comes to building construction, especially for housing, there is no option called product recall, if there is a doubt about the building quality or performance.

3.4.3 Waste reduction

Białko *et al.* (2022) highlighted that technological advancement, a necessary component of social development, has both beneficial and detrimental effects on the environment. As a result, the construction sector has become one of the biggest producers of waste and other unusable materials (Sapuay, 2016). One of the respondents stated that

> 'You do as much as you can in a controlled environment and then you bring a completed building, to a point anyway, to the site. Then you drop it in on a-, you're always going to have to put foundations in of some description. Then you should hopefully just have a much shorter period on site in the finishing phase, rather than having to build in the wet, which is always a problem in this country'.

Considering the complexity of construction projects, efficient planning and scheduling are essential to a successful project completion (Abdullahi and Tembo, 2023). The study reveals the importance of combining lean, off-site and simulation techniques together for housing projects to uplift the process efficiency.

The study reveals that to achieve process efficiency, construction issues need to be identified and categorised by the project's stages. Some respondents stated that 'I'm not sure of the issues related to design, but I think, majorly, the issue when it comes to planning is the supply chain'. To align the efficiency of a housing project, it is essential to address issues arising at each stage of the construction process.

When looking from the off-site manufacturing perspective, several building elements and components often come from different parts of the country or world – for example, the housing project may be in London, UK and the specified off-site manufactured elements may be delivered from Scotland, UK. In that scenario, the constructors need to ensure that, from the manufacturing process through delivery, procedures are properly monitored to achieve process efficiency. Furthermore, some of the respondents stated that

> 'it is quite difficult to properly ensure that the right planning is in place to ensure that all of those elements and components are delivered to site when they are needed. So, I think, when it comes to housing, really, it is an issue of supply chain and logistics. The logistics and supply chain can be quite lengthy, making it challenging to deal with'.

Furthermore, it highlights that waste can be identified at the early stages of construction using LOS and removed accordingly. As a respondent explained, the simulation will address design issues by cutting several of them, which constitute a significant part of the problems in housing delivery.

3.4.4 Process efficiency in design

The study reveals that simulation helps plan and identify design issues early and improves process efficiency by minimising errors during construction. In addition to that, some of the respondents explained that the simulation will support the process flow by combining it with the off-site.

> 'I think it's that the whole process flows through, right from pre-construction to the actual design concept art and the architect. You know, it starts with the client, it starts with the architect. It starts with us as a manufacturer, a designer, an installer'.

One of the respondents explained the impacts of using LOS through a practical scenario, sharing the experience gained from a project he was involved in.

> 'We are using Day Looks technology to conduct quality checks across all factories, as I mentioned before. So, you know, it is a massive project. I think it has got about 94 000 gigabytes of data and 3000 components. I have a statistic I can share with you. Using the clash detection systems within BIM, we successfully resolved approximately 8000 clashes, resulting in approximately seven million pounds in prevented remediation'.

Low productivity and efficiency in construction projects is a result of numerous wasteful activities during the construction process (Klosova and Kozlovská, 2021). As these wastes need to be eliminated, it is necessary to identify the non-value-added activities in housing construction. Having one respondent highlighted that the lean approaches, with the help of the off-site and simulation techniques, will help to understand the exact construction process in a better way, which can eliminate several non-value-added activities that might be involved in a particular construction project. The removal of the identified specific non-value-added activities will significantly reduce waste and project construction duration, addressing one of the prominent drawbacks of housing construction in the UK.

3.4.5 Sustainability and energy-efficient housing

The study reveals that, to achieve sustainability in the built environment, it is significant to construct buildings using energy-efficient methods. Hence, the UK government has initiated a net-zero energy-built environment, considering the environmental and climate changes due to the energy consumption not only in the UK but also in the global context (Hamilton *et al.*, 2013). Similarly, a respondent observed that 'while analysing the current issues with the ongoing housing construction projects, another major issue, of course, is energy efficiency and within that, of course, the larger umbrella of sustainability'. Furthermore, this respondent highlighted the necessity of improving the energy efficiency in the housing buildings, but at the same time, they emphasised that it is a challenging task for the stakeholders in housing construction in the UK.

At the same time, the study found that altogether combining LOS will be able to achieve efficiencies to a considerable extent. Some of the respondents suggested that addressing the current issues with the energy consumption in the housing delivery during and after the completion, which is while using the building itself, is essential – for example, one respondent stated that 'if you are just using off-site manufacture but you are not using digital to enhance that, you are not going to achieve the same level of efficiencies as if you use digital tools to help you with that'. Further evaluation of comments of one respondent shows that proper implementation of off-site construction comes with several benefits in terms of efficiency.

> 'I think one of the issues I identified earlier was in energy efficiency of buildings. And with off-site construction, the chance of you achieving a better energy-efficient building is a lot, lot higher. And that will lure and encourage people to buy into the technology even more'.

The study indicates that achieving energy efficiency in building construction requires interventions at earlier stages of the process, particularly through the use of higher-quality materials and products, which contribute to improved life-cycle performance. It also highlights that delivery and maintenance practices play a significant role in influencing energy consumption. Furthermore, the integration of BIM with off-site construction supports more effective system handover, including the incorporation of maintenance information to enhance long-term energy efficiency.

3.4.6 Innovation

The study reveals that the LOS combination will support innovations in the housing sector, as the benefits that LOS can provide, with numerous methods in detail, can overcome most limitations in housing construction. A respondent highlighted that the two approaches, lean and off-site, are quite transformative and innovative in many ways. But when considering traditional systems, the key issue which is holding back those two approaches, to begin with, is the quite competitive environment. A respondent stated that

> 'So, you know, one of the things that these approaches that we're talking about here, lean and off-site, is that idea that they solidify this integration and so on and I see how they can handle or tackle integration issues. So, a lot of the time this traditional system is still not addressing those and the system, if you like, or the model, the current model that we're using, is quite fragmented and so on'.

3.4.7 Better working conditions

In the construction industry, an employee's performance is greatly influenced by their working environment (Sunday *et al.*, 2023). When there is ongoing housing construction, working conditions have a direct impact on the quality of the final product.

> 'Whereas building houses in a factory off-site in controlled conditions, you're not exposed to the elements, you know, you've got 24/7 manufacturing if you wanted it. You've got the quality, the control, and I think that's the step-change the industry really needs to, sort of, pick up on'.

Alshebani and Wedawatta (2014) pointed out that since construction involves a lot of labour and outdoor work, it is one of the sectors most susceptible to bad weather conditions. Furthermore, the UK has recently experienced extreme climate conditions such as flooding, severe winter weather, snowfall, heavy rainfall, extremely high temperatures and heat waves (Alshebani and Wedawatta, 2014; Kendon *et al.*, 2024).

Since the housing buildings are a lifetime production, their quality needs to be properly monitored, as we discussed earlier. The study reveals that off-site construction is providing solutions to the drawbacks of housing construction as a service rather than a product. Further evaluation of the comments of several respondents indicates that off-site construction is increasingly conceptualised not merely as a product, but as a service-oriented solution that extends beyond factory production to include post-delivery processes and life-cycle considerations. Ajayi *et al.* (2019) explained that off-site construction techniques have been recognised as an appropriate approach to boosting the housing supply and improving productivity in the construction sector.

Additionally, it was identified that using the off-site technique supports tracking the building components while they are being delivered from the manufacturing factory to the site until the installation is completed (Gao *et al.*, 2024;). One respondent stated that

> 'I guess part of my role and remit, going forward with our facilities, would be to track and trace our buildings that leave the factory. So, at every stage point along a production line, we would be doing quality control checks'.

3.4.8 Collaborative working conditions

Additionally, the respondents revealed that LOS approaches support collaborative work. Rahman *et al.* (2014) noted that building projects require collaboration and those involved are realising that one essential component of a beneficial contractual relationship is knowledge and information sharing.

The study also reveals that without practising the management approach, initiating a more collaborative way of working would be helpful, as it would be a bottom-up approach that tackles not only the process and technology but also almost all areas of the housing construction sector in the UK. One respondent explained collaborative working with an example of an aircraft system.

> 'The aircraft has over 2 million parts that work together. And they're manufactured by all different companies. We make excuses in our industry, 'Oh, it's fragmented', 'different', 'different parts'. That's rubbish. Aircraft: some are in Germany, some in America, some in France, some in Japan, yes? And they manufacture their aircraft to electronic precision'.

3.5. Implications for practice and research

The research has identified key factors that influence the use of LOS in housing delivery in the UK and this has further emphasised the need to incorporate LOS approaches into housing in the UK, particularly to address the recurring challenges of cost, time, quality, efficiency and sustainability in the delivery of housing projects. The evidence from this research reveals several practical implications.

(a) *Improved quality and safety*. Off-site construction, when applied with lean construction and simulation, enhances the safety of the working environment and the quality of project output. This finding has been emphasised by Robey and Issa (2015), who stated that prefabrication with lean principles and BIM increased quality, safety of the workforce and productivity.

(b) *Sustainability and energy efficiency*. LOS contributes significantly to the UK's net-zero carbon goals in facilitating the design and production of energy-efficient housing. Furthermore, the simulation tools adopted aid the production of models that indicate the energy performance of the building in the early design stage, while lean and off-site construction help to reduce significantly the waste and environmental impact of construction (Du *et al*., 2023). Overall, this supports the UK Construction 2025 strategy.

(c) *Improved project efficiency and cost management*. The integration of LOS principles facilitates a more predictable project outcome by minimising waste, enhancing accurate planning, reducing defects and rework and maximising the value of the project, which can significantly impact costs. This finding aligns with what was stated, that the combination of BIM, lean and off-site construction improves cost, time and quality.

3.5.1 Implications for research

This study recommends the following areas for further research.

(a) *Quantitative validation of lean off-site simulation benefits*. While this study provides substantial qualitative insights into the benefits of LOS, future research should employ quantitative methods to validate the key findings from the research, specifically in understanding the exact measures of the cost and time savings of LOS for different projects across the UK.

(b) *Barriers to adoption of LOS*. This research has focused on the benefits of LOS; future research should explore the obstacles to the adoption of LOS across different project delivery methods across the UK.

(c) *Integration with emerging technologies*. While the focus of this paper was not to determine the best technology to facilitate LOS, future research should explore how emerging technologies such as artificial intelligence, digital twins and Internet of Things could further advance LOS.
(d) *Post occupancy evaluation and life-cycle performance*. Future research could explore the impact of LOS on different life cycles of the project, especially post occupancy, focusing on the long-term performance of LOS housing.

3.6. Conclusions

This study aims to explore the impact of combining lean off-site and solutions in housing delivery. The study highlights the significant potential of integrating LOS approaches to transform housing delivery in the UK. The findings show that LOS collectively addresses key challenges in cost, time, quality, waste reduction and sustainability, providing a comprehensive solution to the inefficiencies in the housing construction sector. By utilising LOS, housing projects can attain better cost control, improved quality through controlled factory environments and notable reductions in waste and construction time. This integration promotes process efficiency by minimising design errors and optimising supply chain logistics, thereby improving project outcomes.

Furthermore, implementing LOS supports the UK Government's sustainability goals by promoting energy-efficient housing and advancing net-zero targets. Off-site construction and simulation's ability to improve energy performance and simplify maintenance further emphasises their significance in tackling environmental and operational issues. Beyond efficiency improvements, LOS also encourages innovation, collaboration and better working conditions, building a more resilient and flexible housing industry capable of withstanding harsh weather and meeting changing stakeholder needs.

Despite its promise, this study emphasises that the successful adoption of LOS requires following the recommendations.

- A shift in industry practices is required, including investing in skills, technology and collaborative frameworks.
- The findings offer a solid empirical foundation for policymakers, industry leaders and stakeholders to promote LOS implementation and its potential to achieve the UK's ambitious housing delivery goals.

As the construction industry continues to evolve, the integration of LOS proves to be an essential strategy for ensuring sustainable, efficient and high-quality housing delivery in the UK. This study is limited to eighteen qualitative interviews; therefore, the results cannot be generalised to the entire population.

REFERENCES

Abanda FH, Tah JHM and Manjia MB (2022) Modern methods of construction: a review and future trends. *Buildings* **12(5)**: 584, 10.3390/buildings12050584.

Agrawal AK, Zou Y, Chen L, Abdelmegid MA and González VA (2024) Moving toward lean construction through automation: a systematic review. *Developments in the Built Environment* **18**: 100419, 10.1016/j.dibe.2024.100419.

Akinade OO, Oyedele LO, Ajayi SO *et al.* (2020). Design for deconstruction and circular economy: a pathway for offsite construction sustainability. *Journal of Cleaner Production* **258**: 120908.

Awad T, Guardiola J and Fraíz D (2021) Sustainable construction: improving productivity through lean construction. *Sustainability* **13(24)**: 13877, 10.3390/su132413877.

Bajjou MS and Chafi A (2018) Lean construction implementation in the Moroccan construction industry: awareness, benefits and barriers. *Journal of Engineering, Design and Technology* **16(4)**: 533–556.

Biruk S and Rzepecki Ł (2021) A simulation model of construction projects executed in random conditions with the overlapping construction works. *Sustainability* **13(11)**: 5795, 10.3390/su13115795.

Broadhead JP, Daniel EI, Oshodi O and Ahmed S (2023) Exploring offsite construction for the construction sector: a literature review. *Proceedings of the 31st Annual Conference of the International Group for Lean Construction (IGLC31)*, 735–745, 10.24928/2023/0132.

Daniel EI, Babalola A, Oladinrin OT *et al.* (2024) Where lean construction and offsite construction meet: a bibliographic scientometric analysis. *Construction Innovation*, 10.1108/CI-08-2023-0210.

Daniel EI, Konara AN and Oshodi O (2024) Exploring the use of lean, offsite and simulation in housing delivery: evidence from the UK. *Buildings* **14**: 1642, 10.3390/buildings14061642.

Darko A, Chan APC and Adabre MA (2021) Artificial intelligence and digital tools for sustainable construction: a systematic review. *Automation in Construction* **129**: 103775, 10.1016/j.autcon.2021.103775.

Du J, Zhang J, Castro-Lacouture D and Hu Y (2023) Lean manufacturing applications in prefabricated construction projects. *Automation in Construction* **150**: 104790.

Durdyev S, Zavadskas EK, Thurnell D, Banaitis A and Ihtiyar A (2022) Sustainable construction through modern methods: a critical review. *Sustainability* **14(3)**: 1457, 10.3390/su14031457.

Foster TB and Kleit RG (2015) The changing relationship between housing and inequality 1980–2010. *Housing Policy Debate* **25(1)**: 16–40.

Gao MY, Han J, Yang Y *et al.* (2024) BIM-based and IoT-driven smart tracking for precast construction dynamic scheduling. *Journal of Construction Engineering and Management* **150(9)**: 04024117.

Goh M and Goh YM (2019) Lean production theory-based simulation of modular construction processes. *Automation in Construction* **101**: 227–244, 10.1016/j.autcon.2018.12.017.

Gusmao Brissi S, Debs L and Elwakil E (2021) A review on the factors affecting the use of offsite construction in multifamily housing in the United States. *Buildings* **11(1)**: 5, 10.3390/buildings11010005.

Hayward D (2012) Housing construction industry, competition and regulation. *International Encyclopedia of Housing and Home*: 395–403, 10.1016/B978-0-08-047163-1.00229-0.

Heravi G, Rostami M and Kebria MF (2020) Energy consumption and carbon emissions assessment of integrated production and erection of buildings' pre-fabricated steel frames using lean techniques. *Journal of Cleaner Production* **253**: 120045, 10.1016/j.jclepro.2020.120045.

Hosseini MR, Chileshe N and Banihashemi S (2020) Modular construction and occupational health: systematic literature review. *Construction Management and Economics* **38(11)**: 979–995.

Hwang BG, Shan M and Looi KY (2023). Enhancing productivity and sustainability through modern methods of construction: a critical review. *Journal of Cleaner Production* **396**: 136497, 10.1016/j.jclepro.2023.136497.

Hwang BG, Zhao X and See YL (2020). Benefits and challenges of modular integrated construction: sustainability perspective. *Journal of Building Engineering* **32**: 101837.

Jankovic L, Bharadwaj P and Carta S (2021) How can UK housing projects be brought in line with net-zero carbon emission targets? *Frontier of Built Environment* **7**, 10.3389/fbuil.2021.754733.

Kendon M, Doherty A, Hollis D *et al.* (2024). State of the UK Climate 2023. *International Journal of Climatology* **44(S1)**: 1–117, 10.1002/joc.8553.

Krajewska A and Siewczyńska M (2025) Circular economy in the construction sector in materials, processes and case studies: research review. *Sustainability* **17(15)**: 7029, 10.3390/su17157029.

Li J, Zhao X and Xu P (2023). Reused modular steel frames for sustainable housing: a life-cycle carbon evaluation. *Journal of Cleaner Production* **407**: 136791.

Lu Y, Wu Z, Chang R and Li Y (2020) Building information modeling (BIM) and sustainability: simulation-based optimization for energy-efficient building design. *Automation in Construction* **118**: 103309, 10.1016/j.autcon.2020.103309.

Marte Gómez JA, Daniel EI, Fang Y, Oloke D and Gyoh L (2021) Implementation of BIM and Lean construction in offsite housing construction: evidence from the UK, Proc. 29th Annual Conference of the International Group for Lean Construction (IGLC29) (Alarcon LF and González VA (eds)), Lima, Peru, pp. 955–964, 10.24928/2021/0122.

Marzouk M and Abdelkader E (2022) Simulation-based approach for assessing environmental performance and sustainability of construction operations. *Journal of Cleaner Production* **362**: 132336, 10.1016/j.jclepro.2022.132336.

Moradi S and Sormunen P (2024) Integrating lean construction with BIM and sustainability: a comparative study of challenges, enablers, techniques and benefits. *Construction Innovation: Information Process Management* **24(7)**: 188–203, 10.1108/CI-02-2023-0023.

Mostafa S, Chileshe N and Abdelhamid T (2016) Lean and agile integration within offsite construction using discrete event simulation: a systematic literature review. *Construction Innovation* **16(4)**: 483–525, 10.1108/CI-09-2014-0043.

Nesarnobari S, Shahzad WM, Babaeian Jelodar M and Sutrisna M (2024) Offsite construction supply chain management: a scientometric analysis and systematic literature review. *Architectural Engineering and Design Management* **21(3)**: 467–490, 10.1080/17452007.2024.2381497.

O'Connor JT, O'Brien WJ and Choi JO (2015) Industrial project execution planning: modularization versus stick-built. *Practice Periodical on Structural Design and Construction* **21(1)**, 10.1061/(ASCE)SC.1943-5576.0000270.

Pan W and Hon CK (2021). Circular economy potential in modular construction: opportunities for material reuse and resource efficiency. *Resources, Conservation and Recycling* **170**: 105610.

Sacks R, Koskela L, Dave B and Owen R (2017) Interaction of lean and building information modeling in construction. *Journal of Construction Engineering and Management* **143(5)**: 04017009.

Shaqour EN (2022) The impact of adopting lean construction in Egypt: level of knowledge, application and benefits. *Ain Shams Engineering Journal* **13(2)**: 101551, 10.1016/j.asej.2021.07.005.

Sutrisnaa M, Ramnauthb V and Zaman A (2022) Towards adopting off-site construction in housing sectors as a potential source of competitive advantage for builders. *Architectural Engineering and Design Management* **18(3)**: 165–183, 10.1080/17452007.2020.1807306.

Taie AM, Schwerha DJ, Son S, Sullivan G and Ricks M (2024) Exploring the use of lean construction in unusual work conditions: Iraq 2005-2017. *Lean Construction Journal* **2024**: 108–129, 10.60164/hwbzk23xc.

Taiwo R, Hussein M and Zayed T (2022) An integrated approach of simulation and regression analysis for assessing productivity in modular integrated construction projects. *Buildings* **12(11)**: 2018, 10.3390/buildings12112018.

Wu M, Lin J and Zhang X (2022) How human-robot collaboration impacts construction productivity: an agent-based multi-fidelity modeling approach, https://arxiv.org/abs/2203.02393.

Xue H, Zhang S, Su, Y and Wu Z (2018) Capital cost optimization for prefabrication: a factor analysis evaluation model. *Sustainability* **10(1)**: 159.

Yuan Z, Shen L and Wang J (2022) Economic and environmental benefits of prefabrication: evidence from Chinese residential projects. *Journal of Cleaner Production* **367**: 132986.

Zhang XQ (2016) The trends, promises and challenges of urbanisation in the world. *Habitat international* **54**: 241–252.

Emmanuel I Daniel and Fidelis A Emuze
ISBN 978-1-83708-437-1
https://doi.org/10.1108/978-1-83708-436-420261005

Chapter 4
Factors enabling the adoption of modern methods of construction: perspectives of manufacturers and developers

Abimbola Windapo

University of Cape Town, Cape Town, South Africa 0000-0002-3162-6546

Abstract

Low productivity, inefficiency, poor quality and time overruns in conventional construction have stimulated interest in modern methods of construction (MMC), particularly modular and offsite systems. Despite substantial evidence of the benefits of modular and off-site construction, industry-wide adoption, particularly in developing contexts such as South Africa, remains limited, and the factors enabling their broader adoption are not well understood. This chapter investigates the factors driving the adoption of MMC in the construction industry from the perspectives of both manufacturers and developers. A qualitative multiple-case study design was used, drawing on literature, site observations, project documentation and interviews with five representatives of four Agrément-certified MMC systems. The data collected identifies a set of interrelated enablers, including procurement and project delivery models that integrate manufacturers early into the housing delivery process; supportive and predictable regulatory and certification frameworks; access to finance; scalable and collaborative business models; targeted skills development; and the strategic use of digital technologies such as BIM. The chapter argues that changes to the market structure, conventional regulations and support from financial institutions will enable a wider adoption and delivery of MMC in the construction industry.

Keywords: design and build; digital technology; modular construction; offsite construction; regulation; scalability

4.1. Introduction

The construction industry is one of the major pillars contributing to the gross domestic product (GDP) of countries worldwide and thus to economic growth. The South African construction industry, like many globally, contends with low productivity, time and budget overruns, and insufficient housing stock (Ameh and Osegbo, 2011; Hassan *et al*., 2005; Hiyassat *et al*., 2016). Over the years, traditional methods of building, primarily cast-in-situ construction, have dominated. These methods require the transportation of materials, labour and plant to the site, where they are converted into the final product.

Technological innovation, particularly MMC, such as modular and off-site construction, promises substantial improvement in efficiency, sustainability and quality (Windapo *et al*., 2025). Besides project-related benefits, MMC and prefabrication can create more stable employment through factory-based production. However, as framed by diffusion of innovation theory, widespread adoption of MMC in South Africa is hindered by entrenched conservative practices, lack of regulatory support and barriers inherent in the traditional procurement system (Aghimien *et al*., 2019; Chan *et al*., 2018; Windapo *et al*., 2025).

In this chapter, the term innovative building technologies (IBTs) is used broadly, with modular innovative building technologies (MIBTs) referring to modular and prefabricated systems. The chapter addresses the following research question: '*What enabling factors collectively shape the adoption, delivery and scalability of modular/prefabricated building systems in the construction industry?*' The study adopts a qualitative research approach, drawing on literature, field data, case study observations and interviews with manufacturers and developers to identify the enablers of the adoption, delivery and scalability of modular/prefabricated systems.

4.2. Perspectives and concepts

The literature review provides an overview of the state of the construction industry, focusing on productivity challenges, barriers to adopting innovative technologies such as modular and off-site construction methods and the factors that enable their use. Although the primary focus of this chapter is on enablers, the review of barriers is important for understanding the conditions under which MMC can flourish.

4.2.1 State of the construction industry

The construction industry plays a vital role in the economy of South Africa and is one of the largest contributors to economic growth and development (CIDB, 2012). While the industry has shown improvement over the years (CIDB, 2015), it continues to grapple with challenges in adopting innovative technologies that could enhance project and company performance and efficiency.

However, the construction industry is known for conservatism and resistance to change, hindering the adoption of technological innovation (Uusitalo *et al*., 2024; Xue *et al*., 2014; Zuhairy *et al*., 2015). Recent studies have highlighted the need for systemic transformation to effectively address these issues (Chen *et al*., 2022; Rinchen *et al*., 2024). Furthermore, previous research suggests that the ongoing inefficiencies in project delivery, often attributed to siloed approaches among professionals, significantly hamper the industry's progress (Molenaar *et al*., 2016).

4.2.2 Productivity challenges and innovation barriers

Innovation is the successful exploitation of new ideas and is the key business process to compete effectively in the increasingly competitive global environment (Ozorhon *et al.*, 2010). It is the creation and adoption of new knowledge to improve the value of products, processes and services. Although construction has long been a driver of economic activity, it has been criticised for its lower efficiency compared to other industries and its reluctance to innovate (Ozorhon *et al.*, 2010).

Despite its economic importance, there is a low uptake of construction innovation in the South African construction sector, largely due to risk aversion, siloed professional practices and inadequate integration between project participants (CIDB, 2012; Xue *et al.*, 2014). In addition, vested interests and end-user preference for traditional materials and methods limit acceptance of alternative technologies (van Reenen, 2025). These issues are exacerbated by legacy procurement methods such as traditional procurement, which prioritise the lowest capital cost over lifecycle efficiency, environmental impact or end-user needs.

4.2.3 Modular and off-site construction: definitions and benefits

Off-site and modular construction involves the prefabrication and preassembly of substantial building components in controlled environments (MBI, 2019; Smith, 2016), using the same materials and designed to the same codes and standards as conventionally built facilities but in about half the time. Buildings are produced in 'modules' that, when assembled on site, reflect the same design intent and specifications as traditionally built facilities (MBI, 2013). Once joined and sealed, the modules function as an integrated wall, floor and roof assembly (MBI, 2013).

Two primary categories of modular construction are permanent modular construction (PMC) and relocatable buildings (RBs) (MBI, 2013). PMC is an innovative, sustainable construction delivery method utilising off-site, lean manufacturing techniques to prefabricate single or multi-story, whole building solutions in deliverable modules (MBI, 2013). PMC buildings can be constructed of wood, steel, or concrete and manufactured in a safe, controlled setting. The structures are 80-90% complete in a factory-controlled environment before being transported and assembled on site. PMC modules can be integrated into site-built projects or stand alone as a 'turnkey' solution and can be delivered with mechanical, electrical and plumbing (MEP) fixtures and interior finishes (MBI, 2013). RBs are partially or fully assembled buildings constructed in a manufacturing facility using modular processes. RBs are not permanently affixed, but are designed to be reused and can be transported to different building sites (MBI, 2013).

Benefits of modular and off-site construction include rapid project delivery, reduced waste, improved cost prediction, enhanced quality and greater safety (Aghimien *et al.*, 2019; Wang *et al.*, 2020). In traditional on-site construction, materials and personnel must be transported between locations, which has implications for costs, wages and worker well-being. In contrast, factory-based approaches streamline production by relocating key construction activities to controlled environments – for example, modular integrated building technology (MIBT) panels are lightweight, quickly erected and require minimal wet trades, enabling construction speeds of up to five times faster than traditional masonry.

These systems also achieve high compliance with green standards (e.g. IFC EDGE), generate minimal waste and offer improved thermal performance. As a result, modular and off-site construction methods are widely applied in projects such as schools, site offices, medical clinics and sales centres, particularly where relocatable or temporary building solutions are required.

4.2.4 Barriers to adoption: the interplay of procurement and innovation

Table 4.1 summarises the key barriers to the adoption of alternative construction technologies, identified in the literature. These barriers, while negative, also point to the enabling conditions that must be created for wider MMC adoption

Empirical studies in Table 4.1 identify the following key barriers to the use of innovative construction technologies on projects.

4.2.4.1 Lack of client and consultant knowledge/acceptance

Insufficient knowledge among clients regarding the benefits of alternative technologies, are found by Bikitsha and Haupt (2011) and Vernikos *et al.* (2014) as contributing to their limited adoption on construction projects. There are usually acceptance-related challenges that developers face when selling buildings built with non-traditional methods and materials. Because developers follow market demand, lack of end-user understanding of the benefits of innovative building technologies dampens uptake and use.

4.2.4.2 Inadequate training and skills

The current model focus on recruiting unskilled labour, however due to the mobility and cyclical nature of current construction production it becomes difficult to develop and deepen skills. According to Peansupap and Walker (2005), lack of structured training is a critical barrier to the adoption of IBTs. Dupwa (2018) emphasises the need for significant investment in education to facilitate the successful adoption of alternative technologies.

4.2.4.3 Higher perceived upfront costs

Construction has traditionally been a relatively low-barrier-to-entry business (Windapo and Cattell, 2013). Raising capital, building factories and assembly lines and training staff create barriers to the adoption of IBTs. Furthermore, according to Oke *et al.* (2018) and

Table 4.1 Barriers to the use of innovative construction technologies on projects

S/No	Barriers	References
1	Lack of knowledge	Bikitsha and Haupt (2011); Vernikos *et al.* (2014)
2	Training and skills	Peansupap and Walker (2005); Dupwa (2018)
3	Costs	Oke *et al.* (2018); Tam *et al.* (2007); Hossain *et al.* (2013)
4	Legal implications	Mostafa *et al.* (2014); Oke *et al.* (2018); Chen *et al.* (2022)
5	Resistance by consultants	Oke *et al.* (2018); Gambatese and Hallowell (2011); Uusitalo *et al.* (2024)

Source: Author's creation

Tam *et al.* (2007), cost considerations, particularly high training costs and the initial expenditures tied to new technologies, are significant deterrents for many firms.

4.2.4.4 Legal and regulatory complexity

The need for regulatory reviews by professional bodies, planning authorities and funders often introduces delay and uncertainty. Gambatese and Hallowell (2011) state that strict industry standards usually hinder collaboration. Additionally, legal regulations may impede the adoption of innovative technologies (Chen *et al.*, 2022; Mostafa *et al.*, 2014). Chen *et al.* (2022) call for changes in building regulations and operating systems to accommodate new technologies better.

4.2.4.5 Professional resistance and inertia

Oke *et al.* (2018) and Uusitalo *et al.* (2024) found that a significant barrier to adopting alternative technologies is the lack of cooperation among the professional team and broader industry conservatism, which stifles innovation and adoption. Furthermore, the scope of work for some professionals will be reduced when the building elements are factory-designed and installed on-site.

These barriers highlight several critical enablers, including improved knowledge and pilot projects, targeted skills development, supportive regulatory frameworks and collaborative professional cultures. These enabling conditions are further explored below and in the empirical sections of this chapter.

4.2.5 Drivers for the adoption of innovative building technologies

This sub-section provides an overview of the main enablers of the adoption of modular innovative building technologies, grouped under business and economic, social, technological and procurement-related factors.

4.2.5.1 The business and economic environment

The current economic situation has significantly impacted the construction industry. Fewer developments are taking place because spending has been cut. What this entails is fierce competition between construction contractors who, in order to stay in business, devise creative ways to cut their costs and improve their productivity, concurrently, to secure contracts. Modular construction has been proposed as one such strategy, offering unique construction means and methods (Kanaby, 2011).

Gudgel (2011) found that the most important driver of modularisation on projects is its ability to improve productivity. Both contractors and clients indicate that improving productivity is the key driver for using modularisation. Contractors also report that using modularisation on projects provides advantages over competitors (Gudgel, 2011). In addition, clients and end-users demand higher quality at competitive costs/value for money, encouraging the use of factory-based processes where quality control checks can be performed regularly and defects minimised. This controlled environment is not possible on-site. In addition, clients exert pressure on contractors and consultants to keep the project within budget.

4.2.5.2 The social environment

Two main aspects are identified in the social environment, namely site safety, environmental impact and waste reduction.

SITE SAFETY

Much attention has been paid to improving health and safety practices in the construction industry over the years (Jatau *et al.*, 2023). The construction industry is notorious for having poor health and safety records, with accidents and deaths occurring regularly (Mohamed, 2002). Extensive efforts have been and continue to be made to reduce the risk of accidents on construction sites. However, potential dangers on site exist due to the nature of the work. Large machinery, heavy equipment and working at heights are among the potential hazards on construction sites (Lingard and Rowlinson, 2004). New, innovative construction methodologies, such as modularisation, reduce on-site work, shifting more activities to safer, controlled factory environments and thereby lowering exposure to accidents (Nadim and Goulding, 2010).

ENVIRONMENTAL IMPACT AND WASTE REDUCTION

Awareness of environmental issues and waste reduction has increased across industries, including construction. A prominent theme in construction is sustainable development. Sustainable development has been described as development which meets the needs of present generations without compromising the ability of future generations to meet their own needs (Jaillon and Poon, 2008; Pearce *et al.*, 1995). Sustainability is important for designers, engineers and others involved in the construction of new facilities because these facilities have a significant impact on the environment.

Construction activities disturb the environment by generating dust, nose, muddy site run-off and considerable amounts of waste. Specifically, conventional construction methods involve the use of substantial timber formwork and wet trades, which generate significant on-site waste (Jaillon and Poon, 2008). In contrast, off-site manufacturing facilitates reuse and recycling of waste, reduces dust and noise on site and lessens nuisance for neighbouring properties.

4.2.5.3 The technological environment

According to Schoenborn (2012), the advances of technology most associated with buildings include the use of innovative building materials, e-procurement, internet-based purchasing, 3D-computer aided design (3D-CAD), computer aided manufacturing (CAM) tools and computer-aided-engineering (CAE) simulation tools. Building information modelling (BIM), 3D-printing and artificial intelligence (AI) are gaining momentum. 3D-CAD software enables modelling with built-in databases storing product information, assembly groups and dimensions, 4D-CAD software incorporates scheduling and construction sequencing information, and 5D-CAD incorporates cost data. The use of BIM is expected to drive high levels of use of model-driven prefabrication (Gudgel, 2011).

Bady (2011) found that modular construction works well with BIM because off-site construction already integrates systems, materials and labour flow. BIM tools allow schedules and engineering data to be shared seamlessly between consultants, with joint details modelled instantly in 3D. Computer numeric control (CNC) milling machines, 2D laser cutting devices and large-scale 3D printing procedures can remove much handcraft from the construction process/production, reducing errors associated with interpreting drawings with different perspectives by the design and production teams (Schoenborn, 2012).

4.2.5.4 The procurement process and economic structure of the construction industry

The construction industry has significant interactions with other economic sectors as backward and forward linkages. Backward linkages are the relationships between inter-industry purchases and total inputs flowing into construction (materials, labour, equipment and finance), while forward linkages are the outputs flowing from construction to government and society (Wibowo, 2009). The contractor acts as the central integrator of resources. Figure 4.1 shows the areas of the traditional procurement process that change when modular/prefabricated components are used in construction. These include contracting structures, labour patterns and relationships with other industries and can be summarised as follows.

- *Contracting*. In modular construction, two main entities are involved in the production process. A main contractor (with sub-contractors) undertakes on-site works, while a modular building/IBT manufacturer/contractor is responsible for producing the

Figure 4.1 Traditional procurement process and economic structure of the construction industry.

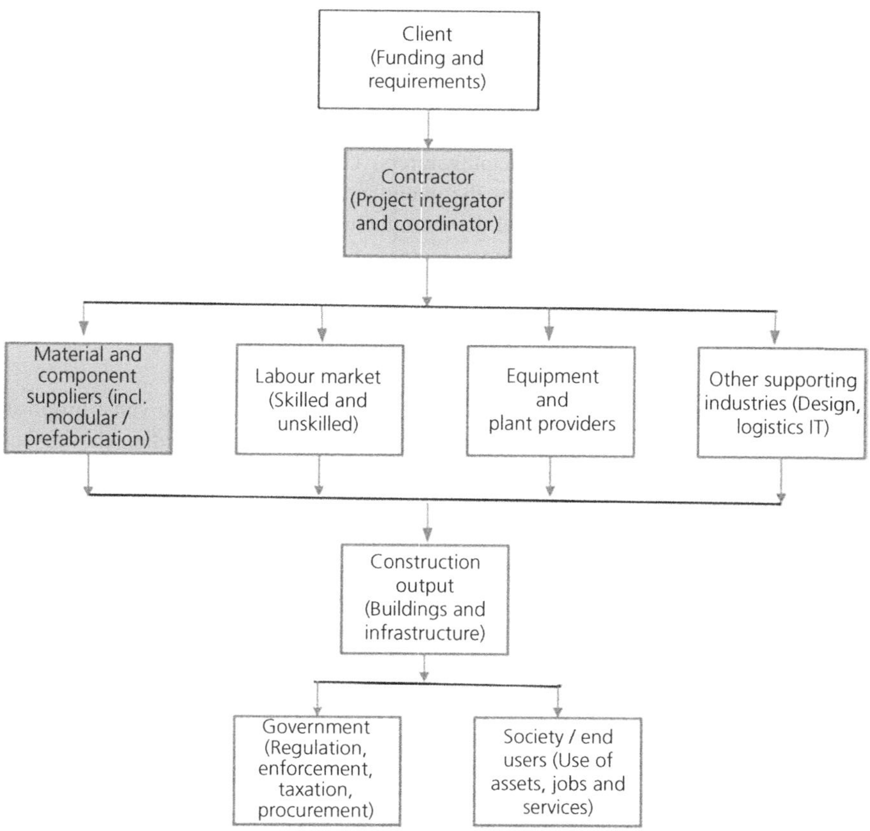

Source: Author's creation

components in the factory. Although the manufacturer performs most of the work, both actors relate with the client. They must comply with regulations, use resources (machines, materials, money and labour) and engage with other industries, such as banking and insurance. The traditional procurement method, in which the design is completed before bidding, is often unsuitable because design decisions may not align with manufacturing and assembly requirements.

- *Labour*. Modular off-site construction typically requires less labour on site and more labour in factories during production. Wibowo (2009) shows that labour accounts for approximately one-third of a contractor's budget in traditional construction. However, through modular off-site construction, this will be significantly reduced because fewer site workers are required.
- *Other industries, such as the banking and insurance industry*. These sectors will continue to play a role in the construction of modular and off-site construction technologies similar to those in traditional construction, except that modular and off-site construction technologies allow for units to be produced in larger numbers in factories. Financing and insurance arrangements, therefore, differ from the stage-based, incremental model in traditional construction. More capital is therefore required upfront to cover the high level of work-in-progress and the components produced but not yet installed.

The construction of modular off-site and innovative building technologies is inherently collaborative. In contrast to traditional construction where management and organisation occur primarily on site, in modular and off-site IBTs, a significant part of the building production process is undertaken in factories by manufacturers. The contractors work closely and collaborate with manufacturers, except that the manufacturer is also the contractor. The construction programme is prepared around the manufacturer's schedules so that the work can proceed in sequence as needed. Furthermore, increased supervision and control are exercised in the factory compared with on-site production. Also, because design changes are difficult once factory production begins, the design process becomes tightly integrated with production, requiring, to a large extent, early decision-making and client certainty.

Despite growing interest, there are gaps in knowledge demonstrating the value proposition of modular IBTs to improve their acceptance and use and in understanding the impact of procurement methods on their adoption and scalability. The traditional procurement (design–bid–build) model separates design and construction responsibility and often awards contracts to the lowest bidder, increasing perceived risk around unfamiliar technologies. In contrast, integrated models such as design and build or construction management approaches bring designers, constructors and supply chain partners together earlier, promoting collaboration, risk-sharing and the uptake of innovation (Molenaar *et al.*, 2016). Furthermore, for off-site construction/prefabrication decisions about the production process need to be taken early in the design development process, so that the project is built around that methodology to avoid redesigning.

4.3. Methodological approach

The research adopts a qualitative case study approach, which is recognised as suitable for investigating complex real-world phenomena such as the use of modular off-site innovative building technology (IBT) in the construction industry (Yin, 2014). A purposive sampling

Table 4.2 Summary of case studies

Innovative Building Technology	Technology Type	Housing Elements
MIBT panels	Lightweight interlocking panels	External/Internal walls
PERI Monolithic systems	Aluminium formwork systems; one-time pour	Integrated construction
KWIKFRAME	Welded steel wire	External/Internal walls
Saint-Gobain light-guage steel frame	Light steel framework; Saint-Gobain's fibre cement boards	External/Internal walls

Source: Author's creation

technique was employed, emphasising homogeneity among participants to maximise the depth and relevance of insights (Saunders *et al.*, 2023). The purposive sampling technique ensures respondents share core professional characteristics and enhances the validity of cross-case comparisons. All sampled participants were representatives of manufacturers specialising in IBTs, registered with Agrément South Africa. The sample included five respondents and four IBTs shown in Table 4.2, meeting the recommended minimum threshold for robust qualitative case study investigation (Creswell and Poth, 2018). Three of the investigated cases focused on walling systems, with the fourth comprising a monolithic construction method.

Data were collected using a multi-method qualitative approach, combining interviews and self-reports from key informants, focusing on perceived enablers of MMC/MIBT adoption and a review of case project documentation, building specifications and other materials provided by respondents to corroborate and extend insights from the interviews. The analysis followed a thematic approach (Braun and Clarke, 2006), suitable for identifying and interpreting recurrent patterns both within and across case studies.

The data collected was coded inductively and deductively, first to surface emergent themes and then to align and contrast them with established constructs from the literature on construction innovation, procurement and adoption of MMC. Specific steps included review of documentation, initial coding to identify themes related to enablers, procurement, technology and stakeholder perspectives; comparative analysis between cases; triangulation of perspectives across respondents and comparison to document evidence for corroboration; and synthesising and situating the findings in the literature.

4.4. Results and discussion

4.4.1 Case context

The study evaluated four types of IBT – MIBT panels, PERI Monolithic construction technique, KWIKFRAME, and Saint-Gobain's light gauge steel frame presented in the following sub-sections and shown in Figure 4.2.

4.4.1.1 MIBT panels

The MIBT panel is an industrial building technology that offers an alternative to traditional brick-and-mortar construction. This technology is characterised by the off-site fabrication

Figure 4.2 Examples of modular innovative building technologies (a) MIBT panels and (b) KWIKFRAME

(a) (b)

Sources: Courtesy of (a) MIBT, 2025 (b) KwikFrame, 2025

and onsite assembly of lightweight composite panels. The materials used in the production of MIBT panels include cement, cellulose fibre cement sheets, expanded polystyrene and additives that enhance their strength and insulation, resulting in panels that weigh 54 kg/m^2, lighter than conventional masonry walls.

The manufacturing process of MIBT panels involves mixing the raw materials, vertically pouring it into moulds and an exothermic curing phase that takes about 14 h. After curing, panels are lightly watered and left to stand upright for two days. Assembly onsite requires limited skilled labour and is often completed within 12–14 days. Its advantages include speed and efficiency, thermal insulation, and enhanced structural strength – being 60% stronger than double clay bricks and increased internal space, due to their reduced weight and design.

4.4.1.2 PERI Monolithic construction technique

The PERI Monolithic construction technique adopts an industrialised production system, utilising in-situ reinforced concrete combined with reusable aluminium formwork. This process allows for the single-pour creation of integrated structural shells – including walls, floors and slabs to form a seamless, solid unit. One key feature of this system is the use of slab drop-heads, which enable the early removal and reuse of formwork.

Among the advantages of the PERI system are speed and efficiency; it is possible to complete an entire housing unit within three days. The method enables rational thinner slab designs, reducing the overall dead load and foundation size required. The resulting surfaces are smooth and durable, and the integrated placement of window and door openings, as well as electrical and plumbing services, enhances construction precision and airtightness. The system provides savings in material and labour and supports sustainability goals through reduced carbon emissions, lower resource usage and recyclable formwork. Socially, it empowers local communities by enabling the use of mostly unskilled or semi-skilled labour, who can be easily upskilled through training.

4.4.1.3 KWIKFRAME system

The KWIKFRAME system is an innovative, non-brick prefabricated factory-made walling solution designed to provide rapid and durable construction in a variety of environments, including areas with unstable soils and extreme weather conditions. Central to its technology are 2 mm-dia. welded steel-wire panels, assembled into a rigid three-dimensional framework. These pre-manufactured panels are produced off site to standard or custom dimensions and then transported to the building site for quick, efficient assembly.

Once installed on site, the panels are filled with dry river sand or aggregate dust to provide mass and stability, after which both sides are coated with a cementitious plaster up to 80 mm thick. For enhanced thermal efficiency, polyurethane insulation is added. The advantages of the KWIKFRAME system include speed of installation, up to 70% faster than conventional brick and mortar methods. A team can erect and complete 100 m^2 of walling in 8 h. It is lightweight at only a quarter of the mass of double brick walls, therefore reducing foundation requirements. It is strong, supporting up to 17.8 t/m^2, and has good thermal and acoustic properties.

4.4.1.4 Saint-Gobain's light-gauge steel frame (LSF) and fibre cement board system

Saint-Gobain's light-gauge steel frame (LSF) is a three-dimensional skeleton of precision-manufactured, light-gauge galvanised steel that serves as the primary structural support for a range of building projects. Both the exterior and interior wall surfaces are then clad with proprietary fibre cement boards. These panels are joined, rendered or painted to achieve desired finishes, while glass wool or similar insulation materials are installed between wall layers, contributing to acoustic and thermal efficiency.

The steel frames and fibre cement boards are fabricated off site to standards and delivered ready to assemble. On site, the lightweight frames are anchored to simpler foundations than those required for brick-and-concrete construction, accelerating the build process and reducing overall resource consumption. Skilled or semi-skilled teams can fix insulation materials and fibre cement boards and apply finishes. The advantages of Saint-Gobain's LSF include its certification for fire, acoustic and durability performance by leading national authorities; it is engineered to withstand harsh weather, resist pests and maintain structural and aesthetic quality over time. The use of local manufacturing minimises transport emissions and supports regional economies, while water use and building waste are lower than those of brick-and-mortar construction. Although the material cost per m^2 is high, this is balanced by foundation savings.

4.4.1.5 Commonalities across the identified technologies

A cross-case analysis of the selected construction technologies reveals several patterns in their approach to delivering modern building solutions. Foremost among these is a shared emphasis on industrialisation and prefabrication. Across all four technologies, a significant portion of the prefabrication process is relocated from the traditional building site to controlled factory environments. This off-site manufacturing not only applies to major structural elements such as panels, steel frames, or specialised formwork, but also results in lean, largely dry onsite assembly processes.

Closely related to this is the enhanced speed and efficiency of the production process. All case studies indicate substantially reduced construction time, with some systems (MIBT panels and KWIKFRAME) achieving up to five times the speed of traditional brickwork, with the PERI Monolithic system and Saint-Gobain LSF enabling completion months ahead of conventional timelines. The industrialised nature of these methods also reduces the need for skilled labour, making workforce training and upskilling in these technologies more accessible and rapid.

A further recurring pattern is evident in materials innovation. Each system incorporates alternatives or advanced materials ranging from composite cementitious panels to modular steel or aluminium frameworks, welded steel meshes and proprietary fibre cement boards. Additionally, these technologies often integrate high-performance insulation materials, such as expanded polystyrene, polyurethane or glass wool to improve thermal and acoustic comfort.

4.4.2 Perspectives of the enablers of the adoption of off-site manufactured innovative building technologies in construction

The study sought to know the key enablers of these innovative construction systems. Information was sourced from five manufacturer representatives using a convenient sampling technique that combined interviews and self-reported questionnaires, as shown in Table 4.3.

The following subsections summarise the respondents' views on the key enablers, or, where absent, the constraints to the adoption of these innovative construction systems. For each theme, direct quotations are followed by interpretive commentary highlighting what they reveal about industry perspectives.

Table 4.3 Respondents profile

Respondent Code	Innovative Building Technology	Method of Data Collection
RP1	MIBT panels	Self-report
RP2	PERI monolithic systems	Self-report
RP3	KWIKFRAME	Interview
RP4	Multiplex	Self-report
RP5	Multiple products sales representative	Self-report

Source: Author's creation

4.4.2.1 Transaction and project delivery models

The relationship between developers/clients and IBT manufacturers exhibits both standardisation and customisation, influenced by market sector and procurement culture; as observed by one respondent.

> 'Most modular rooms are a standard size, to work for transport, but are then either standard or customised finishes and layout'.

This comment highlights how logistics and transport constraints encourage standardisation at the module level, while market expectations drive customisation of finishes. The ability to balance these considerations is an important enabler of adoption.

Similarly, another respondent highlighted that, in South Africa,

> 'the procurement model … tends to be very traditional … most contractors do not have the management skills required to orchestrate large design and build contracts, and generally the architects will dictate the design and specification in the commercial and residential world'.

In traditional procurement, project roles are rigid and fragmented, and this environment encourages risk aversion and reliance on familiar conventional materials and methods. This environment constrains the uptake of off-the-shelf standardised modular solutions and favours bespoke systems driven by architects' preferences. By contrast, where manufacturers have direct relationships with developers, such as those serving mining or long-term leasing markets, standardised modular technologies gain more traction. From an industry perspective, therefore, integrated project delivery models (e.g. design and build and construction management) that involve manufacturers early and promote shared risk-taking, collaborative problem-solving and greater willingness to pilot modular systems are seen as enablers of MMC adoption (Howell and Higgins, 1990; Molenaar *et al.*, 2016).

4.4.2.2 Financial barriers and access to lending

Financial barriers are among the most cited challenges. One respondent noted that

> 'the main barriers with respect to obtaining funding are for the manufacturers to be able to secure the sales required to underwrite the capital investment needed to prefabricate'.

The volatility of the construction sector makes planning challenging, with most manufacturers hesitant to invest in expensive innovation and instead preferring to 'stay very close to components that are compatible with traditional build solutions'. Meanwhile another respondent echoed frustration with demand, stating that the 'hit rate to date is 1% and only in student accommodation'.

This indicates that from the manufacturers' perspectives, long-term pipelines or anchor clients are key enablers, as they can underwrite investment.

Financial institutions have stringent and sometimes inconsistent requirements for non-traditional systems. According to one respondent,

> 'X bank's requirements for evaluation of alternative building technology (ABT/IBT) involve an extensive list of documentation, including full certificates SABS (South African Bureau of Standards) or Agrément-approved test results, NHBRC registration, evidence of supply capacity, durability and quality control'.

This exhaustive process can dissuade manufacturers and developers alike. This respondent further lamented, 'I don't have any of these issues from any of the other banks', pointing to the highly variable institutional practices inhibiting market growth.

These comments suggest that consistent, transparent lending criteria and recognition of certified IBTs would materially enable adoption by reducing transaction costs and uncertainty.

4.4.2.3 Business and compliance management

Regulatory approval processes are often onerous. According to one,

> 'South African standards and building codes have not maintained pace with improvement in technology, materials and design and as such innovation is avoided as regulatory approvals require more time and often negate the time savings and proposed innovation'.

This perspective indicates that outdated or ambiguous codes can neutralise the time-saving advantages of MMC/IBT, discouraging both manufactures and clients from innovating. At the same time, respondents describe rigorous internal quality control and traceability as both barriers and enablers. As another explained,

> 'We have a detailed construction methodology statement that is 94 pages. We also have a detailed quality production management system'.

From the industry viewpoint, therefore, compliance and traceability are double-edged; they enhance safety by engagement of construction professionals at every stage but can be burdensome for smaller or under-resourced manufacturers or new entrants.

4.4.2.4 Technology and component sourcing

On technological innovation, respondents note that many specialised components cannot be sourced through conventional retail channels. One respondent observed, 'I would expect that very little is procured from a regular hardware store'.

Similarly, one respondent noted that

> 'very few unconventional components are used in the building system and when they are, they would be part of the investment required to prefabricate and be sourced from local industry component manufacturers'.

These statements show that reliable local supply chains and dedicated component manufacturers are essential enablers. Where unique or patented components are required, manufacturers must invest in custom equipment and exclusive supply agreements, further increasing the importance of scale and long-term demand.

4.4.2.5 Intellectual property and licensing

Issues of intellectual property (IP), licensing and compliance frameworks also influence industry dynamics. One respndent stated that

> 'We are the third oldest ABT certificate holder for building systems in the country (South Africa)....but our biggest challenge is simply to get authorities to sit down and listen'.

This indicates that even long-established, certified systems struggle for recognition and market penetration. Another respondent offered a recommendation that 'I would suggest reaching out to the legal fraternity to answer this question', implying that navigating IP is complex and case specific. From an industry perspective, it can be deduced that clear, accessible guidance on IP and standardised licensing models would help lower transaction costs and support the wider use of proven systems.

4.4.2.6 Collaboration and industry culture

The findings show that industry culture enables the adoption of off-site innovative building technology and knowledge-sharing. As one respondent reflected,

> 'The South African construction industry fears its own shadow. The top-tier stakeholders do not exchange information readily, as this may be seen as collusive conduct, and our industry bodies...are weak'.

However, examples of successful collaboration emerge. However, examples of successful collaboration can be identified. Peikko (Finland), Cape Concrete and the architectural designer at A-I-R Inc. (USA) collaborated to deliver what is considered to be one of the first luxury homes in South Africa constructed using precast concrete elements.

These contracting accounts suggest that stronger industry platforms and collaborative networks, particularly those clarifying what constitutes lawful collaboration, are important enablers of learning and innovation diffusion.

4.4.2.7 Skills and training

Several respondents highlight the importance of skills, training and ease of adoption. One respondent noted that

> 'Building with Mi Panel can be taught to anyone, who can then build with it within a day – no special skill required. We have done this many times with the projects we have completed'.

Similarly, another described a practical approach where

> 'we train you, bring your first project to us, or you build a toilet ... or you go ... to a site where they are building and you train for four days there. But we don't take on contractors unless they are licensed'.

These comments show that manufacturers view accessible training programmes and licensing schemes as key enablers, ensuring quality while expanding the pool of competent installers. They also highlight that MMC can create opportunities for rapid upskilling of semi-skilled labour.

4.4.2.8 Social acceptance and perception

A common thread is a negative perception among end-users and professionals, often manifesting as a preference for brick-and-mortar. One observeed that 'there is a negative perception with respect to modular housing in Africa as traditional brick and mortar is favoured in contrast to modular construction'. Similarly, another attributed limited affordability and access to 'bribery and corruption', asserting that 'the moment you put money into it, it becomes an unaffordable system'.

These views underline that public awareness, trust in non-traditional systems and transparent procurement are essential enabling conditions. Without them, even technically superior systems may fail to achieve scale.

4.4.2.9 Scale and supply chain

Commercial viability depends heavily on scale, as one respondent observed that 'scale is required to be commercially viable'. For manufacturers, having a large demand is necessary for scale. Fragmented demand and limited market penetration block the realisation of economies of scale. Another detailed their industry experience.

> 'If you take all of those (56) innovative systems registered with Agrément together, we probably have anything between 1% and 4% of the market ... all 56! Cement and bricks, traditional building ... forms the rest of the market'.

From manufacturers' perspectives, this limited demand limits economies of scale and discourages investment. Enablers, therefore, include aggregated demand programmes, standardisation across public projects and long-term framework agreements.

4.4.2.10 Digital technologies

The sector is unevenly developed in terms of digital enablers such as BIM and 3D modelling. One noted that

> 'Advanced technologies like BIM, robotics, 3D modelling tools play a pivotal role in construction for efficiency and sustainability. Companies with existing technological expertise have a competitive edge, making it harder for new entrants without such capabilities'.

Another similarly affirmed that 'in commercial construction, BIM plays a significant role'. These views suggest that digital capability, particularly BIM integration, is an enabling factor, enabling better coordination, clash detection and optimisation of modular designs.

4.4.2.11 Synthesis of industry perspectives

It emerged from the findings that the use and adoption of off-site modular and prefabricated IBTs are fundamentally shaped by a complex interplay of regulatory, financial, technical and socio-cultural factors. Key enablers identified include

- the predictable certification and compliance processes
- integrated and collaborative procurement models

- access to finance aligned with prefabrication business models
- pilot projects and awareness campaigns
- investment in structured training and licensing
- supportive industry networks
- digital capability.

Conversely, adoption of the IBTs is constrained by regulatory inertia, fragmented procurement, inconsistent lending criteria, negative perception of the systems and insufficient market demand.

A strategic approach involving government support, regulatory reform and education is essential for unlocking the potential of modular innovation in the construction industry. As one respondent aptly concluded,

> 'Sustainability is a misunderstood word in the built environment ... we are the most EDGE compliant and LEED dominant building system right now ... MI Panel hits all the right notes when it comes to meeting 14 of the 17 UN development goals and is ideally suited for ESG (environmental, social, governance) investing and specifically embraces the critical factors of impact investments'.

However, according to respondents, financial institutions and many clients lack awareness of these benefits, indicating a gap between technical performance and market valuation that must be bridged through policy and procurement reform. Traditional procurement models often fail to account for ESG benefits or embodied energy savings. In contrast, responsible procurement frameworks explicitly incorporate ESG metrics into prequalification and selection, aligning project delivery with broader investor and societal goals.

4.4.3 Discussion of findings

The findings from this study support and expand upon existing knowledge about the enablers to the adoption of IBTs in the construction sector, particularly in South Africa.

4.4.3.1 Industrialisation, prefabrication and productivity

A pattern across all four case technologies – MIBT Panels, PERI Monolithic construction, KWIKFRAME and Saint-Gobain LSF – was the centrality of industrialised off-site prefabrication. Each system leverages controlled factory environments to produce major structural components, resulting in reduced site work, improved productivity and higher quality control. In line with literature (MBI, 2019; Smith, 2016), such approaches result in build times up to five times faster than traditional brickwork, with added benefits of minimal waste and reduced dependency on skilled on-site labour. This supports Gudgel's (2011) finding that productivity improvement and cost-time savings are key enablers for the adoption of modular building systems.

4.4.3.2 Innovation in materials and performance

The use of innovative materials such as composite cements, steel meshes, aluminium frameworks, high-performance insulation further differentiates these systems from traditional techniques. These materials enable thermal, structural and acoustic performance, aligned to prior studies that cite improved quality and sustainability as major benefits of modularisation (Aghimien *et al.*, 2019; Wang *et al.*, 2020). The lower environmental footprint and alignment with certification systems such as EDGE and LEED reflect the industry's increasing emphasis on sustainability (Pearce *et al.*, 1995).

4.4.3.3 Regulatory, financial and socio-cultural barriers

It emerged that regulatory, financial and socio-cultural barriers constrain the uptake of IBTs. Respondents cited the onerous approval processes that delay projects and discourage innovation, confirming previous studies on regulatory complexity as a barrier to the adoption of IBTs (Mostafa *et al.*, 2014; Chen *et al.*, 2022). Financial institutions' inconsistent requirements and a lack of recognition for lifecycle and ESG benefits limit access to capital, in line with the findings of Oke *et al.* (2018) and Tam *et al.* (2007) on cost-related barriers. Negative perceptions among professionals and end users, rooted in a preference for brick-and-mortar and resistance to unfamiliar methods, hamper demand and market penetration (Bikitsha and Haupt, 2011). Furthermore, this cultural inertia aligns with the literature on industry conservatism and the slow diffusion of innovation (Xue *et al.*, 2014; Uusitalo *et al.*, 2024).

At the same time, these constraints highlight enabling opportunities such as updating codes to be performance based, standardising bank requirements for certified IBTs, and running targeted awareness and demonstration programmes to build trust in modular solutions.

4.4.3.4 Role of procurement and collaboration

The findings support the argument (see Molenaar *et al.*, 2016) that traditional procurement processes, characterised by rigid roles and integration, impede IBT adoption, whereas integrated construction models, such as design and build or management-oriented procurement, such as construction management, with early contractor and supply-chain involvement, foster collaboration and shared risk-taking, thus enabling modular innovation. This integration is vital to overcoming the fragmented, project-by-project approach that stifles system-wide learning and scalability. These integrated approaches – design and build and management-oriented procurement – are evident in successful IBT applications and therefore serve as key enablers.

4.4.3.5 Skills, training and knowledge gaps

While the modular systems evaluated are designed to be accessible to semi-skilled or unskilled labour, there is a broader need for structured training, upskilling and knowledge dissemination about the technologies (Peansupap and Walker, 2005). Additionally, the integration of digital enablers such as BIM is currently limited to larger, well-resourced players, presenting competitive disadvantages for smaller manufacturers and impeding industry-wide uptake (Schoenborn, 2012; Gudgel, 2011). Addressing these gaps through training, accreditation and digital capacity-building would significantly enable MMC adoption.

4.4.3.6 Systemic enablers

The study's findings suggest that realising the productivity, environmental and social benefits of off-site modular innovation requires a systemic rather than project-by-project approach. The findings indicate that maximising the productivity, efficiency and sustainability gains of IBTs in construction will require a multi-pronged approach involving regulatory reform, transformation of the procurement process, education, investment in training and digital capacitation and improved demand.

According to Chen *et al.* (2022), government and industry bodies must update standards and streamline approval processes to better accommodate performance-based alternative technologies. Transitioning from traditional procurement to more integrated and collaborative procurement models can leverage the benefits of modular systems (Molenaar *et al.*, 2016). Parallel investment in training, digital capability and local supply chains will drive demand-led, scalable adoption of MMCs/IBTs

4.5. Implications of the findings

The findings have implications for how the South African construction industry and similar developing countries can unlock the potential of modular and off-site MMC. The findings suggest the need for systemic rather than project-level responses, involving coordinated changes in regulation, procurement, finance, business models, skills development and digital capability.

4.5.1 Implications for industry practice and procurement

The results suggest that procurement and project delivery models are central levers for MMC adoption. Traditional design-bid-build, with fragmented responsibilities and lowest-price tendering, is structurally misaligned with factory-based production that requires early design-freeze, integration with manufacturing processes and collaborative planning. The following are key requirements to achieve successful adoption.

- Developers and contractors interested in using MMC will need to adopt more integrated forms of delivery – such as design and build, construction management or framework agreements – that bring manufacturers into projects from the concept stage, share risks and support standardisation at the design/module level while allowing customisation at the finish level.
- Construction professionals/teams should recognise that MMC is not a substitute product but a different production logic and adjust the briefing, design development, programming and contract strategies accordingly.

At company level, the findings suggest that scale and repeatability are preconditions for commercial viability. Individual, one-off projects are unlikely to justify the capital investment required in factories, specialised equipment and trained staff. Aggregating demand through programme-based procurement (e.g. standardised schools, clinics, housing typologies) and long-term framework contracts is therefore essential to allow manufacturers to plan capacity and lower unit costs.

4.5.2 Implications for policy, regulation and certification

The perspectives of manufacturers highlight that current regulatory and certification systems, while necessary for quality assurance, are often slow, opaque and poorly aligned with innovation cycles. The following are the implications of the findings.

- Regulators and standards bodies should prioritise updating the national building regulations and standards to be more performance-based and technology-neutral. Clear technical criteria for structural, fire, acoustic and energy performance would allow certified MMC systems to be accepted more readily in approvals, rather than requiring case-by-case justifications.

- Agrément SA and the NHBRC can play a catalytic role by streamlining certification processes for proven systems and by publishing guidance notes for building control officers. Predictable and transparent procedures would reduce transaction costs and perceived risk for both innovators and clients.
- Government housing and infrastructure programmes should recognise certified MMC/IBT systems as compliant solutions and improve demand through policy targets for their use in appropriate building types. This would help create a clearer market signal and help overcome conservative defaults to masonry construction.

The findings indicate that regulatory frameworks must shift from being predominantly gate-keeping mechanisms to becoming enablers of responsive innovation.

4.5.3 Implications for financial institutions and investment

Manufacturers' experiences with banks show that inconsistent, often onerous requirements for non-traditional systems constitute a major barrier, although MMC systems provide quantifiable benefits in speed, quality and environmental performance. Key considerations are as follows.

- Banks and other financiers should develop due diligence templates for evaluating MMC/IBT systems, anchored in Agrément/NHBRC certification, track record and performance data, rather than treating each proposal as a risk.
- Lending criteria and valuation practices should incorporate life-cycle cost and ESG metrics, which recognise rapid delivery, reduced defects and superior energy performance as having the potential to increase asset value and reduce long-term credit risk.
- Public finance institutions and DFIs could pilot guarantee schemes or concessional finance for MMC projects in priority sectors such as affordable housing, thereby underwriting early pipelines and de-risking private investment in factories and tooling.

If such financial adaptation is not used, the capital-intensive nature of prefabrication will continue to discourage manufacturers from scaling and will limit the market to niche applications.

4.5.4 Implications for skills development and construction industry capacity

The case studies show that MMC systems are intentionally designed to enable semi-skilled workers to be trained quickly; however, their adoption is constrained by broader knowledge and skills gaps. Potential solutions include the following.

- The Construction Education Training Authority (CETA) and manufacturers should create structured, modular training and licensing programmes in MMC installation, quality control and maintenance. These should be integrated into existing artisan and contractor development routes.
- Universities and TVET colleges should incorporate MMC concepts, digital design (e.g. BIM) and factory-based construction into curricula for built-environment disciplines, so that future professionals are familiar with off-site production processes and logics, and can design effectively for them.

4.5.5 Implications for digital transformation

Off-site production relies on accurate, information-rich models. Therefore, digital transformation and the adoption of MMC are interrelated. As a result, policy initiatives that promote the use of BIM implementation in public projects will indirectly support MMC uptake. Key considerations include the following.

- For large contractors and manufacturers, further investment in BIM, 3D/4D/5D modelling and digital fabrication interfaces can enhance coordination between design and production, reduce errors and support mass customisation.
- For SMEs and new entrants, support from institutions such as the Construction Industry Development Board (cidb) and industry associations will help lower the barriers to assessing digital tools (e.g. open licences, shared licences, training hubs, open libraries of shared details).

4.5.6 Implications for sustainable development and housing policy

The findings have direct implications for housing and sustainability policy. The case studies demonstrate that MMC/IBTs can reduce construction time, embodied waste and in some cases, operational energy use, while creating local employment in factories. Therefore, promoters such as the Department of Human Settlements could explicitly frame MMC as a tool to address housing backlogs, improve quality and reduce environmental impacts, aligning it with national climate commitments and sustainable development goals.

The study's findings imply that the wider adoption of MMC in South Africa will not emerge spontaneously from individual firm initiatives. It requires coordinated action across regulation, procurement, finance, skills and digital infrastructure, guided by a clear policy vision that recognises modular and off-site construction as strategic enablers of a more productive, sustainable and inclusive construction sector.

4.6. Conclusions and recommendations

The study sought to identify the enabling factors that shape the adoption, delivery and scalability of off-site modular prefabricated building systems in the construction industry. Using a case study of available modular technologies and interviews in data collection, the study reveals that while IBTs offer transformative potential for the South African construction sector, including reduced production time, improved quality, enhanced sustainability and increased productivity, their adoption remains hindered by entrenched challenges.

Key among these are regulatory inertia, procurement fragmentation, limited client and professional knowledge, inconsistent access to finance and socio-cultural preference for traditional methods of construction. These barriers resonate with longstanding issues identified in the literature, such as conservatism, low uptake of innovation and rigid procurement systems. Based on these findings, the study concludes that the constraints stifle the sector's capacity to address critical priorities such as delivering affordable housing and advancing environmental sustainability using MMCs/IBTs.

The study recommends the following pathways for scaling the adoption of off-site Modular IBTs.

- Policymakers, SABs, NHBRC must update building codes and approval processes to accommodate performance-based modular solutions.
- Agrément SA must introduce expedited, transparent certification mechanisms for compliant IBTs to reduce time to market and uncertainty for innovators and clients/developers.
- Clients/developers must shift from traditional, fragmented procurement methods toward integrated approaches such as design and build or construction management.
- Clients should embed value-based criteria including ESG performance, lifecycle cost and productivity in tender evaluation to incentivise innovation.
- Manufacturers should launch targeted awareness campaigns for clients, consultants and financiers and the public to dispel the myths and highlight the track record of IBTs in speed, quality and sustainability.
- Manufacturers should also promote demonstration projects and knowledge-sharing platforms to showcase lived successes and best practice.
- The NHBRC and manufacturers should develop structured, modular upskilling programmes in IBT assembly, digital design tools (e.g. BIM) and procurement methods for the existing and future workforce.
- The cidb should support SMEs to access digital construction technologies and networking opportunities with IBT manufacturers.
- The National Government should provide targeted initiatives, such as tax breaks or grants, for early adopters, especially for projects in affordable or sustainable housing development.

Finally, there is a need for a joint roundtable meeting with all the relevant stakeholders over three to six months, during which they should sign off on some achievable goals or changes, with a commitment from the Government and stakeholders that the changes will be adopted or enacted timeously.

Acknowledgements

This research was supported by the National Research Foundation of South Africa CPRR Award (448168). The views expressed are those of the author and should not be attributed to the NRF. The author would also want to acknowledge the contribution of the respondents, especially Mr. Graham Teede OAM FAIB FNZIOB, for providing the necessary information used in the chapter.

REFERENCES

Aghimien D, Aigbavboa C and Matabane K (2019) Impediments of the fourth industrial revolution in the South African Construction Industry. In *Proceedings of the 11th International Conference on Construction in the 21st Century*.

Ameh OJ. and Osegbo EE (2011) Study of relationship between time overrun and productivity on construction sites. *International Journal of Construction Supply Chain Management* **1**: 56–67.

Bady S (2011) *Five Award-Winning Modular Innovations*. Building Design and Construction. https://www.bdcnetwork.com/five-award-winning-modular-innovations (accessed 23/04/26).

Bikitsha L and Haupt T (2011) Impact of prefabrication on construction site health and safety: Perceptions of designers and contractors. In: Proceedings of the 6th Built Environment Conference. Jere, N. (ed.). Johannesburg, South Africa, 31 July – 2 August 2011, pp. 191–204.

Braun V and Clarke V (2006) Using thematic analysis in psychology. *Qualitative Research in Psychology* **3(2)**: 77–101.

Chan APC, Darko A, Olanipekun AO and Ameyaw EE (2018) Critical barriers to green building technologies adoption in developing countries: the case of Ghana. *Journal of Cleaner Production* **172**: 1067–1079.

Chen X, Chang-Richards AY, Pelosi A *et al.* (2022) Implementation of technologies in the construction industry: a systematic review. *Engineering, Construction and Architectural Management* **29(8)**: 3181–3209.

Construction Industry Development Board (CIDB) (2012) *The Construction Industry as a Vehicle for Contractor Development and Transformation*. CIDB, Pretoria, South Africa.

Construction Industry Development Board (CIDB) (2015) *Construction Industry Indicators*. CIDB, Pretoria, South Africa.

Creswell JW and Poth CN (2018) *Qualitative Inquiry and Research Design: Choosing Among Five Approaches*, 4th ed. SAGE Publications.

Dupwa RP (2018) *Investigation of the Utilisation of Modular Construction in South Africa*. University of the Witwatersrand, Johannesburg, South Africa.

Gambatese JA and Hallowell M (2011) Enabling and measuring innovation in the construction industry. *Construction Management and Economics* **29(6)**: 553–567.

Gudgel J (2011) *Everything Old Has Become New Again – Not Your Grandpa's Prefab!*. McGraw-Hill Construction, USA.

Hassan MM, Doheim R and Bahgat M (2005) Delay causes in major Egyptian construction projects. *Journal of Construction Engineering and Management* **131(6)**: 617–623.

Hiyassat MA, Hiyari MA and Sweis GJ (2016) Factors affecting construction labour productivity: a case study of Jordan. *International Journal of Construction Management* **16(2)**: 138–149.

Hossain MK, Munns A and Rahman MM (2013) Enhancing team integration in building information modelling (BIM) projects. *BIM Management and Interoperability* **1**: 78–92.

Howell JM and Higgins CA (1990) Champions of technological innovation. *Administrative Science Quarterly* **35(2)**: 317–341.

IFC (International Finance Corporation) (various years). *Excellence in Design for Greater Efficiencies (EDGE)*. International Finance Corporation, Washington DC, USA.

Jaillon L and Poon CS (2008) Sustainable construction aspects of using prefabrication in dense urban environment: a Hong Kong case study. *Construction management and Economics* **26(9)**: 953–966.

Jatau T, Emuze F and Smallwood J (2023) *Moving the Construction Safety Climate Forward in Developing Countries*. Routledge.

Kanaby M (2011) Utilizing modular construction as a strategy. *FMI Quarterly* **1(2)**: 59–65.

Lingard H and Rowlinson S (2004) *Occupational Health and Safety in Construction Project Management*. Routledge.

MBI (2013) *Why build Modular?* [Online] http://www.modular.org/HtmlPage.aspx?name=why_modular (accessed 13/8/2013).

MBI (2019) *Permanent Modular Construction Report*. Modular Building Institute, Charlottesville, United States.

Mohamed S (2002) Safety climate in construction site environments. *Journal of Construction Engineering and Management* **128(5)**: 375–384.

Molenaar K, Leicht R, Messner J, Franz B and Esmaeili B (2016) *Maximizing Success in Integrated Projects: An Owner's Guide*. Charles Pankow Foundation. https://www.pankowfoundation.org/site/assets/files/1829/research_report_-_maximizing_success_in_integrated_projects.pdf (accessed 23/04/26)

Mostafa S, Dumrak J, Chileshe N and Zuo J (2014) Offsite manufacturing in developing countries: current situation and opportunities. In *Proceedings of the 5th International Conference on Engineering, Project and Production Management*.

Nadim W and Goulding JS (2010) Offsite production in the UK: the way forward? A UK construction industry perspective. *Construction Innovation* **10(2)**: 181–202.

Oke A, Aghimien D, Aigbavboa C and Koloko N (2018) Challenges of digital collaboration in the South African Construction Industry. In *Proceedings of the International Conference on Industrial Engineering and Operations Management*, pp. 6–8.

Ozorhon B, Abbott C, Aouad G and Powell J (2010) Innovation in construction: a project life cycle approach. *Salford Centre for Research and Innovation in the Built Environment (SCRI) Research Report* **4**: 903–1012.

Peansupap V and Walker DH (2005) Factors enabling information and communication technology diffusion and actual implementation in construction organisations. *Journal of Information Technology in Construction* **10(14)**: 193–218.

Pearce AR, Hastak M and Vanegas JA (1995, November). A decision support system for construction materials selection using sustainability as a criterion. *Proceedings of the NCSBCS Conference on Building Codes and Standards* **1(4)**.

Rinchen S, Banihashemi S and Alkilani S (2024) Driving digital transformation in construction: strategic insights into building information modelling adoption in developing countries. *Project Leadership and Society* 5: 100138.

Saunders M, Lewis P and Thornhill A (2023) *Research Methods for Business Students*. 9th edn. Pearson, Harlow, UK.

Schoenborn J (2012) *A Case Study Approach to Identifying the Constraints and Barriers to Design Innovation for Modular Construction*. Doctoral dissertation, Virginia Tech.

Smith RE (2016) *Offsite and Modular Construction Explained*. Offsite Construction Council, National Institute of Building Sciences.

Tam VWY, Tam CM, Zeng SX and Ng WCY (2007) Towards adoption of prefabrication in construction. *Building and Environment* **42(10)**: 3642–3654.

Uusitalo P, Peltokorpi A, Seppänen O and Alhava O (2024) Towards systemic transformation in the construction industry: a complex adaptive systems perspective. *Construction Innovation* **24(7)**: 341–368.

van Reenen C (2025) Innovative building technologies in the human settlements sector. Presentation made at the Online IBT Summit of the Department of Housing, 15 November, 2025 (MS Teams).

Vernikos V, Goodier C, Broyd T, Robery P and Gibb A (2014) Building information modelling and its effect on offsite in civil engineering. *Proceedings of the Institution of Civil Engineers: Management, Procurement and Law* **167**: 152–159.

Wang M, Wang C, Sepasgozar S and Zlatanova S (2020) A systematic review of digital technology adoption in offsite construction: current status and future direction towards industry 4.0. *Buildings* **10(11)**: 204.
Wibowo A (2009) *The Contribution of the Construction Industry to the Economy of Indonesia: A Systemic Approach*. Diponegoro University, Indonesia.
Windapo AO and Cattell KS (2013) The South African construction industry: perceptions of key challenges facing its performance, development and growth. *Journal of Construction in Developing Countries* **18(2)**: 65–79.
Windapo A, Lefoka M, Hall J and Narker T (2025) Use of modular and offsite technologies in construction: barriers and pathways to adoption. *CIB Conferences* **1(1)**, 177.
Xue X, Zhang R, Yang R and Dai J (2014) Innovation in construction: a critical review and future research. *International Journal of Innovation Science* **6(2)**: 111–126.
Yin RK (2014) Case Study Research: Design and Methods. 5th edn. Thousand Oaks, CA: Sage Publications.
Zuhairy M, Ibrahim H and Ismail N (2015) Relationship between innovation and organizational performance in construction industry in Malaysia. *Universal Journal of Industrial and Business Management* **3**: 87–99.

Emmanuel I Daniel and Fidelis A Emuze
ISBN 978-1-83708-437-1
https://doi.org/10.1108/978-1-83708-436-420261006

Chapter 5
Barriers to building information modelling adoption in cost management in modern methods of construction

Christiana Ada Adah[1], Henry Odeyinka[2] and Emmanuel I. Daniel[3]

[1]*Baze University, Abuja, Federal Capital Territory, Nigeria* 0000-0003-1562-7749
[2]*Obafemi Awolowo University, Ile-Ife, Nigeria* 0000-0002-6516-4167
[3]*University of Wolverhampton, Wolverhampton, United Kingdom* 0000-0002-5675-1845

Abstract

The construction of buildings using modern methods of construction (MMC) is gaining traction globally as it promotes sustainability, climate action and the circular economy for waste minimisation. One of the digital technologies MMC utilises in database management is building information modelling (BIM). Despite the significance of BIM in MMC and construction activities generally, its adoption in developing countries, including Nigeria, is reportedly low, unlike in developed countries. Studies show that some cost management firms (CMFs) are already adopting BIM, but its adoption is fraught with numerous challenges. Using a Likert-type survey instrument, the study utilised a two-round Delphi technique to obtain the opinions of experts from 18 CMFs that have adopted BIM on the barriers to BIM adoption among CMFs. The experts' responses were analysed using descriptive and inferential statistical tools. The study found that the prominent barriers to BIM adoption among CMFs are the cost of BIM software and setup, the cost of BIM training, the lack of adequately trained BIM experts, inadequate BIM education and the lack of clear BIM benefits evaluation. The study will assist both professional and regulatory bodies of cost management practice in devising measures to mitigate the barriers to BIM adoption among CMFs.

Keywords: adoption; barriers; building information modelling; cost management firms; digital technologies; modern methods of construction

5.1. Introduction

Growing concerns about sustainability, alternative energy and lifestyle changes have led to an increase in buildings utilising MMC (Sánchez-Garrido *et al.*, 2023). MMC are off-site manufacturing and on-site techniques that utilise innovative construction processes to deliver quality and sustainable housing (Construction Industry Federation, 2021; Sánchez-Garrido *et al.*, 2023). Under MMC, construction often takes place in a factory setting, on-site and in 3D units or modules (Sweeney, 2024). MMC promotes sustainability, climate action and the circular economy to minimise waste (Construction Industry Federation, 2021; Sweeney, 2024).

Three key elements that drive MMC are digitisation, manufacturing and performance (Davies, 2018). Digitisation is reported to enhance the production processes of MMC by enabling precision assembly and creating a dynamic database that spans the entire life cycle of a building. The continuous monitoring of performance makes digitisation an essential component of MMC (Davies, 2018). Moreover, Davies (2018) emphasises that performance is accurately monitored and analysed through digitisation and manufacturing processes. One of the digital technologies and innovations that MMC leverages on is BIM (CIF, 2021; Davies, 2018).

BIM and other digital tools such as property technology (PropTech), big data, artificial intelligence (AI), the Internet of Things (IoT), remote inspection, robotics and data analytics, allow for performance tracking throughout the lifespan of the building, creating a continuous feedback loop that involves design, manufacturing, construction, management and, ultimately, feeding back into the design process (Davies, 2018). BIM enhances MMC processes, serving as a database for MMC, tracking activities such as design and quality control and evaluating the performance of buildings throughout their life cycle. BIM also simplifies the integration of modular units, delivery systems and personnel. In MMC, BIM is utilised to assist with clash detection, helping to identify potential dimensional issues and the use of tolerances to control critical features in an assembly (Davies, 2018).

BIM has transformed construction business practices globally, transformed the traditional professional boundaries including those of cost managers within the architectural, engineering and construction (AEC) industry, as BIM positively influences every stage of a project, including the stakeholders involved, with outcomes including project delivery and organisational performance (Abubakar *et al.*, 2014; Adesi *et al.*, 2018; Babatunde *et al.*, 2020; Kulasekara *et al.*, 2016). BIM is used for visualisation, fabrication, cost estimating, construction sequencing, collision detection and facilities management (Azhar, 2011; Ibrahim and Abdullahi, 2016). BIM has been accepted as the standard for building design and documentation in the AEC industry globally due to its capabilities to improve efficiencies, increase productivity, enhance collaboration among project participants and give organisations a competitive edge over their rivals (Aranda-Mena *et al.*, 2008; Arayici *et al.*, 2011). AEC industry stakeholders, particularly organisations cannot afford to ignore BIM because of its potential (Aranda-Mena *et al.*, 2008).

Among AEC stakeholders, CMFs are key beneficiaries whose traditional functions has been transformed by BIM, as it offers capabilities that enhances CMFs' performance and practice (Makenya and Ally, 2018; Zainon *et al.*, 2018). In the AEC industry, cost managers are responsible for managing resources during project execution to deliver value for money

(Osunsanwo and Dada, 2019). Some of their traditional functions include cost plans production, cost estimates, project document preparations such as bills of quantities (BOQ), financial statements and the resolution of contract claims (Ashworth *et al.*, 2013). However, these activities are reportedly tedious, time-consuming and prone to human errors and inaccuracies (Babatunde *et al.*, 2020; Ismail *et al.*, 2018; Wong *et al.*, 2015). BIM has transformed the traditional roles of cost managers, enabling them to generate reliable cost estimates, rapid and accurate automated quantity take-off, enhance their project planning ability, facilitate cost planning and BOQ production (Ismail *et al.*, 2019; Makenya and Ally, 2018; Marsh, 2017).

Cost managers need BIM to efficiently execute their various tasks, especially in their cost management role during the execution of construction projects by way of traditional construction methods as well as MMC. BIM possesses capabilities that improve cost managers' service delivery and tackles prominent challenges such as low productivity, low profit margin, lack of innovation and inefficient service delivery (Alutwela *et al.*, 2016; Osunsanwo and Dada, 2019).

There has been an exponential demand for BIM in recent years due to the benefits to the construction industry globally (Kaduma, 2022; Safour *et al.*, 2021). Developed countries have made progress with BIM adoption and implementation of its tools, while developing countries are still struggling with BIM adoption (Olawumi and Chan, 2019; Saka *et al.*, 2019). Countries like Finland, USA, Norway, Denmark and Austria, have made progress in BIM implementation introducing BIM mandate, standards and policies in their respective countries, leading to BIM becoming part of the requirements for construction contracts and practice (Kaduma, 2022; Safour *et al.*, 2021).

The most significant stride made towards BIM adoption was in 2016 when the UK Government mandated BIM level 2 for all key projects costing more than £5 million in their Construction Strategy. This initiative made BIM compulsory for securing government contracts and led to accelerated adoption among contractors in the UK (Hamma-Adama, 2020; Kaduma, 2022; Mtya, 2019).

BIM adoption in African countries like Nigeria is still at infancy stage phase, with little to no initiatives towards improving its adoption reported in literature (Kaduma, 2022; Olawumi and Chan, 2019). Evidences abound that BIM adoption among CMFs is low, as most of them are yet to fully embrace BIM (Akinsiku, 2016; Babatunde *et al.*, 2020; Ismail *et al.*, 2019; Ogunseiju, 2019). Olugboyega and Aina (2016) reported that among AEC professionals, cost managers had the lowest percentage of BIM usage in comparison to architects and engineers. Some studies reported that a few CMFs that manage project costs in some Nigerian cities are already adopting BIM (Aka *et al.*, 2020; Babatunde *et al.*, 2020). However, the challenges of BIM adoption among CMFs need studying due to the paucity of research on BIM adopters in Nigeria. Generally, there is a dearth of studies that focus on BIM adoption among AEC organisations, particularly CMFs. There is the need to fill this research gap by researching on BIM adopters in the Nigerian AEC.

From the foregoing, this study assessed the barriers to the adoption of BIM adoption among CMFs in Nigeria by focusing on two study objectives: *(a)* determine the barriers to BIM adoption and *(b)* determine the prominent barriers to BIM adoption, with the aim of creating awareness among relevant stakeholders in the AEC industry – particularly firms, professional bodies and regulatory bodies – and to develop initiatives that will promote the adoption of BIM among firms in the Nigerian AEC industry.

5.2. Perspectives and concepts

5.2.1 Overview of building information modelling

BIM has several definitions. One of the definitions of BIM is that 'BIM is a repository of digital information that facilitates the efficient management of project information from conception by way of simplifying and presenting a real-world simulation of a preconceived project facility' (Olawumi and Chan, 2019, p. 1225).

BIM revolves around three components – people, processes and digital technology (Badiru *et al.*, 2015). BIM uses intelligent objects to support the design, construction and operation of building and engineering projects within a shared digital environment that enables the integration, management and exchange of project information (Ibrahim and Abdullahi, 2016). It is a successor to AutoCAD, offering several applications that were lacking in 2D and 3D CAD technologies (Wong *et al.*, 2011). BIM is adopted as a platform, tool and collaborative information environment that supports coordination among project stakeholders (Ibrahim and Abdullahi, 2016).

Herr and Fischer (2017) posited that BIM forecasts and manages construction processes and building performance to reduce labour, materials and energy costs. BIM as a 3D modelling tool, provides project details to all project stakeholders throughout the project's life cycle (Sheth and Malsane, 2014).

5.2.2 Cost benefits

BIM is reported to offer several cost benefits to construction projects. The cost benefits that BIM offers may be tied to its capabilities – for example, the 5D BIM which is the BIM dimension associated with cost management functions connected to construction costs and having project management features (Kissoon *et al.*, 2015; Raji *et al.*, 2020). This enables cost analysis and the monitoring of the project budget to be done concurrently (Mesaros *et al.*, 2019), allowing for adjustments in costs when project designs change and enabling clients to assess the contractor's bill for a project at any given time (Kjartansdóttir *et al.*, 2017), enbling proper management of project costs throughout the project duration.

With BIM, the project team can visualise the development of project activities and the project's associated expenses over time by way of the 5D BIM, which improves project cost accuracy and predictability (Marsh, 2017). 5D BIM enables cost managers to optimise their workflows and deliver high-quality service (Harrison and Thurnell, 2015), leading to accurate cost estimation and forecast of projects. In the same vein, 5D BIM enables specification data and quantities to be extracted and assigned unit costs (Mesaros *et al.*, 2019; Stanley and Thurnell, 2014), thereby promoting cost efficiency in construction projects. Additionally, with the appropriate BIM library and coding system, the capabilities of 5D BIM also assist in offering precise and trustworthy guidance on choosing suitable construction materials (Kim and Park, 2016).

Conversely, BIM has been used in projects worldwide to enhance cost and time efficiency. According to Azhar (2011), BIM was utilised in several major projects in Australia, resulting in an 80% reduction in cost estimation time, a 10% decrease in project cost and a 7% decrease in project duration. Ibem *et al.* (2018) also reported that BIM usage in projects can lead to 15–40% in cost savings. Although the statistics looks good and positive, there is a need to expatiate how these were achieved. These statistics indicates the cost benefits of BIM when utilised in construction projects and its impact on cost management functions. During tender documentation, BIM tools are mostly used for cost estimation and the production of bills of quantities (Ismail *et al.*, 2019). Some of the BIM tools utilised in cost management practice include Autodesk Quantity Takeoff, Buildsoft Takeoff, Cubicost, Navisworks Manage and Revit (Babatunde *et al.*, 2020; Ismail *et al.*, 2019).

5.2.3 Collaboration

One of the major positive highlights of BIM in construction projects is its ability to enhance collaboration among project participants (Aranda-Mena *et al.*, 2008; Arayici *et al.*, 2011). This capacity of BIM is significant because most problems in construction projects are tied to lack of collaboration among project participants often leading to disputes, division, delays and project failures. BIM improves collaboration among project team members through shared ownership of design and construction processes, enabling the design team to work together on a project and addressing interdisciplinary problems in construction projects (Aibinu and Venkatesh, 2012; Harrison and Thurnell, 2015; National Building Specification [NBS], 2016).

Collaboration among project participants is achieved through BIM capability that enables project information to be created, stored and shared in a format superior to paper-based techniques (Sawhney *et al.*, 2017). BIM uses a single platform for information storage and sharing and offering an enabling platform for collaboration among project stakeholders. This allows all project participants to have access to project information and keep every participant abreast with changes, corrections and updates.

5.2.4 Adoption barriers

The AEC industry has adopted BIM and is being transformed as a result of BIM's many advantages and ability to address the various problems that the sector is facing, including cost overrun, poor project quality and project complexity (Hamma-Adama and Kouider, 2018; Saka *et al.*, 2019; WEF, 2018).

Although adopting BIM in the AEC sector has numerous advantages, several factors make adoption difficult (Gibbs *et al.*, 2015). It would be easier to create strategies for successful BIM adoption if these factors were understood (Abubakar *et al.*, 2014; Ezeokoli *et al.*, 2016). It is noteworthy that most research on BIM in developing countries like Nigeria focuses on the lack of BIM implementation and barriers to BIM adoption (see Aka *et al.*, 2020; Babatunde *et al.*, 2021; Bouhmoud and Loudyi, 2020; Makenya and Ally, 2018; Ogunmakinde and Umeh, 2018; Olugboyega and Windapo, 2021; Sahil, 2016; Udomdech *et al.*, 2021). In contrast, broader studies across developing countries tend to focus more on assessing the current level of BIM implementation, with comparatively less emphasis on identifying and analysing adoption barriers (Hooper *et al.*, 2020; Mesaros *et al.*, 2019; Pučko *et al.*, 2014; Sawhney, 2014; Sawhney *et al.*, 2017; WEF, 2018).

Research indicates that the barriers to BIM adoption, especially in developing countries, are largely similar (Desbalo and Bargstadt, 2020; Saka and Chan, 2019). The barriers are classified as people/process related, economic related and technology related (Saka and Chan, 2019).The economic-related variables are mainly in the area of costs of BIM setup and BIM training (Saka and Chan, 2019). At the same time, software, hardware and networks are among the technology-related elements of BIM adoption; leadership, infrastructure, human resources and products/services are among the process-related elements of BIM adoption. While contracts, regulations and education/research are all part of the policy-related aspect of BIM adoption (Mtya, 2019).

Saka and Chan (2019) reported that people/process-related factors were the major inhibitors of BIM adoption among stakeholders in the AEC industry. Saka and Chan (2019) further noted that lack of BIM training, lack of BIM awareness and lack of BIM demand from clients are the prominent people/process-related BIM adoption barriers in developing regions like Africa.

One of the major obstacles impeding the adoption of BIM in the industry is the lack of client demand for BIM, as clients' requests for BIM is requisite for advancing BIM adoption among firms and its usage on projects (Aibinu and Venkatesh, 2014; Olugboyega and Windapo, 2021). Bouhmoud and Loudyi (2020) and Akerele and Etiene (2016) also noted that another barrier to the adoption of BIM is the absence of government support for BIM adoption in developing nations like Nigeria. According to Hamma-Adama and Kouider (2018), BIM has not received any national attention in Nigeria, nor has the government provided any support to increase BIM adoption at the industry level. The outcome is that most AEC stakeholders in Nigeria are left in limbo regarding the adoption of BIM.

The absence of policy (rules and procedures) governing the use of BIM is another factor impeding its deployment. Significantly, meaningful BIM adoption cannot occur without the availability of rules and norms. Thus, legal support and national adoption and implementation standards and guidelines are important variables influencing the implementation of BIM (Ezeokoli *et al.*, 2016). Hamma-Adama and Kouider (2018) reported that BIM adoption lacks government involvement, particularly in some developing countries like Nigeria. Hence, the lack of government involvement in promoting BIM may translate to the absence of policy and regulation to guide BIM adoption in countries that are reported to have no government involvement in BIM promotion.

The lack of support from top management is another significant barrier to BIM adoption in the AEC industry (Awodele *et al.*, 2022). Another barrier to BIM adoption among stakeholders in the AEC sector is the lack of knowledge about BIM, which makes stakeholders ignorant about its benefits, thereby affecting its adoption (Ezeokoli *et al.*, 2016). When stakeholders do not know what BIM is about and its relevance, the drive to adopt it may not exist. The lack of collaboration among key stakeholders of the AEC industry also inhibits BIM adoption. To enhance BIM adoption, collaboration is required between construction firms and technology firms (Awodele *et al.* 2022; Gilchrist *et al.*, 2021).

The expensive cost of hardware and software needed to set up BIM is another significant barrier to its wider adoption in the AEC industry (Awodele *et al.*, 2022). According to

Babatunde *et al.* (2021), setting up BIM software and hardware requires a significant amount of upfront investment from firms intending to adopt it (Babatunde *et al.*, 2021). Reluctance or resistance to change is also a barrier to BIM adoption among stakeholders in the AEC sector. Most people in the AEC industry are reported to be comfortable with the status quo in their professional practice and are often reluctant to accept novel ideas or innovations in their practice (Awodele *et al.*, 2022; Babatunde *et al.*, 2021).

Poor infrastructure in areas of variable internet connectivity and power supply also hinders BIM adoption in the AEC sector, particularly in African countries (Bouhmoud and Loudyi, 2020; Ezeokoli *et al.*, 2016). The availability and efficiency of infrastructure are quite significant for BIM adoption as it requires the necessary software, hardware and network to be functional (Mtya, 2019).

The barriers to the adoption of BIM identified in the literature are presented in Table 5.1.

Table 5.1 Barriers to building information modelling adoption (continued on next page)

S/N	Factors	Literature Sources
	People/Process-related factors	
1	Inadequate BIM education	Bouhmoud and Loudyi (2020), Badiru *et al.* (2015)
2	Insufficient BIM demand from clients	Bouhmoud and Loudyi (2020)
3	Ignorance about BIM	Bouhmoud and Loudyi (2020), Desbalo and Bargstadt (2020)
4	Fear of BIM by clients and other stakeholders	Saka and Chan (2019), Ezeokoli *et al.* (2016)
5	Lack of BIM training	Babatunde *et al.* (2021), Bouhmoud and Loudyi (2020)
6	Lack of adequately trained BIM experts	Aka *et al.* (2020), Desbalo and Bargstadt (2020)
7	Lack of familiarity with BIM by stakeholders	Desbalo and Bargstadt (2020), Saka and Chan (2019)
8	Lack of government support	Desbalo and Bargstadt (2020), Saka and Chan (2019)
9	Cultural resistance to change	Awodele *et al.* (2022), Bouhmoud and Loudyi (2020)
10	Lack of policy for BIM adoption	Abubakar *et al.* (2014); Hamma-Adama and Kouider (2018)
11	Lack of support from senior leadership or top management for BIM adoption	Awodele *et al.* (2022), Babatunde *et al.* (2021), Desbalo and Bargstadt (2020)
12	Absence of new or modified contract terms that incorporate BIM	Abubakar *et al.* (2014)

Table 5.1 Continued

S/N	Factors	Literature Sources
13	Lack of industry push	Babatunde *et al.* (2021)
14	Lack of comprehensive framework and implementation plan for BIM	Desbalo and Bargstadt (2020)
	Economic-related factors	
15	Cost of BIM software and set-up	Awodele *et al.* (2022), Babatunde *et al.* (2021),
16	Cost of BIM training	Saka and Chan (2019), Ezeokoli *et al.* (2016)
17	Lack of clear BIM benefits evaluation	Saka and Chan (2019), Desbalo and Bargstadt (2020)
	Technology-related factors	
18	Lack of IT infrastructure	Bouhmoud and Loudyi (2020), Desbalo and Bargstadt (2020), Saka and Chan (2019)
19	Inadequate BIM standards and guidelines	Babatunde *et al.* (2021), Bouhmoud and Loudyi (2020), Desbalo and Bargstadt (2020)
20	BIM software complexity	Awodele *et al.* (2022), Bouhmoud and Loudyi (2020)
21	The longer time it takes to learn BIM software	
22	Longer time required to implement BIM	Desbalo and Bargstadt (2020)
23	ICT illiteracy	Desbalo and Bargstadt (2020), Saka and Chan (2019)
24	Legal issues around ownership and copyrights	Bouhmoud and Loudyi (2020), Desbalo and Bargstadt (2020)

Source: Synthesis from literature reviewed

5.3. Methodological approach

The study's population comprised cost management firms located in Lagos and Abuja, Nigeria, and registered with the Quantity Surveyors Registration Board of Nigeria (QSRBN), the regulatory body overseeing CMFs in Nigeria. A total of 205 CMFs (118 in Abuja and 87 in Lagos) were registered with QSRBN in 2023 (QSRBN, 2023). The sampling frame comprised CMFs in Lagos and Abuja that have adopted BIM in their practices. The non-probability purposive sampling technique was adopted, starting with a pilot survey followed by the snowball sampling method, due to the lack of information on CMFs that have implemented BIM in the study areas. Eighteen companies, twelve located in Lagos and six in Abuja, were identified as having adopted BIM, forming the sample size for this research.

As part of the study on assessing the capability maturity of CMFs for BIM adoption, primary data were collected using the Delphi technique. The Delphi technique was adopted due to the sample size anticipated for this study. It is a technique used to collect information from respondents who are competent in their field of expertise or an area, using a series of questionnaires to achieve consensus among the experts (Aghimien *et al.*, 2020; Evans and Farrell, 2021). BIM experts from 18 CMFs that have adopted BIM were used for the Delphi survey. The number was sufficient, as most construction management studies utilised between 5 and 20 experts in their Delphi studies (Aghimien *et al.*, 2020; Evans and Farrell, 2021).

The Delphi process consists of developing the questionnaire, choosing the panel of experts, gathering initial data (first round) and analysing the data garnered from the experts. The next step involves distributing the analysed data retrieved from the first round of data collection so that the panellists can compare their responses to the group mean and decide whether to review or confirm their responses. This process is repeated until a consensus is reached among the panellists (Saunders *et al.*, 2019). Most studies in construction management establish that consensus can be reached in 2 to 3 rounds of the Delphi survey (Ameyaw *et al.*, 2016; Evans and Farrell, 2021).

Since the selection of respondents to form the expert panel is key to the successful outcome of the Delphi survey (Ameyaw *et al.*, 2016; Evans and Farrell, 2021), some of the criteria the respondents for this study must fulfil include *(a)* work in a cost management firm that has adopted and possesses a thorough grasp of BIM concepts; *(b)* possess a bachelor's degree, at the very least in construction, management and engineering, or a closely connected profession; *(c)* be an affiliate of a professional body in cost management, in this case, Nigerian Institute of Quantity Surveyors (NIQS).

The questions on barriers to BIM adoption among CMFs were based on existing literature. The barriers were to be rated by the panellists on a 6-point Likert scale ranging from 'no importance' to 'very high importance'. However, to enhance the reliability and suitability of the questions, and refine where necessary (Saunders *et al.*, 2019), the Delphi survey instrument was first pilot-tested with six panellists (four from firms in Abuja and two from firms in Lagos). During the first round of the Delphi survey, 18 experts received both closed-ended and open-ended questions electronically, along with instructions on how to respond. The responses were subsequently collected and analysed. The data from the first round was sent to the experts for the second round, where their responses confirmed a consensus had been reached. Importantly, the experts did not review their initial responses during this second round.

The data collected from each round of the Delphi study were analysed using mean score analysis. Additionally, Chi-Square and Kendall's coefficient of concordance – two commonly used consensus tools in Delphi studies – were employed to assess the level of agreement among the experts (Aghimien *et al.*, 2020; Ameyaw *et al.*, 2016). The Cronbach's alpha test and the Shapiro-Wilk test were used to assess the reliability and normality of the data collected, as they are commonly utilised for evaluating data in Delphi studies (Aghimien *et al.*, 2020; Evans and Farrell, 2021). The Kruskal-Wallis H test determines if significant differences exist among three or more groups of respondents (Paul *et al.*, 2021). The data collected was analysed using SPSS version.

5.4. Results and discussion

5.4.1 Profile of experts

Eighteen experts participated in the Delphi survey, achieving a 100% response rate. Their demographic information is important as it indicates their expertise in BIM adoption among CMFs. Analysis showed that 44.4% of the experts held a bachelor's degree (the highest number), master's degree qualification (38.9%) ranked next and, lastly, Higher National Diploma qualification. The results indicated that a majority of experts (55.6%) were members of the Nigerian Institute of Quantity Surveyors (NIQS), the organisation overseeing CMFs in Nigeria, while fellows (FNIQS) of the organisation ranked next (22.2%). Additionally, the average number of years these experts have worked in the construction industry was 12. The experts in the Delphi survey were well-educated, experienced in the industry and mostly professionally qualified. As a result, their input can be considered reliable for this study.

5.4.2 Barriers to BIM adoption among cost management firms

The barriers to BIM adoption among CMFs are outlined in Table 5.2. Previous studies (Moyanga and Agboola, 2020) indicated that various factors, such as financial capacity, the age of the firm, its size and annual turnover, influence CMFs' ability to adopt new technologies. In this study, the primary distinguishing factor examined was the age of the CMFs. As a result, the factors promoting BIM adoption among CMFs were categorised into three types of firms: those that have existed for 1–10 years (young firms), 11–20 years (middle-aged firms) and more than 20 years (established firms).

5.4.2.1 Consensus among experts

The consensus among the experts was measured using the Chi-Square (χ^2) and Kendall's coefficient of concordance (W^a) is presented in Table 5.2. The computed Chi-Square (χ^2) value was 57.024, while its critical Chi-Square ($\chi^2\alpha$) value was 35.172 ($p=0.05$) at a degree of freedom of 23 (obtained from the statistical table). The χ^2 value being higher than its $\chi^2\alpha$ value indicates that consensus was reached among the experts (Aghimien *et al.*, 2020; Evans and Farrell, 2021). The result of the Kendall's coefficient of concordance (W^a) as shown in Table 2 was 0.138, indicating weak agreement among the experts, as Moslem *et al.* (2019) reported that a coefficient of concordance value of 0.10 indicates weak agreement, 0.30 indicates moderate agreement and 0.60 and above indicates strong agreement.

5.4.2.2 Reliability and normality of data

The reliability of the data collection instrument was measured using the Cronbach's alpha test. The Cronbach's alpha was 0.933, as shown in Table 5.2. This value is considerably higher than the 0.700 minimum value threshold set for confirming the reliability of the research instrument for data collection, indicating that the data collected were reliable (Paul *et al.*, 2021). The normality of the data collected was also tested using Shapiro-Wilk's test of normality. The data were not normally distributed, with 0.000 as the significance level (*p*-value), which is less than the 0.05 value threshold ($p<0.05$), as shown in Table 5.2. The result indicated that non-parametric statistical approaches should be used for subsequent analyses. Moreover, since the number of respondents who participated in the Delphi survey was less than 30, the assumption of a normal population does not apply.

Table 5.2 Barriers to BIM adoption among cost management firms (continued on next page)

N	Factors	All			Firms 1–11 Yrs			Firms 11–20 Yrs			>20 Yrs			K-W	
		MS	SD	R	MS	SD	R	MS	SD	R	MS	SD	R	χ^2	Sig.
1	Cost of BIM software and set-up	4.44	0.705	1	5.00	0.000	1	3.75	0.500	2	4.55	0.688	1	6.810	0.033**
2	Cost of BIM training	4.11	0.758	2	4.33	0.577	4	3.50	0.577	4	4.27	0.786	2	3.336	0.189
3	Lack of adequately trained BIM experts	4.06	1.259	3	5.00	0.000	1	3.50	1.000	6	4.00	1.414	8	4.657	0.097
4	Inadequate BIM education	4.00	1.372	4	4.33	1.155	6	3.50	1.291	9	4.09	1.514	7	1.470	0.480
5	Lack of clear BIM benefits evaluation	3.94	0.802	5	3.67	0.577	9	3.50	0.577	4	4.18	0.874	4	3.912	0.141
6	Lack of IT infrastructure	3.94	0.873	6	4.33	0.577	4	3.25	0.957	14	4.09	0.831	6	2.882	0.237
7	Lack of government support	3.78	1.396	7	3.67	2.309	12	4.00	1.414	1	3.73	1.272	15	0.445	0.801
8	BIM software complexity	3.72	0.826	8	3.67	1.155	10	3.25	0.500	10	3.91	0.831	9	2.659	0.265
9	Lack of support from senior leadership or top management for BIM adoption	3.72	0.895	9	3.00	1.000	18	3.25	0.957	14	4.09	0.701	5	4.530	0.104
10	Lack of BIM training	3.72	1.406	10	3.33	1.528	15	3.50	1.000	6	3.91	1.578	11	1.295	0.523
11	Ignorance about BIM	3.67	1.188	11	2.33	1.528	21	3.25	1.258	18	4.18	0.751	3	5.130	0.077
12	Lack of familiarity with BIM by stakeholders	3.67	1.237	12	4.00	1.000	7	3.25	0.957	14	3.73	1.421	16	1.441	0.487
13	Lack of policy for BIM adoption	3.67	1.414	13	3.33	2.082	16	3.50	1.000	6	3.82	1.471	12	0.846	0.655
14	Inadequate BIM standards and guidelines	3.61	0.850	14	3.00	1.000	18	3.25	0.500	10	3.91	0.831	9	4.262	0.119
15	Insufficient BIM demand from clients	3.56	1.199	15	3.67	1.528	11	3.25	1.258	18	3.64	1.206	17	0.512	0.774
16	Cultural resistance to change	3.56	1.247	16	4.00	1.000	7	3.00	1.633	21	3.64	1.206	17	1.099	0.577
17	Lack of comprehensive framework and implementation plan for BIM	3.56	1.338	17	4.67	0.577	3	3.25	0.500	10	3.36	1.567	23	4.513	0.105

Table 5.2 Continued

N	Factors	All			Firms 1–11 Yrs			Firms 11–20 Yrs			>20 Yrs			K-W	
		MS	SD	R	MS	SD	R	MS	SD	R	MS	SD	R	χ^2	Sig.
18	Lack of industry push	3.50	1.249	18	3.33	0.577	13	3.75	0.500	2	3.45	1.572	21	0.415	0.813
19	Absence of new or modified contract terms that incorporate BIM	3.44	1.464	19	3.33	2.887	17	3.25	0.957	14	3.55	1.293	19	0.591	0.744
20	Legal issues around ownership and copyrights	3.39	1.243	20	3.00	1.732	20	2.75	1.258	22	3.73	1.104	14	2.091	0.351
21	ICT illiteracy	3.33	1.188	21	3.33	0.577	13	3.25	0.500	10	3.36	1.502	22	0.600	0.741
22	Longer time required to implement BIM	3.17	1.249	22	1.67	0.577	24	2.75	1.893	24	3.73	0.647	13	6.887	0.032**
23	The longer time it takes to learn BIM software	3.06	1.162	23	2.00	0.000	23	2.75	1.258	22	3.45	1.128	20	5.040	0.080
24	Fear of BIM by clients and other stakeholders	2.56	1.464	24	2.33	1.528	21	3.00	1.414	20	2.45	1.572	24	0.444	0.801
	Cronbach's α reliability coefficient		0.933												
	Number of respondents (N)		18												
	Kendall's coefficient of concordance (W^a)		0.138												
	Computed χ^2		57.024												
	χ^2 critical value from statistical table ($p = 0.05$)		35.172												
	Degree of freedom (df)		23												
	Significance level (p)		0.000												

Note. **Significant at $p = 0.05$; MS = Mean Score; SD = Standard Deviation; R = Ranking; χ^2 = Chi-Square; K-W = Kruskal Wal
Source: Authors' creation

5.4.2.3 Barriers to BIM adoption

The barriers to BIM adoption among CMFs based on their age band (that is, 1-10 years, 11-20 years and above 20 years) and the overall mean scores of the barriers to BIM adoption among CMFs are presented in Table 5.2. The Kruskal-Wallis' test was first conducted on the data collected to compare the responses of the panellists across the three age categories. The result of the data analysis revealed that there was no significant difference in the panellists' scoring of twenty-two out of the twenty-four (24) barriers to BIM adoption assessed for 1–10 years, 11–20 years and above 20 years categories of firms. However, there was a difference in the scoring of the panellists on two of the barriers (the cost of BIM software and set-up and the longer time required to implement BIM). This is because a significance level of less than 0.05 was derived for the two barriers.

This implies that the panellists viewed the two barriers differently. The reason for this difference in view could be that the three categories of CMFs have goals, capabilities and resources in line with their age band – for instance, established firms (11–20 years) may have more financial resources than young firms (1–10 years), and both categories of firms may not view the cost of BIM software and set-up as a BIM-adoption inhibiting factor in the same way. The results in Table 5.2 show that the scores for all the barriers by the panellist had mean values (MS) above 3.00. The mean score of the 24 ranked barriers ranged from an overall MS of 2.56 for fear of BIM by clients and other stakeholders to an overall MS of 4.44 for the cost of BIM software and set-up ($2.56 \leq MS \leq 4.44$). This indicates that the scores of the barriers to BIM adoption among CMFs were moderate to high.

For the three categories of CMFs, the results in Table 5.2 shows that the barriers to BIM adoption are the cost of BIM software and set-up and the cost of BIM training are significant barriers to BIM adoption that cuts across all categories of CMFs. While lack of adequately trained BIM experts and lack of IT infrastructure were top barriers to BIM adoption among young firms, they were not significant barriers to established firms. While lack of clear BIM benefits evaluation was a significant barrier to BIM adoption among established firms, it was among the top barriers to BIM adoption reported by young firms.

Established firms reported lack of support from senior leadership or top management for BIM adoption and ignorance about BIM as major barriers to BIM adoption, while these factors were not reported as significant barriers by young firms. Meanwhile, younger firms reported lack of comprehensive framework and implementation plan for BIM as a significant barrier to BIM adoption. However, this factor was not considered significant by established firms.

The findings of the study, as highlighted in Table 5.2, show that for CMFs in all three categories, the cost of BIM software and set-up and the cost of BIM training were among the major barriers to BIM adoption. These findings indicate that the cost of BIM implementation and the training of staff on BIM is a major concern among all CMFs, irrespective of their years of existence. Thus, strategies need to be deployed to mitigate the effects of these factors. The overall result revealed that the top barriers to BIM adoption among CMFs are the cost of BIM software and set-up (MS = 4.44), the cost of BIM training (MS = 4.11), the lack of adequately trained BIM experts (MS = 4.06), inadequate BIM education (MS = 4.00) and the lack of clear BIM benefits evaluation (MS = 3.94).

From the findings, it is noteworthy that the barriers to BIM adoption among CMFs were mostly economic-related and people/process-related variables. The major concerns of CMFs are the cost of implementation and training staff on BIM, inadequate BIM experts and BIM education and the lack of a clear evaluation of BIM benefits. The findings of this study partly agree with the findings of Saka and Chan (2019) that people/process-related factors were the major inhibitors of BIM adoption among stakeholders in the AEC industry. The findings of this study also align with those of Saka and Chan (2019) that the lack of BIM training was among the major barriers to BIM adoption in developing regions like Africa.

The findings of this study agree with the findings of Ezeokoli *et al.* (2016) that the cost of training staff on BIM is quite significant and is a major barrier to BIM adoption among organisations in the AEC sector. The findings of this study further align with those of Awodele *et al.* (2022), who established that the expensive cost of hardware and software needed to set up a BIM system is a significant barrier to the wider adoption of BIM in the AEC industry. The study's findings also agree with Abubakar *et al.* (2014), who reported that the lack of skilled personnel with BIM knowledge and skills is among the major barriers to BIM adoption among AEC industry stakeholders.

The study's findings are in tandem with those of Hamma-Adama and Kouider (2018), that the lack of government support for BIM adoption in developing nations like Nigeria inhibits BIM adoption among stakeholders; although relevant, it is not among the top-ranked barriers in this study. This study's findings align with those of Awodele *et al.* (2022), who reported that the lack of support from top management is a significant barrier to BIM adoption in the AEC industry. Although this factor is significant, it is not among the top-ranked barriers in this study.

However, the study's findings are divergent from those of Saka and Chan (2019) that the lack of BIM awareness and the lack of BIM demand from clients were among the prominent barriers to BIM adoption. This is because the two barriers were not among the highly ranked barriers to BIM adoption among CMFs in the study areas. The results contradict those of Olugboyega and Windapo (2021), who reported the lack of client demand for BIM as one of the major barriers to BIM adoption. However, this barrier was not among the highly ranked barriers to BIM adoption among CMFs. The findings of this study were divergent from those of Awodele *et al.* (2022) and Babatunde *et al.* (2021), who reported that resistance to change is a significant barrier to BIM adoption among AEC stakeholders; interestingly, this barrier was not a major factor in this study.

5.5. Implications of the study

The findings of this study present several important implications for the construction industry and its stakeholders, including firms, professional and regulatory bodies, policymakers and technology developers. Given the critical role of digitisation in implementing MMC, technology developers need to subsidise the costs associated with digital tools like BIM. This would enhance accessibility for various types of firms within the construction sector.

Additionally, there is a pressing need for more comprehensive education and training on BIM and other relevant digital tools. Offering these programs at subsidised costs could significantly increase BIM adoption among construction professionals and enhance its application in MMC, ultimately resulting in a greater number of BIM experts in the industry. Policymakers and industry leaders should focus on creating strategies that support young and middle-aged firms in adopting technologies like BIM. Advocating for widespread adoption through policies that encourage BIM integration, establish necessary IT infrastructure and provide government support will be crucial.

Moreover, a clear evaluation of the benefits of BIM is necessary, particularly from the perspective of firms that have already incorporated it into their processes. Implementing these strategies will not only enhance the utilisation of BIM in MMC but also lay the groundwork for future research on BIM integration in MMC.

5.6. Conclusions

The study set out to identify and prioritise the most significant barriers to BIM adoption among CMFs. It utilised the Delphi approach to solicit expert opinions on BIM adoption among CMFs in Abuja and Lagos, Nigeria. The study identified key barriers, including the cost of BIM software and set-up, the cost of training, the lack of adequately trained BIM experts, inadequate BIM education and the absence of clear methods for evaluating BIM benefits. Overall, these barriers were primarily cost related and manpower related.

This study contributes to the body of existing knowledge by revealing the barriers to BIM adoption among CMFs in a developing country like Nigeria, where the adoption of BIM is still low and MMC is still nascent. The study is significant as it draws on the experiences of BIM-adopters, rather than the generic opinions of AEC stakeholders. It is believed that the findings will assist both cost management professionals and regulatory bodies in devising measures to mitigate the barriers to BIM adoption among CMFs and foster MMC implementation.

While this study significantly contributes to the body of knowledge, care must be taken in generalising its findings – for example, while the study was conducted in two major states of the country, future studies would benefit more from broader coverage, which would lead to a larger sample size than obtained in the study. This study utilised the Delphi approach; future research could explore other methods, such as interviews and focus group discussions, for a deeper understanding of the topic.

REFERENCES

Abubakar M, Ibrahim YMY, Kado D and Bala K (2014) Contractors perception of the factors affecting building information modelling (BIM) adoption in the Nigerian Construction Industry. *Computing in Civil and Building Engineering*, 167–178.

Adesi M, Murphy R and Kehily D (2018) The role of digitisation in the strategic planning process of Irish Quantity Surveying (QS) practices. In *Proceeding of the 34th Annual ARCOM Conference* (Gorse C and Neilson CJ (eds)). Association of Researchers in Construction Management, Belfast, UK, pp. 250–259.

Aghimien DO, Aigbavboa CO and Oke AE (2020) Critical success factors for digital partnering of construction organisations – a Delphi study. *Engineering, Construction and Architectural Management* **27(10)**: 3171–3188, 10.1108/ECAM-11-2019-0602.

Aibinu A and Venkatesh S (2014) Status of BIM adoption and the BIM experience of cost consultants in Australia. *Journal of Professional Issues in Engineering Education and Practice* **140(3)**: 1–10, 10.1061/(ASCE)EI.1943-5541.0000193.

Aibinu AA and Venkatesh S (2012) The rocky road to BIM adoption: quantity surveyors perspectives. *CIB Joint International Symposium on Management of Construction: Research to Practice*, pp. 539–554.

Aka A, Iji J, Isa RB and Bamgbade AAA (2020) Assessing the relationships between underlying strategies for effective Building Information Modeling (BIM) implementation in Nigeria construction industry. *Architectural Engineering and Design Management* **17(5–6)**: 1–13, 10.1080/17452007.2020.1847026.

Akerele AO and Etiene M (2016) Assessment of the level of awareness and limitations on the use of building information modeling in Lagos State. *International Journal of Scientific and Research Publications* **6(2)**: 229–234.

Akinsiku OE (2016) The impacts of Building Information Modelling (BIM) on Quantity Surveying Practice. *Lagos Journal of Environmental Studies* **8(1)**: 85–93.

Alutwela A, Perera BAKS, Aluthwela A and Perera KKS (2016) Characteristics of quantity surveying firms and their impact on performance. *Proceedings of Institute of Quantity Surveyors Sri Lanka Annual Technical Sessions: Social Responsibility of Young Quantity Surveyors*, Colombo, Sri Lanka.

Ameyaw EE, Hu Y, Shan M, Chan APC and Le Y (2016) Application of Delphi method in construction engineering and management research: a quantitative perspective. *Journal of Civil Engineering and Management* **22(8)**: 991–1000, 10.3846/13923730.2014.945953.

Aranda-Mena G, Crawford J, Chevez A and Froese T (2008) BIM demystified does it make business sense to adopt BIM. *Proceedings of the International Conference on Information Science and Technology in Construction*, Santiago, Chile.

Arayici Y, Coates P, Kiviniemi A, Koskela L and Kagioglou M (2011) BIM implementation and adoption process for an architectural practice. *Structural Survey* **29(1)**: 7–25.

Ashworth A, Hogg K and Higgs C (2013) *Willis's Practice and Procedure for the Quantity Surveyor*, 13th ed., John Wiley and Sons Ltd, West Sussex, United Kingdom.

Awodele IA, Mewomo MC and Eze EC (2022) Inhibitors to the adoption of building information modelling in modular construction: a case study of the Nigerian Construction Industry. *Journal of Construction in Developing Countries*, 10.21315/jcdc-01-22-0004.

Azhar S (2011) Building Information Modeling (BIM): trends, benefits, risks and challenges for the AEC Industry. *Leadership and Management in Engineering* **11(3)**: 241–252.

Babatunde SO, Perera S, Ekundayo D and Adeleye TE (2020) An investigation into BIM-based detailed cost estimating and drivers to the adoption of BIM in quantity surveying practices. *Journal of Financial Management of Property and Construction* **25(1)**: 61–81, 10.1108/JFMPC-05-2019-0042.

Babatunde SO, Udeaja C and Adekunle AO (2021) Barriers to BIM implementation and ways forward to improve its adoption in the Nigerian AEC firms. *International Journal of Building Pathology and Adaptation* **39(1)**: 135–154. https://doi.org/10.1108/IJBPA-05-2019-0047

Badiru YY, Kherun NA, Muhammad RE and Abdul'azeez, AD (2015) Education at the confluence of effective Building Information Modelling Integration to Building Industry in Malaysia. In, *Proceedings of the 6th West Africa Built Environment Research (WABER) Conference* (Laryea S and Leiringer R (eds)), WABER, Accra, Ghana, pp. 357–371.

Bouhmoud H and Loudyi D (2020) Building Information Modeling (BIM) barriers in Africa versus global challenges. *The Proceedings of the 6th IEEE Congress on Information Science and Technology (CiSt)*, 5-12 June, Institute of Electrical and Electronics Engineers Inc., Agadir - Essaouira; Morocco, pp. 495–501, 10.1109/CiSt49399.2021.9357248.

Construction Industry Federation (2021) *Modern Methods of Construction*, Construction Industry Federation, Dublin, Eire.

Davies A (2018), *Modern Methods of Construction: A Forward-Thinking Solution to the Housing Crisis?*, London, UK.

Desbalo MT and Bargstadt IHJ (2020) Perceived benefits and barriers of Building Information Modeling (BIM) adoption in the AEC Sectors of Ethiopia. *International Journal of Engineering Research and Technology (IJERT)* **9(10)**: 777–789.

Evans M and Farrell P (2021) Barriers to integrating building information modelling (BIM) and lean construction practices on construction mega-projects: a Delphi study. *Benchmarking* **28(2)**: 652–669, 10.1108/BIJ-04-2020-0169.

Ezeokoli FO, Okoye PU and Nkeleme E (2016) Factors affecting the adaptability of building information modelling (BIM) for construction projects in Anambra State Nigeria. *Journal of Scientific Research and Reports* **11(5)**: 1–10, 10.9734/JSRR/2016/26563.

Gibbs DJ, Emmitt SBA, Lord W, *et al.* (2015) BIM and construction contracts – CPC 2013's approach. *Proceedings of Institution of Civil Engineers: Management, Procurement and Law* **168**: 285–293, 10.1680/jmapl.14.00045.

Gilchrist CR, Cumberlege R and Allen C (2021) Lack of implementing building information modelling in the quantity surveying profession. *IOP Conference Series: Earth and Environmental Science* **654**: 1–11, 10.1088/1755-1315/654/1/012025.

Hamma-Adama M (2020) *Framework for Macro Building Information Modelling (BIM) Adoption in Nigeria*, Thesis, Robert Gordon University, Aberdeen.

Hamma-Adama M and Kouider T (2018) A review on building information modelling in Nigeria and its potentials. *World Academy of Science, Engineering and Technology International Journal of Civil and Environmental Engineering* **12(11)**: 1113–1119, 10.5281/zenodo.2021965.

Harrison C and Thurnell D (2015) BIM implementation in a New Zealand consulting quantity surveying practice. *International Journal of Construction Supply Chain Management, School of Engineering, Auckland University of Technology* **5(1)**: 1–15, 10.14424/ijcscm 501015-01-15.

Herr CM and Fischer T (2017) Challenges to the adoption of BIM in the Chinese AEC Industries: an extended BIM adoption model. In *22nd International Conference of the Association for Computer-Aided Architectural Design Research in Asia: Protocols, Flows and Glitches* (Janssen P, Loh P, Raonic A and Schnabel MA (eds)), The Association for Computer-Aided Architectural Design Research in Asia (CAADRIA), Hong Kong, pp. 179–187.

Hooper E, Sinclair D, Bain D, Gelder J and Turpin M (2020) *10th Annual BIM Report*, Newcastle Upon Tyne.

Ibem EO, Uwakonye UO, Akpoiroro G, Somtochukwu M and Oke CA (2018) Building Information Modeling (BIM) adoption in architectural firms in Lagos, Nigeria. *International Journal of Civil Engineering and Technology (IJCIET)* **9(9)**: 902–915.

Ibrahim YM and Abdullahi M (2016) Introduction to Building Information Modelling. *Proceedings of the 3-Day Workshop/Annual General Meeting of the Nigerian Institute of Quantity Surveyors*, Lagos, Nigeria.

Ismail NAA, Adnan H and Bakhary NA (2019) Building Information Modelling (BIM) adoption by quantity surveyors: a preliminary survey from Malaysia. *IOP Conference Series: Earth and Environmental Science* **267**, 10.1088/1755-1315/267/5/052041.

Ismail NAA, Idris NH, Ramli H, Sahamir SR and Rooshdi RRRM (2018) Sustainable BIM-based cost estimating for quantity surveyors. *Chemical Engineering Transactions, Italian Association of Chemical Engineering - AIDIC* **63**: 235–240, 10.3303/CET1863040.

Kaduma L (2022) BIM Adoption Across the World: A Global Outlook. https://medium.com/specter-automation-insights/bim-adoption-across-the-world-a-global-outlook-1b3879f23bc6 (accessed 21/11/2024).

Kim KP and Park KS (2016) Implication of quantity surveying practice in a BIM-enabled environment. *Proceedings of Pacific Association of Quantity Surveyors Congress (PAQS), Christchurch*, pp. 1–11.

Kissoon C, Willis C, Regmi T, Zhang J and Cheung F (2015), *Utilization of BIM in Construction Cost and Project Management Practices: North America, China and the UK, RICS Report*, London, UK.

Kjartansdóttir IB, Mordue S, Nowak P, Philip D and Snæbjörnsson JT (2017) *Building Information Modelling* (Książek M and Rolson J (eds)), Civil Engineering Faculty of Warsaw University of Technology, Warsaw, Warsaw, Poland, 10.4324/9781315628011.

Kulasekara G, Jayasena HS and Ariyachandra MRMF (2016) Comparative effectiveness of quantity surveying in BIM implementation. In *Proceedings of the 9th International Conference of Faculty of Architecture Research Unit (FARU)* (Rajapaksha U *et al.* (eds.)). University of Moratuwa, Sri Lanka, pp. 317–328.

Makenya AR and Ally AA (2018) Practical application of building information modeling for quantity surveying profession in Tanzania. *International Research Journal of Advanced Engineering and Science* **3(1)**: 170–176.

Marsh DB (2017) *Building Information Modelling (BIM) and the UK Quantity Surveying Organisation: A Framework for Value Creation*. Liverpool John Moores University.

Mesaros P, Smetankova J and Mandicak T (2019) The fifth dimension of BIM – implementation survey. *IOP Conference Series: Earth and Environmental Science* **222**: 1–9, 10.1088/1755-1315/222/1/012003.

Moslem S, Ghorbanzadeh O, Blaschke T and Duleba S (2019) Analysing stakeholder consensus for a sustainable transport development decision by the fuzzy AHP and interval AHP. *Sustainability* **11(3271)**: 1–22, 10.3390/su11123271.

Moyanga DT and Agboola AA (2020) The growth status of quantity surveying firms in Nigeria: management executive's perspective. *Journal of Surveying, Construction and Property* **11(1)**: 74–82, 10.22452/jscp.vol11no1.7.

Mtya A (2019) *Evaluation of Building Information Modelling (BIM) Adoption, Capability and Maturity Within South African Consulting and Construction Firms*. University of Cape Town.

NBS (2016) *National BIM Report*. RIBA Enterprises Ltd.

Ogunmakinde OE and Umeh S (2018) *Adoption of BIM in the Nigerian Architecture Engineering and Construction (AEC) Industry*. A Paper Presented at the 42nd Australasian Universities Building Education Association (AUBEA).

Ogunseiju OR (2019) *Assessment of Capability Readiness of Quantity Surveying Firms in Building Information Modelling Usage in Lagos State, Nigeria*. Obafemi Awolowo University.

Olawumi TOTO and Chan DWM (2019) Development of a benchmarking model for BIM implementation in developing countries. *Benchmarking: An International Journal* **26(4)**: 1210–1232, 10.1108/BIJ-05-2018-0138.

Olugboyega O and Aina OO (2016) Analysis of building information modelling usage indices and facilitators in the Nigerian Construction Industry. *Journal of Logistics, Informatics and Service Sciences* **3(2)**: 1–36.

Olugboyega O and Windapo A (2021) Modelling the indicators of a reduction in BIM adoption barriers in a developing country. *International Journal of Construction Management*: 1–11, 10.1080/15623599.2021.1988196.

Osunsanwo HF and Dada JO (2019) Evaluating quantity surveying firms' performance: an application of balanced scorecard technique. *International Journal of Productivity and Performance Management*, 10.1108/IJPPM-06-2018-0209.

Paul CA, Aghimien DO, Ibrahim AD and Ibrahim YM (2021) Measures for curbing unethical practices among construction industry professionals: quantity surveyors' perspective. *Construction Economics and Building* **21(2)**: 1–17, 10.5130/AJCEB.v21i2.7134.

Pučko Z, Nataša N and Klanšek U (2014) Building information modeling based time and cost planning in construction projects. *Organization, Technology and Management in Construction: An International Journal* **6(1)**: 958–971, 10.5592/otmcj.2014.1.6.

Raji M, Ogunsemi DR and Olanipekun AO (2020) Assessing the willigness of quantity surveyors to pay for 5D BIM. In *5th Research Conference of the Nigerian Institute of Quantity Surveyors (RECON 5)* (Oyewobi LO, Shittu AA, Mohammed YD and Oke AA (eds)). Nigerian Institute of Quantity Surveyors, Federal University of Technology, Minna, pp. 176–188.

Safour R, Ahmed S and Zaarour B (2021) BIM adoption around the world. *International Journal of BIM and Engineering Science (IJBES)* **4(2)**: 49–63, 10.54216/ijbes.040203.

Sahil AQ (2016) *Adoption of Building Information Modeling in Developing Countries: A Phenomenological Perspective*. Colorado State University.

Saka AB and Chan DWM (2019) A scientometric review and meta synthesis of building information modelling (BIM) research in Africa. *Buildings* **9(85)**: 1–21, 10.3390/buildings9040085.

Saka AB, Chan DWM and Olawumi TO (2019) A systematic literature review of building information modelling in the architecture, engineering and construction industry - the case of Nigeria. *The Proceedings of Environmental Design and Management International Conference*, pp. 728–738.

Saka AB, Chan DWM and Siu FMF (2019) Adoption of building information modelling in small and medium-sized enterprises in developing countries: a system dynamics approach. *CIB World Building Congress*, Hong Kong, China.

Sánchez-Garrido AJ, Navarro IJ, García J and Yepes V (2023) A systematic literature review on modern methods of construction in building: an integrated approach using machine learning. *Journal of Building Engineering* **73**, 10.1016/j.jobe.2023.106725.

Saunders MNK, Lewis P and Thornhill A (2019) *Research Methods for Business Students*, 5th ed., Pearson Education Limited, Harlow.

Sawhney A (2014) *International BIM Implementation Guide*, 1st ed., Royal Institution of Chartered Surveyors, London, UK.

Sawhney A, Khanzode AR and Tiwari S (2017) *Building Information Modelling for Project Managers*, London, UK.

Sheth AZ and Malsane SM (2014) Building information modelling, a tool for green built environment. *Proceeding of All India Seminar on Innovation in Green Building Technology: Green Build*, The Institute of Engineers (India), Nagpur, India.

Stanley R and Thurnell D (2014) The benefits of and barriers to, implementation of 5D BIM for quantity surveying in New Zealand. *Australasian Journal of Construction Economics and Building* **14(1)**: 105–117.

Sweeney R (2024) *Modern Methods of Construction: Barriers and Benefits for Irish Housing*, TASC (Think-tank for Action on Social Change) and the Chartered Institute of Building (CIOB), Dublin, Eire.

Udomdech P, Papadonikolaki E and Davies A (2021) Digital transformation in developing countries: a study into BIM adoption in Thai design and engineering SMEs. *World Academy of Science, Engineering and Technology International Journal of Structural and Construction Engineering* **15(8)**: 367–374.

WEF (2018) *An Action Plan to Accelerate Building Information Modeling (BIM) Adoption*, Clony/Geneva, Switzerland.

Wong K-DA, Wong KF and Nadeem A (2011) Building information modelling for tertiary construction education in Hong Kong. *Journal of Information Technology in Construction* **16**: 467–476.

Wong PF, Salleh H and Rahim FAM (2015) A relationship framework for building information modeling (BIM) capability in quantity surveying practice and project performance. *Informes de La Construcción* **67(540)**: e119, 10.3989/ic.15.007.

Zainon N, Mohd-rahim FA, Aziz NM, Kamaruzzanman SN and Puidin S (2018) Catching up with building information modeling: challenges and opportunities for quantity surveyors. *Journal of Surveying, Construction and Property (JSCP)* **9(1)**: 19–31.

Modern Methods of Construction: Foundations, Processes and Applications

Emmanuel I Daniel and Fidelis A Emuze
ISBN 978-1-83708-437-1
https://doi.org/10.1108/978-1-83708-436-420261007

Chapter 6
Enhancing construction safety through modern methods of construction: a risk mitigation approach

Wakisa Simukonda[1], Fidelis A. Emuze[2] and Emmanuel I. Daniel[3]

[1]*School of Built Environment, Malawi University of Business and Applied Sciences, Malawi*
0000-0003-1940-0840
[2]*Department of Engineering Technology, Texas State University, Texas, USA*
0000-0001-7714-4457
[3]*School of Architecture and Built Environment, University of Wolverhampton, UK*
0000-0002-5675-1845

Abstract

Modern methods of construction (MMC) are increasingly recognised for their potential to improve safety in the construction industry; however, empirical evidence on their risk reduction remains limited. This chapter synthesises secondary evidence to examine MMC as a proactive approach to managing risk. Using a desktop research method within an interpretivist framework, six MMC and traditional construction case studies were analysed across four aspects: comparative safety performance, factory-controlled environments, technological integration and regulatory frameworks. Findings suggest that MMC can significantly reduce accident rates compared to traditional methods by integrating prevention, precision and control throughout design and production processes. Smart technologies (STs), including building information modelling (BIM), the internet of things (IoT) and artificial intelligence (AI), facilitate predictive hazard detection, real-time monitoring and proactive risk management. Regulatory frameworks and standards institutionalise safety through compliance, quality assurance and standardised workflows. A conceptual model of MMC's risk mitigation pathways shows how off-site prefabrication, factory-controlled processes, technological integration and regulatory alignment transform construction safety from reactive, site-based management to proactive, design-driven risk mitigation. The study highlights the importance for policymakers, industry practitioners and researchers to prioritise MMC adoption, incorporate STs and strengthen regulatory frameworks to deliver safer, more efficient and resilient construction practices worldwide.

Keywords: case-study; construction; mitigation; off-site; risk; safety

6.1. Introduction

In recent years, the global construction industry has been undergoing a quiet but transformative revolution, moving away from traditional, labour-intensive on-site building methods toward modern methods of construction (MMC) (Rafa and Khalid, 2024). This transition is driven by government and industry efforts to address persistent productivity challenges, improve project performance and meet growing housing demands (Payne and Serin, 2023). The shift reflects an ongoing quest to modernise construction practices, echoing early reform calls from landmark reports such as Latham's Constructing the Team (1994) and Egan's Rethinking Construction (1998), which emphasised integration, collaboration and efficiency as keys to sectoral transformation.

In response, several governments have developed strategic policies to scale up MMC adoption – for instance, the UK Construction 2025 Strategy and Hong Kong's 2017 Modular Integrated Construction Policy Agenda both target a more diversified and sustainable construction market (Payne and Serin, 2023; Government of Hong Kong, 2017). As a result of supportive policy interventions, the global MMC sector has reportedly grown from USD 104.65 billion in 2024 to USD 112.54 billion in 2025 and is projected to reach USD 214.76 billion by 2034 (Precedence Research, 2025).

Other independent market analyses provide more conservative estimates. Grand View Research (2025) reports that the market was valued at USD 103.55 billion in 2024 and is expected to reach USD 162.42 billion by 2030, with residential applications accounting for 53.2% of the market and steel-based modular construction representing 41.3% of material share, while MarketsandMarkets (2025) projects growth from USD 104.1 billion in 2024 to USD 140.8 billion by 2029, reflecting a compound annual growth rate CAGR of 6.2%. The market highlights suggest that, while strong growth is anticipated, the precise trajectory depends on technological innovation and regulatory alignment (Grand View Research, 2025; MarketsandMarkets, 2025).

The UK's Ministry of Housing, Communities and Local Government (MHCLG, 2019) categorises MMC into seven distinct types: *(a)* 3D structural systems such as bathroom pods, *(b)* 2D structural systems like cross-laminated timber, *(c)* non-systemised structural elements, *(d)* additive manufacturing, *(e)* non-structural assemblies, *(f)* productivity-enhancing site processes, and *(g)* innovative technological site applications. Among these, 3D volumetric components, manufactured up to 95% completion off-site, offer significant safety benefits, as they reduce on-site labour and hazardous operations (Ayinla *et al.*, 2019; Smith, 2016).

While MMC have been widely promoted for their time, cost and quality (TCQ) advantages (HM Government, 2013), their potential as a proactive safety enhancement and risk mitigation strategy has not received equivalent attention – for instance, the UK Construction 2025 Industrial Strategy: Government and Industry in Partnership prioritises reductions in cost by 33%, time by 50% and greenhouse gas emissions by 50%. However, an equally critical dimension of project sustainability (i.e. safety) remains underemphasised. Masood (2024) found that while prefabricated construction transfers activities to a more controlled off-site environment, it continues to cause injuries such as laceration and fractures, caused by strain, carrying and puncture. Similarly, Meacham (2022) argues that, despite MMC being frequently promoted as a safer construction approach, additional regulatory guidance, performance monitoring and oversight mechanisms are required, revealing that the safety implications of MMC remain underexplored.

Furthermore, although digital technologies such as BIM, IoT and AI are increasingly cited as enablers of construction safety, studies of their impact on MMC-based risk management remain limited. Emerging studies demonstrate that BIM can support hazard identification and safety risk management during modular construction processes (Chatzimichailidou and Ma, 2022), while IoT technologies enhance real-time monitoring and situational awareness through sensor-enabled safety tracking (Zhai *et al.*, 2019). Similarly, AI and machine learning applications show promise for automating hazard prediction, analysing safety data and improving decision support (Li, 2024; Pan and Zhang, 2023). However, most of this research remains confined to simulation and modelling, with limited empirical validation under real-world MMC project conditions (Chatzimichailidou and Ma, 2022; Daniel *et al.*, 2025; Pereira, 2025). Rafa and Khalid (2024) further confirm that the social and safety benefits of MMC, particularly for workers, are largely overlooked in academic and industry research, leaving a thin evidence base to substantiate claimed accident reductions.

Collectively, these studies highlight a significant gap in the literature. Although MMCs are promoted to improve safety outcomes, there is limited evidence explaining how and under what contextual conditions these improvements take place. This underscores the imperative for rigorous, evidence-based research drawing on existing secondary data to examine how factory environments, technological advancements and regulatory practices contribute to risk mitigation in MMC processes. Therefore, the main aim of this study is to explore how MMC can be strategically utilised to reduce risks and improve construction safety. The book chapter aims to answer the overarching research question: How can MMC be leveraged to mitigate risks and enhance construction safety? To operationalise this inquiry, the study addresses four specific sub-questions.

(a) *Comparative safety performance.* How do incident rates in MMC projects compare with those in traditional construction projects?
(b) *Impact of factory-controlled environments.* In what ways do off-site, factory-controlled construction environments influence worker safety outcomes?
(c) *Technological contributions.* How do digital technologies such as BIM, IoT and AI contribute to accident prevention in MMC projects?
(d) *Regulatory influence.* What role do building regulations and standards play in shaping safety outcomes in MMC versus traditional construction?

6.2. Perspectives on enhancing construction safety through MMC

This section synthesises theoretical, technological and regulatory perspectives on MMC in relation to enhancing safety and mitigating risks. It is structured around the four research sub-questions mentioned in Section 1.

6.2.1 Comparative safety performance: MMC vs traditional construction

The construction industry, although employing about 7% of the global workforce, accounts for a disproportionately high share of occupational fatalities, estimated at 30–40% (Sunindijo and Zou, 2012). Many of these accidents are caused by unsafe worker behaviours, often occurring within traditional, on-site construction methods (Gao *et al.*, 2019; Meng *et al.*, 2021). In contrast, MMCs, which use off-site prefabricated components, are generally regarded as safer due to less dependence on on-site labour and lower exposure to high-risk tasks (Masood, 2024). Studies show that modular construction projects have significantly fewer injuries, with reductions of up to 60% compared to traditional methods. This improvement mainly results from reducing working-at-height activities and mechanising repetitive manual tasks (Gan *et al.*, 2021; Gao *et al.*, 2022). From a theoretical viewpoint, these results support the principles of design for safety (DfS), where risk mitigation is integrated during the design and manufacturing phases rather than addressed reactively on-site (Hasanzadeh *et al.*, 2018).

6.2.2 Impact of factory-controlled environments

Off-site, factory-controlled construction environments constitute a central mechanism for hazard mitigation in MMC. By relocating high-risk activities such as structural assembly, mechanical installation and finishing works to controlled production settings, MMC significantly reduces exposure to environmental hazards, site congestion and unsafe acts (Johnsson and Rudberg, 2015). Lean production principles, standardised workflows and mechanised handling in factory environments further limit manual lifting, repetitive strain and slip/fall incidents (Blismas and Wakefield, 2009). Organisationally, these environments encourage process standardisation, more explicit role allocation and better supervision, which collectively enhance safety outcomes (Ayinla *et al.*, 2019).

6.2.3 Technological contributions to safety

Smart technologies (STs) play a critical role in accident prevention within MMC projects. BIM, IoT-enabled sensors and AI-driven predictive analytics allow for early identification of potential hazards, simulation of complex construction sequences and continuous monitoring of worker safety conditions (Daniel *et al.*, 2025) – for example, BIM facilitates clash detection and virtual assembly, reducing on-site errors that can lead to accidents. At the same time, IoT and wearable devices provide real-time feedback to workers and supervisors regarding unsafe conditions. This integration of technology operationalises the concept of proactive safety management, shifting the industry away from reactive, compliance-based approaches toward anticipatory, data-driven interventions (Gan *et al.*, 2021).

6.2.4 Regulatory influence on safety outcomes

Policy and regulatory frameworks shape the safety landscape for both MMC and traditional construction. Developed countries such as the UK, Singapore and Hong Kong have introduced standards, prequalification schemes and digital reporting requirements that support safe adoption of off-site construction methods (Chan *et al.*, 2020; HM Government, 2013). These regulations ensure that MMC manufacturers comply with stringent quality, safety and operational protocols. In contrast, developing regions often face fragmented regulatory oversight, weak enforcement and limited institutional capacity, which can compromise the safety potential of MMC adoption (Kamanga and Steyn, 2022). Therefore, regulatory alignment, building codes and standardisation are critical for realising the risk mitigation benefits of MMC, reinforcing the need for evidence-based policy frameworks and industry-academia collaboration.

6.2.5 Integrative perspective

Synthesising these insights, MMC enhances construction safety through four interconnected pathways.

(a) *Off-site prefabrication*. Relocation of high-risk activities to off-site facilities.
(b) *Factory-controlled environments*. Reduced on-site exposure to risks through standardisation, stable conditions and mechanised workflows.
(c) *Technological integration*. Leveraging BIM, IoT, AI and other digital tools for proactive hazard management.
(d) *Regulatory support*. Policies, standards and enforcement mechanisms that institutionalise safe practices.

Collectively, these perspectives highlight MMC as a multidimensional strategy for risk mitigation, combining process innovation, digital technology, organisational adaptation and policy alignment to deliver safer, more sustainable construction projects.

6.3. Methodological approach

This study uses a desktop research methodology grounded in an interpretivist epistemological framework to explore how MMC can be employed to mitigate safety risks. Desktop research was chosen for its ability to synthesise evidence from various secondary sources such as journal articles, case studies and industry reports (Karimi *et al.*, 2024; Scolobig *et al.*, 2023).

The interpretivist paradigm highlights understanding meaning, context and complexity in social phenomena rather than aiming for universal laws through statistical generalisation (Creswell and Creswell, 2023). Knowledge is regarded as socially constructed through interpreting human experiences, institutional practices and project interactions. In this study, interpretivism offers a framework for analysing how MMC processes, technologies and regulatory contexts impact safety and risk mitigation outcomes across different construction environments. The interpretivist approach was operationalised through qualitative content analysis of published case studies. Each case was examined to identify causal and contextual relationships between MMC implementation and safety outcomes. Coding focused on four thematic areas aligned with the research questions: *(a)* comparative safety performance, *(b)* factory-controlled environments, *(c)* technological contributions, and *(d)* regulatory impact. This facilitated the extraction of both quantitative data (e.g. reductions in accident rates) and qualitative insights (e.g. managerial practices and organisational safety behaviour).

The use of interpretivism is justified on three grounds. Firstly, the study's aim is understanding rather than generalisation, consistent with interpretivism's emphasis on depth of inquiry over breadth (Lincoln *et al.*, 2018). Secondly, MMC's safety impacts emerge from complex social, technological and organisational interactions that cannot be fully captured by positivist or quantitative methods (Bogna *et al.*, 2020; Easterby-Smith *et al.*, 2021). Thirdly, interpretivism allows for context-sensitive synthesis of heterogeneous data, integrating academic, industrial and policy perspectives to provide a complete understanding of MMC as a safety-enhancing construction approach (Denzin and Lincoln, 2023). This orientation facilitates insight into why and how MMC improves safety, an essential perspective in construction management research where behaviour, regulation and organisational culture interact dynamically (Fellows and Liu, 2022).

Data were drawn from secondary sources, including peer-reviewed publications, institutional safety datasets, government documents and industry reports produced between 2015 and 2025. Selection emphasised credibility, recency and relevance to ensure interpretive reliability (Karimi *et al.*, 2024). MMC project cases were identified through academic databases and industry websites. In contrast, traditional construction cases were sourced from the US Occupational Safety and Health Administration (OSHA) database, which contains comprehensive, verified records of accident investigations. Only projects with detailed safety outcomes were retained to ensure analytical robustness.

Representative MMC cases included the following.

- New Grange Hospital in the UK (MMCUK1), developed by Laing O'Rourke in 2020, utilised extensive off-site prefabrication of 1200 precast walls, 821 columns, 661 service modules and 243 bathroom pods. This approach reduced on-site work by 237 099 hours and 60 workers, cut hazard exposure by 33%, minimised falls and heavy lifting, and enhanced ergonomics and workflow predictability (Banks *et al.*, 2018).
- The Sekisui House residential projects in Japan (MMCJP) (Gao *et al.*, 2022) employed factory-controlled modular fabrication, achieving zero major on-site accidents, reducing exposure to high-risk tasks, enhancing quality control and decreasing injuries by 60% compared to conventional construction through minimised manual handling and the mitigation of fall hazards (Sekisui House, 2020).
- Project Two Fifty-One in London (MMCUK2), which includes a 40-storey residential tower and an eight-storey commercial block made with modular rooms and precast components, achieved a 50% reduction in on-site labour, a 23% decrease in hazard exposure and fewer reported injuries. BIM was used for design, sequencing and modular assembly, enabling virtual hazard detection and reducing clashes and assembly errors (Banks *et al.*, 2018).

Representative traditional construction cases included the following.

- At 3 Thirty-Three Downtown in Florida, USA (TCUSA1), a 20-storey multi-family residential building experienced a fatal fall when a formwork table collapsed in 2016. Contributing factors included inadequate fall protection, high-elevation work and an increased volume of on-site tasks.
- At the Imperial Pacific International Hotel in Saipan Island in 2020 (TCSI), a scaffold collapse caused three injuries, attributed to defective equipment and improper work procedures, highlighting the high prevalence of equipment-related and manual handling accidents in traditional construction.
- On October 19, 2021, during the construction of a one-storey Dollar General store in Orange City, Florida (TCUSA2), a timber roof structure collapsed due to out-of-plane truss buckling, resulting in one fatality and two injuries caused by improper work methods, poor fall protection, noncompliance with industry practices and adverse weather conditions.

A cross-case comparative analysis was then employed to identify recurring safety patterns and contextual differences between MMC and traditional projects. Evidence was triangulated across multiple sources to enhance credibility (Flick, 2023). The analysis followed a structured process involving

(a) thematic coding of safety outcomes
(b) identification of causal links between MMC features and risk reduction
(c) interpretive synthesis to construct a conceptual model of mmc as a systemic risk mitigation framework.

Through this interpretivist methodology, the study integrates empirical evidence, theoretical reflection and contextual interpretation to illuminate how MMC transforms construction safety from a reactive, site-based activity into a proactive, design-embedded risk mitigation approach (Fellows and Liu, 2022; Goh and Loosemore, 2017).

6.4. Results and discussion

6.4.1 MMC as a risk mitigation approach

MMC represent a transformative shift in how construction projects are designed and constructed, particularly in relation to safety and risk mitigation. The findings from the interpretivist analysis demonstrate that MMC functions not merely as a technological innovation but as a systemic safety paradigm that embeds prevention, precision and performance into all phases of project delivery. This section synthesises evidence from secondary sources, structured around the study's four sub-questions, to illustrate MMC's risk mitigation pathways, contributing to safer construction environments.

6.4.2 Comparative safety performance: MMC vs traditional construction

Comparative analysis of MMC and traditional construction consistently reveals substantial safety advantages linked to off-site, factory-controlled production. Evidence from case studies such as the MMCUK1(Laing O'Rourke, 2020) and MMCJP (Gao *et al.*, 2022) demonstrates how transferring construction activities to controlled factory settings significantly reduces on-site risks. At MMCUK1, prefabrication of walls, columns and bathroom pods cut on-site labour by 60% and reduced working hours by over 237 000, resulting in a 33% decline in hazard exposure and the near elimination of working-at-height incidents. Likewise, MMCJP reported approximately 60% fewer injuries, attributed to reduced manual handling, standardised workflows and enhanced environmental control within factory settings.

These results are reinforced across the literature. Gao *et al.* (2022) and Sekisui House (2020) confirmed that modular production minimises manual tasks and stabilises working conditions, directly lowering accident rates. Similarly, Laing O'Rourke's outcomes at MMCUK1 (Ayinla *et al.*, 2019) exemplify how prefabrication embeds risk mitigation within the construction process itself.

Further, digital integration amplifies these benefits. MMCUK2 contributed to a 23% reduction in on-site hazard exposure through BIM-led modular sequencing (Banks *et al.*, 2018), while studies by Gan *et al.* (2021) found an additional 40–50% decrease in fall-related and manual operation injuries through the application of automation and artificial intelligence-based safety monitoring systems. The Building and Construction Authority (BCA) (2020) in Singapore recorded approximately a 70% reduction in reportable site accidents for modular projects. Likewise, the UK Health and Safety Executive (HSE, 2019) and HM Government (2013) noted 60–80% safety improvements through design for manufacture and assembly (DfMA) and off-site fabrication.

Conversely, traditional construction projects such as 3 Thirty-Three Downtown, Florida (OSHA, 2016) and the Imperial Pacific International Hotel, Saipan (2020) exhibit high incident rates arising from falls, scaffold collapses and equipment failures, typical of labour-intensive, site-based operations. Comparative statistics (see Table 6.1) show that MMC significantly reduces accident frequencies across key categories: falls from height (5–10% compared to 35–40%), struck-by incidents (3–5% compared to 20–25%), manual handling injuries (2–5% compared to 15–20%), electrocutions (<1% compared to 5–10%) and equipment failures (2–4% compared to 10–15%) (Banks *et al.*, 2018; Gao *et al.*, 2022; OSHA, 2016).

Synthesising evidence from academic, industrial and governmental sources, it is evident that MMC displaces, digitises and regulates high-risk operations, resulting into a substantial reduction in accident occurrence relative to traditional construction. This quantitative advantage underscores MMC's role as an active safety management system rather than a reactive site-control measure.

6.4.3 Factory-controlled environments: minimising on-site exposure

Factory-controlled settings in MMC not only transfer physical work from unpredictable construction sites to managed indoor environments but also incorporate advanced technologies and systematic controls that significantly improve safety performance. The highly regulated

Table 6.1 Comparative analysis of accident types and hazard frequency between MMC and traditional construction

Accident Type / Hazard	Traditional Construction	Modern Methods of Construction	Source
Falls from height	35–40%	5–10%	Gao *et al.*, 2022; OSHA 2016
Struck by / moving objects	20–25%	3–5%	Jeong *et al.*, 2021; Fard *et al.*, 2017
Manual handling / musculoskeletal injury	15–20%	2–5%	Banks *et al.*, 2018
Electrocution / contact with utilities	5–10%	<1%	Jeong *et al.*, 2021; Gao *et al.*, 2022
Equipment / scaffold failure	10–15%	2–4%	OSHA (2016–2021)
Other / miscellaneous	5–10%	1–2%	Fard *et al.*, 2017

Source: Authors' creation

production environment ensures strict compliance with occupational safety standards, continuous monitoring and immediate action in case of deviations. At MMCUK1, for example, extensive prefabrication within a controlled factory environment eliminated many high-risk tasks such as formwork, rebar tying and high-level installations, resulting in a 33% reduction in hazard exposure and improved ergonomic conditions for workers (Banks *et al.*, 2018). The use of precision-engineered modules and mechanical handling systems further decreased strain injuries and human error. Similarly, MMCJP shows how automated fabrication, robotic lifting and structured task sequencing can virtually eliminate falls and manual handling injuries, contributing to a 60% decline in total injuries and no major on-site incidents (Gao *et al.*, 2022; Sekisui House, 2020).

Furthermore, factory settings promote improved supervision and data-driven safety monitoring systems through sensors and BIM, which enable real-time detection of potential hazards before assembly. This was exemplified at MMCUK2, where digital integration and modular sequencing allowed virtual hazard detection and cut on-site labour by half, contributing to a 23% reduction in hazard exposure (Banks *et al.*, 2018). Overall, these examples demonstrate how MMC's factory-controlled environments embed safety through engineering precision, automated handling and continuous oversight, replacing the variability and reactive safety culture of traditional construction with a proactive, systematic risk management approach.

6.4.4 Technological pathways: BIM, IoT, AI and automation

The complexity of MMC operations has heightened the need for smart technologies (STs) capable of collecting and processing data in real time to support proactive decision-making. Central to this digital transformation is the integration of BIM with the IoT and AI, which enables a continuous flow of data between the digital and physical environments. Integrating IoT connectivity and AI algorithms (e.g. deep learning) with BIM's extensive project data has the potential to enhance MMC safety impact. This progression aligns with Deng *et al.*'s (2021) five-level taxonomy of BIM evolution, where Levels 3 and 4 represent BIM integration with IoT and AI, respectively.

The adoption of these technologies is exemplified in several MMC case studies. At MMCUK1, BIM and digital twin technologies were deployed to coordinate the design, sequencing and modular assembly of over 1200 precast walls, 821 columns, 661 service modules and 243 bathroom pods (Laing O'Rourke, 2020), This digital integration reduced on-site work by 237 099 hours and workforce exposure by 33%, as IoT-enabled tracking systems and automated quality checks provided real-time oversight of component handling and installation (Banks *et al.*, 2018).

Similarly, at MMCJP IoT sensors and robotics were embedded within the modular production lines to monitor ergonomic conditions, control environmental variables and optimise machine-assisted lifting. This factory-controlled system, enhanced by AI-driven safety analytics, achieved zero major on-site accidents and contributed to a reduction in total injuries by 60% (Gao *et al.*, 2022; Sekisui House, 2020). On MMCUK2, BIM was employed for design coordination, modular sequencing and clash detection, supported by digital simulations that predicted assembly risks before installation. This use of virtual hazard mapping and workflow automation contributed to a 50% reduction in on-site labour and a 23% decrease in hazard exposure (Banks *et al.*, 2018).

Beyond these projects, empirical studies further demonstrate BIM-IoT and BIM-AI applications for safety management. Wang *et al.* (2020) developed the Intelligent Construction System for Prefabricated Buildings (ICSPB-IoT), which combines IoT-based sensors with a BIM-integrated safety database to identify unsafe practices and issue corrective alerts in real time. Zhai *et al.* (2019) introduced an IoT-enabled BIM platform that reduced abnormal events by 85% in production and 80% in transportation and assembly by monitoring modular components through sensor data. Likewise, Tan *et al.* (2020) integrated BIM and LiDAR technologies for automated geometric inspections, identifying defective modules before installation and eliminating struck-by risks.

BIM-AI applications have also advanced predictive safety management, with Awolusi *et al.* 2018) and Sidani *et al.* (2018) detailing BIM-linked physiological and environmental monitoring systems and Zhang *et al.* (2015) developing a BIM-based fall hazard identification model. Collectively, these developments illustrate how the BIM-IoT-AI system supports risk prediction, real-time supervision and intelligent decision-making throughout the MMC processes. This integrated digital environment transforms MMC into a proactive, data-driven and precision-engineered construction paradigm, where safety performance is continuously optimised through automation and analytics.

6.4.5 Regulations and standards in shaping safety outcomes in MMC

Regulations and building standards played a decisive role in shaping the positive safety outcomes observed in all three MMC case studies by embedding safety into design, planning and production processes rather than relying solely on reactive site control.

At MMCUK1, Laing O'Rourke's use of extensive off-site prefabrication was strongly guided by the Construction (Design and Management) Regulations 2015 (CDM, 2015) and the Health and Safety at Work Act (1974), which place a legal duty on designers and contractors to eliminate or reduce risks 'as far as reasonably practicable' at the design stage (Health and Safety Executive [HSE], 2015). These regulations encouraged early risk assessment and collaborative design coordination, leading to the relocation of high-risk activities such as working at height and manual lifting to controlled factory environments. Compliance with BS 5975:2019 (Temporary Works Procedures) and BS EN 1090 (Execution of Steel and Aluminium Structures) further ensured that modular components were safely designed, tested and installed (British Standards Institution [BSI], 2019). As a result, the project achieved a 33% reduction in hazard exposure and improved worker ergonomics (Banks *et al.*, 2018). Here, regulation acted not as a constraint but as a design philosophy that guided safe innovation.

MMCJP exemplifies how national safety and quality laws can institutionalise a culture of prevention. The company's operations were underpinned by Japan's Industrial Safety and Health Act (1972) and the Building Standards Act (1950, revised 2015), both of which mandate manufacturers to ensure factory safety, structural reliability and consistent product testing (Ministry of Health, Labour and Welfare [Japan], 2015). These frameworks incentivised Sekisui House to adopt robotic lifting systems, automated inspection technologies and ergonomic workflows within its production facilities, thereby reducing the likelihood of falls, strains and human error. Consequently, the firm reported zero major on-site accidents and a 60% drop in total injuries compared to traditional construction methods (Gao *et al.*, 2022;

Sekisui House, 2020). In this case, regulatory compliance did not merely enforce minimum safety; it fostered technological innovation that redefined safe construction practices.

Similarly, MMCUK2 demonstrates how digital compliance BIM can help teams meet the requirements of UK Building Regulations Part A (Structural Safety) and Part C (Site Preparation and Resistance to Contaminants) while also aligning with CDM 2015's emphasis on design-stage risk management (BSI, 2018; HSE, 2015). BIM's capacity for clash detection and virtual hazard simulation ensured that design flaws were addressed before on-site assembly, contributing to a 50% reduction in on-site labour and a 23% drop in hazard exposure (Banks *et al.*, 2018).

Overall, across these cases, regulations and standards served as enablers of safety innovation rather than bureaucratic barriers. By enforcing accountability, encouraging safe design integration and promoting digital and factory-based controls, they transformed MMC projects into construction environments where safety is inherently designed into the process rather than merely inspected after execution.

6.4.6 Conceptual model of MMC risk mitigation approach

Figure 6.1 illustrates how MMC function as an integrated risk mitigation system that enhances both construction safety and project performance. This system operates through four interconnected pathways: off-site prefabrication, which transfers high-risk on-site activities to controlled off-site facilities, reducing exposure to hazards; factory-controlled environments, which provide standardised workflows, mechanisation, quality assurance and safer working conditions, thereby minimising manual handling and fall risks; digital technologies such as BIM, IoT and AI, which support hazard prediction, real-time monitoring and process optimisation; and regulatory frameworks, which institutionalise safety standards and enforce compliance mechanisms (summarised in Table 6.2).

Together, these pathways not only reduce accident frequency and severity but also foster proactive risk management and promote safer, more sustainable construction practices. Based on the evidence presented in this chapter, the combination of off-site factory environments, digital technologies and policy-driven safety systems enables the early detection and elimination of risks. It improves compliance with established safety standards, thereby significantly reducing accident frequency by at least 70% in MMC operations compared to traditional construction.

6.5. Implications of the study

The findings of this study have profound implications for both theory and practice in the construction industry, particularly regarding how MMC can be institutionalised as a proactive risk mitigation strategy. Firstly, the evidence confirms that MMC fundamentally redefines construction safety management by shifting risk away from dynamic, uncontrolled sites to predictable, factory-controlled environments, thereby transforming safety from a reactive compliance activity into a design-embedded process. This shift has theoretical significance for extending the DfS and prevention through design (PtD) frameworks, as it demonstrates that hazard elimination can be achieved through structural and procedural innovation rather than post-incident intervention.

Figure 6.1 A conceptual model of MMC's risk mitigation pathways in construction

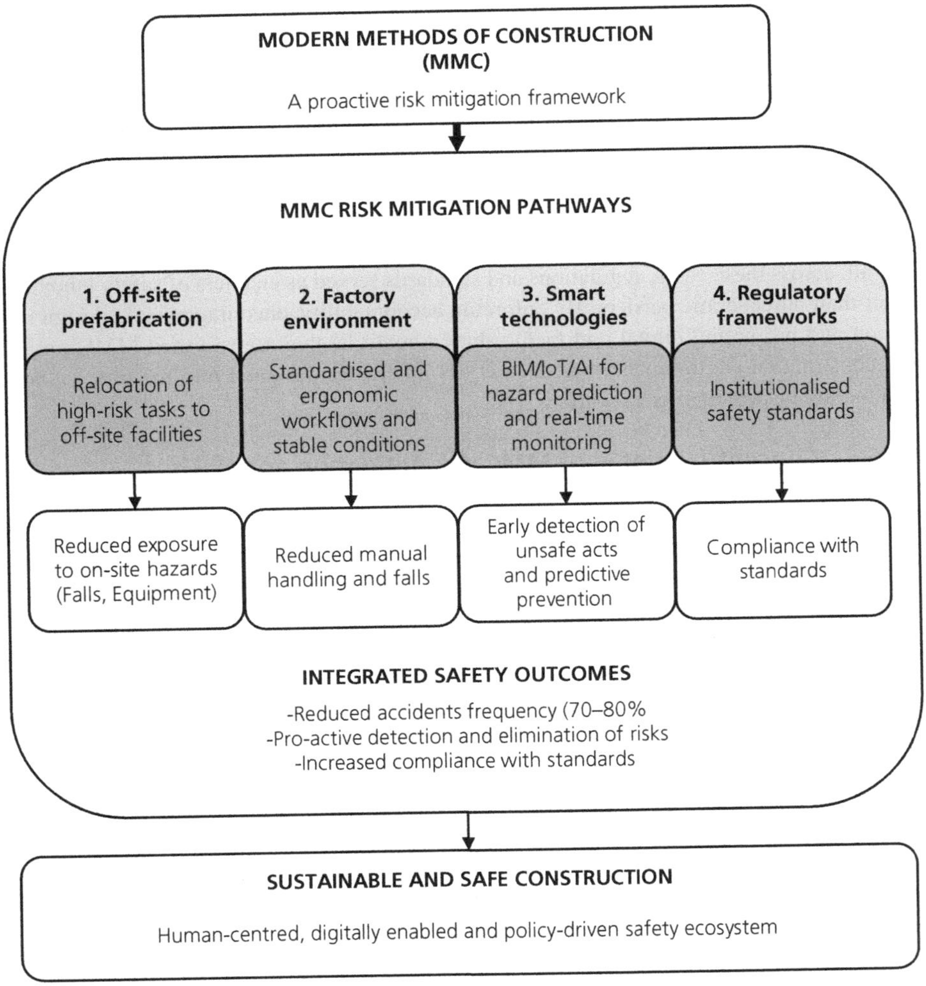

Source: Authors' creation

Secondly, the integration of digital technologies, such as BIM, IoT and AI, illustrates a paradigm of data-driven safety governance. By enabling real-time monitoring, predictive analytics and simulation-based risk forecasting, these STs allow project teams to anticipate and control hazards before they occur, reinforcing the transition toward predictive, evidence-based safety systems.

From a practical perspective, the study underscores the need for policy realignment and institutional capacity building to maximise MMC's safety potential. Governments and regulatory

Table 6.2 MMC's risk mitigation pathways

MMC Pathway	Direct Mitigation Process	Primary Risk Reduced	Supporting Evidence / Case Study
Off-site prefabrication	Transfers high-risk tasks (e.g. assembly, lifting) from site to controlled factory	Reduces falls from height, manual handling injuries, equipment failure	(Banks *et al.* 2018; Gao *et al.*, 2022)
Factory-controlled environment	Standardised workflows, ergonomic workstations, environmental stability	Mitigates climatic hazards, fatigue and on-site variability risks	(Gan *et al.*, 2021)
Digital design and planning (BIM)	Enables virtual hazard detection, clash analysis and sequencing optimisation	Prevents assembly errors, structural failures and rework-related injuries	(Banks *et al.*, 2018);
IoT-based monitoring	Tracks real-time worker and equipment safety data	Reduces unsafe behaviour and equipment misuse	(Hasanzadeh *et al.*, 2018)
AI and predictive analytics	Identifies high-risk scenarios before accidents occur	Prevents falls and manual operation errors	
Automation and robotics	Replaces manual lifting, cutting and repetitive tasks	Reduces musculoskeletal disorders and contact injuries	(Gan *et al.*, 2021)
Regulatory and policy integration	Ensures compliance through design standardisation, audits and prequalification	Reduces noncompliance accidents and poor practice risks	(BCA, 2020; Chan *et al.*, 2020; HM Government, 2013)

Source: Authors' creation

bodies must integrate MMC-specific safety standards, factory accreditation systems and digital compliance monitoring into national construction regulations. This approach, already demonstrated in Singapore and the UK, could be replicated in developing countries, where fragmented regulatory systems and limited enforcement capacity hinder MMC adoption.

Economically, the reduction in accident occurrence in MMC has significant implications for productivity, insurance costs and project life-cycle performance. Safer work environments contribute to fewer disruptions, reduced compensation claims and enhanced stakeholder confidence, factors that cumulatively support sustainable growth in the construction sector. Environmentally, MMC's precision manufacturing and reduced site exposure align with green construction objectives, minimising waste, emissions and resource inefficiency.

Ultimately, this study positions MMC not merely as a construction innovation but as a systemic safety transformation model, integrating design, technology and regulation to advance the global agenda for sustainable, resilient and safe built environments.

6.6. Conclusions

The primary aim of this study was to explore how MMC can be strategically utilised to reduce risks and enhance construction safety. Based on a comprehensive review of secondary data, the research demonstrates that MMC functions as a systematic risk mitigation strategy, fundamentally transforming safety management across the construction life cycle. By relocating high-risk, labour-intensive tasks from unpredictable on-site environments to controlled factory settings, MMC substantially decreases exposure to critical hazards such as falls from height, manual handling injuries, electrocutions and equipment failures. Its strength lies in proactive risk elimination rather than reactive management, achieved through standardisation, mechanisation and precision manufacturing.

The integration of digital technologies such as BIM, IoT and AI further embeds predictive safety controls, enabling real-time hazard detection and informed decision-making before incidents occur. Evidence from industrial case studies consistently shows that MMC reduces accident rates compared to traditional construction methods, confirming its role as a systematic safety engineering solution. Supported by DfMA principles, strict quality assurance and regulatory compliance, MMC institutionalises safety from design through delivery. Beyond improving safety, MMC enhances productivity, resource efficiency and environmental sustainability, positioning it as a strategic, integrated framework that places prevention, precision and performance at the heart of modern construction practice.

MMC face several challenges despite their benefits, including high initial costs, skills shortages, limited availability of trained personnel, design inflexibility, coordination issues between off-site and on-site activities, regulatory barriers, limited supply chains and resistance to change within the industry. However, the future of MMC is promising as these barriers are increasingly being addressed: investment in training and upskilling can overcome skills gaps; technological advancements in design and modular construction enhance flexibility and coordination; policy support and financial incentives ease regulatory and cost constraints; and a growing awareness of sustainability and efficiency leads to cultural acceptance. As supply chains become more robust and standardised, MMC is poised to become a mainstream approach for large-scale and high-demand construction projects.

To guide the strategic implementation of MMC and to strengthen empirical understanding, contextual relevance and the integration of technology, policy and practice, the following knowledge gaps need to be addressed.

- *Empirical validation of safety outcomes in MMC*. There remains limited empirical evidence quantifying the causal relationship between MMC adoption and reduced accident rates across diverse contexts. Future research should employ longitudinal and mixed-method studies to validate MMC's safety benefits beyond anecdotal and case-based evidence.

- *Comparative risk modelling frameworks*. Development of standardised safety and risk assessment models that quantitatively compare MMC and traditional construction methods across project types and regions remains a research priority.
- *Contextual and geographical variability*. Most studies originate from developed economies (e.g. the UK, Japan, Singapore). Further research is needed to examine how MMC safety and risk mitigation mechanisms perform in developing countries, where regulatory systems, skills and resources differ significantly.
- *Digital technology integration and efficacy*. While BIM, IoT and AI are frequently cited as enablers of MMC safety, few studies assess their actual impact in real projects. Future research should focus on real-world implementation for safety monitoring and risk forecasting impact of construction workers' hazard identification skills on their visual attention.
- *Regulatory and institutional frameworks for MMC safety*. Research should evaluate how policy instruments, certification systems and enforcement mechanisms shape MMC safety performance and compliance, particularly in contexts with weak regulatory oversight.

Addressing these gaps is essential to advance future research on MMC, encompassing empirical studies, comparative modelling, contextual adaptation, technology validation and regulatory evaluation.

REFERENCES

Awolusi I, Marks E and Hallowell M (2018) Wearable technology for personalized construction safety monitoring and trending: review of applicable devices. *Automation in Construction* **85**: 96–106.

Ayinla KO, Cheung F and Tawil A (2019) Offsite manufacturing and construction: drivers, barriers and future directions. *Construction Innovation* **19(2)**: 239–260.

Banks C, Kagioglou M and Aouad G (2018) Design for manufacture and assembly in modular high-rise buildings: the case of Project Two Fifty-One, London. *Journal of Construction Engineering and Management* **144(12)**: 1–12.

BCA (Building and Construction Authority) (2020) *Modular Construction and Design for Manufacturing and Assembly (DfMA) Implementation Report 2020*. https://www1.bca.gov.sg/buildsg/productivity/buildability-buildable-design-and-constructability/building-control-regulations?utm_source=chatgpt.com (accessed 12/10/2025).

Blismas N and Wakefield R (2009) Drivers, constraints and the future of offsite manufacture in Australia. *Construction Innovation* **9(1)**: 72–83.

Bogna D, Koskela L and Rooke J (2020) *Research Methods in Built Environment: Exploring the Interpretivist Paradigm*. Routledge, London, UK.

BSI (British Standards Institution) (2018) UK building regulations: approved documents A and C. BSI, London, UK. https://www.gov.uk/government/collections/approved-documents (accessed 12/10/2025).

BSI (British Standards Institution) (2019) BS EN 1090 – Execution of Steel and Aluminium Structures. BSI, London, UK. https://knowledge.bsigroup.com/ (search for BS EN 1090) (accessed 20/08/2025).

CDM (Construction (Design and Management) Regulations) (2015) *The Construction (Design and Management) Regulations 2015*. The Stationery Office, London, UK. https://www.hse.gov.uk/pubns/books/l153.htm (accessed 11/10/2025).

Chan DWM, Lu W and Kwan JC (2020) Modular integrated construction in Hong Kong: implementation strategy and policy framework. *Built Environment Project and Asset Management* **10(4)**: 566–581.

Chatzimichailidou M and Ma Y (2022) Using BIM in the safety risk management of modular construction. *Safety Science* **154**: 105852.

Creswell JW and Creswell JD (2023) *Research Design: Qualitative, Quantitative and Mixed Methods Approaches*, 6th ed. SAGE, Thousand Oaks, CA, USA.

Daniel EI, Oshodi OS, Nwankwo NI, Emuze FA and Chinyio E (2025) The use of digital technologies in construction safety: A systematic review. *Buildings* **15**: 1386.

Deng M, Menassa CC and Kmat VR (2021) From BIM to digital twins: a systematic review of the evolution of intelligent building representations in the AEC-FM industry. *Journal of Information Technology in Construction* **22(2021)**: 58–83.

Denzin N and Lincoln YS (2023) *The SAGE Handbook of Qualitative Research*, 6th ed. SAGE, Thousand Oaks, CA, USA.

Easterby-Smith M, Thorpe R and Jackson P (2021) *Management and Business Research*, 7th ed. SAGE, London, UK.

Egan J (1998) Rethinking Construction: The Report of the Construction Task Force. https://constructingexcellence.org.uk/wp-content/uploads/2014/10/rethinking_construction_report.pdf (accessed 10/07/2025).

Fardi MM, Terouhid SA, Kibert CJ and Hakim H (2017) Safety concerns related to modular/prefabricated building construction. *Journal of injury control and safety promotion*, **24(1)**:10-23. doi: 10.1080/17457300.2015.104786

Fellows R and Liu A (2022) *Research Methods for Construction*, 5th ed. Wiley-Blackwell, Oxford, UK.

Flick U (2023) *An Introduction to Qualitative Research*, 7th ed. SAGE, London, UK.

Gan X, Chang R and Wen T (2021) Factory-based production and its impact on safety in modular construction. *Engineering, Construction and Architectural Management* **28(10)**: 3147–3166.

Gao S, Jin R and Lu, W (2022) Safety performance in modular integrated construction: evidence from Japan and China. *Safety Science* **150**: 105713.

Gao YF, Gonzalez VA and Yiu TW (2019) The effectiveness of traditional tools and computer-aided technologies for health and safety training in the construction sector: a systematic review. *Computers and Education* **138**: 101–115.

Goh YM and Loosemore M (2017) The impact of lean production on safety performance in construction. *Safety Science* **98**: 285–295.

Government of Hong Kong (2017) *The Chief Executive's 2017 Policy Address: Policy Agenda*. Hong Kong. https://www.policyaddress.gov.hk/2017/eng/pdf/Agenda.pdf (accessed 17/10/2025).

Grand View Research (2025) *Modular Construction Market Size Report, 2025–2034*. Grand View Research, San Francisco, CA, USA. https://www.grandviewresearch.com/industry-analysis/modular-construction-market (accessed 24/10/2025).

Hasanzadeh S, Esmaeili B and Dodd M (2018) Safety by design: leveraging technology to reduce risk in construction. *Safety Science* **109**: 186–197.

Health and Safety at Work etc. Act (1974) Health and Safety at Work etc. Act 1974. https://www.legislation.gov.uk/ukpga/1974/37/contents (accessed 05/09/2025).

HM Government (2013) *Construction 2025: Industrial Strategy for Construction*. Department for Business, Innovation and Skills, London, UK. https://www.gov.uk/government/publications/construction-2025-strategy (accessed 03/11/2025).

HSE (Health and Safety Executive) (2015) Managing health and safety in construction: construction (design and management) regulations 2015 – Guidance on Regulations (L153). HSE Books, London, UK. https://www.hse.gov.uk/pubns/books/l153.htm (accessed 17/08/2025).

HSE (Health and Safety Executive) (2019) HSE science and research publications 2019. https://solutions.hse.gov.uk/resources/hse-science-and-research-publications-2019 (accessed 12/10/2025).

Jeong G, Kim H, Lee HS, Park M and Hyun H (2022). Analysis of safety risk factors of modular construction to identify accident trends. *Journal of Asian Architecture and Building Engineering*, **21(3)**: 1040–1052.

Johnsson H and Rudberg M (2015) Coordinating off-site production and on-site assembly through value chain integration. *Construction Management and Economics* **33(7)**: 565–579.

Kamanga T and Steyn W (2022) Challenges in the adoption of modern construction methods in Sub-Saharan Africa. *Journal of Construction in Developing Countries* **27(1)**: 67–88.

Karimi F, Zhang G and Skitmore M (2024) Qualitative synthesis in construction management: integrating desktop research and interpretivism. *International Journal of Project Management* **42(3)**: 234–246.

Laing O'Rourke (2020) *New Grange Hospital Project Safety Report*. Laing O'Rourke, London, UK. https://www.laingorourke.com/thinking/grange-university-hospital/ (accessed 12/09/2025).

Latham M (1994) Constructing the Team: Final Report of the Government/Industry Review of Procurement and Contractual Arrangements in the United Kingdom Construction Industry. https://constructingexcellence.org.uk/wp-content/uploads/2014/10/Constructing-the-team-The-Latham-Report.pdf (accessed 10/07/2025).

Li J (2024) The relationship between artificial intelligence (AI) and building information modelling (BIM) technologies for sustainable building in the context of smart cities. *Sustainability* **16(24)**:10848.

Lincoln YS, Lynham SA and Guba EG (2018) Paradigmatic controversies, contradictions and emerging confluences. In *The SAGE Handbook of Qualitative Research*, 5th ed. (Denzin NK. and Lincoln YS (eds)). SAGE, Thousand Oaks, CA, USA.

MarketsandMarkets (2025) *Global Modular Construction Market Report 2025–2034*. MarketsandMarkets, Pune, India. https://www.marketsandmarkets.com/Market-Reports/modular-construction-market-11812894.html (accessed 24/10/2025).

Masood R (2024) Modelling prefabricated construction safety. *Applied Sciences* **14(4)**:1629.

Meacham BJ (2022) Fire performance and regulatory considerations with modern methods of construction. *Buildings and Cities* **3(1)**:464–487.

Meng Q, Liu W, Li Z and Hu X (2021) Influencing factors, mechanism and prevention of construction workers' unsafe behaviours: a systematic literature review. *International Journal of Environmental Research and Public Health* **18**: 2644.

MHCLG (Ministry of Housing, Communities and Local Government) (2019) *Modern Methods of Construction: Category Definitions*. MHCLG, London, UK. https://www.london.gov.uk/sites/default/files/cast_-_mmc_-_december_2020.pdf (accessed 04/09/2025).

Ministry of Health, Labour and Welfare [Japan] (2015) *Industrial Safety and Health Act (1972) and Building Standards Act (1950, revised 2015)*. Government of Japan, Tokyo, Japan.

OSHA (Occupational Safety and Health Administration) (2016) *Construction Accident Investigation Reports 2010–2016*. U.S. Department of Labour, Washington, DC, USA. https://www.osha.gov/construction/engineering (accessed 17/09/2025).

OSHA (2020) *Investigation of the January 10, 2020, Scaffold Collapse in Saipan, Commonwealth of the Northern Mariana Islands*. Engineering Report No. 2020-07. Department of Labor Washington DC, USA.

Pan Y and Zhang L, (2023) Integrating BIM and AI for smart construction management: current status and future directions. *Archives of Computational Methods in Engineering* **30**:1081–1110.

Payne J and Serin B (2023) *Modern Methods of Construction in the UK Housing Sector: Challenges and Prospects*. UK Collaborative Centre for Housing Evidence, London, UK.

Pereira F (2025) BIM for safety: applying real-time monitoring technologies to reduce construction accidents. *Applied Sciences* **15(4)**: 2218.

Precedence Research (2025) *Global Modular Construction Market Forecast 2024–2034*. Precedence Research, Ottawa, Canada. https://www.precedenceresearch.com/modular-construction-market (accessed 24/10/2025).

Rafa N and Khalid R (2024) *Modern Methods of Construction (MMC) for Net Zero Housing: Implications from the Social Sciences and Humanities*. UK Energy Research Centre (UKERC) Working Paper, 21 March. https://ukerc.ac.uk/publications/modern-methods-of-construction-mmc-for-net-zero-housing-implications-from-the-social-sciences-and-humanities/ (accessed 01/11/2025).

Scolobig A, Komendantova N and Patt A (2023) Desktop research and the interpretivist approach in disaster risk reduction. *International Journal of Disaster Risk Science* **14(5)**: 891–905.

Sekisui House (2020) *Annual Report 2020: Safety and Quality in Modular Housing*. Sekisui House Ltd, Osaka, Japan. https://www.sekisuihouse.co.jp/library/english/company/sustainable/2020/all-A4.pdf?utm_source=chatgpt.com (accessed 10/10/2025).

Sidani A, Duarte J, Dinis F *et al.* (2018) Impact of BIM-based virtual and augmented reality interfaces on health and safety in construction projects: protocol for a systematic review. *International Journal of Occupational and Environmental Safety* **2(1)**: 67–74.

Smith RE (2016) *Offsite Architecture: Constructing the Future*. Routledge, New York, NY, USA.

Sunindijo RY and Zou PXW (2012) How project managers' skills may influence the development of safety climate in construction projects. *International Journal of Project Organisation and Management* **4(3)**: 286–301.

Tan Y, Li S and Wang Q (2020) Automated geometric quality inspection of prefabricated housing units using BIM and LiDAR. *Remote Sensing* **12(15)**: 1–24.

Wang X, Wang S, Song X and Han Y (2020) IoT-based intelligent construction system for prefabricated buildings: study of operating mechanism and implementation in China. *Applied Sciences* **10(18)**: 1–18.

Zhai Y, Chen K, Zhou JX and *et al.* (2019) An Internet of Things-enabled BIM platform for modular integrated construction: a case study in Hong Kong. *Advanced Engineering Informatics* **42**: 100997.

Zhang S, Sulankivi K, Kiviniemi M and *et al.* (2015) BIM-based fall hazard identification and prevention in construction safety planning. *Safety Science* **72(2015)**: 31–45.

Part II

Performance impact and benefits of MMCs

Emmanuel I Daniel and Fidelis A Emuze
ISBN 978-1-83708-437-1
https://doi.org/10.1108/978-1-83708-436-420261008

Chapter 7
Modern methods of construction and quality management

Olugbenga Timo Oladinrin[1,2]

[1]*School of Art, Design and Architecture, Faculty of Arts, Humanities and Business, University of Plymouth, Plymouth PL4 8AA, UK* 0000-0003-3555-2430
[2]*Cidb Centre of Excellence, Faculty of Engineering and the Built Environment, University of Johannesburg, Johannesburg 2006, South Africa*

Abstract

The construction sector is undergoing a revolution in the form of modern methods of construction (MMC), which aims to improve efficiency, sustainability and quality performance. In this chapter, the science of MMC is assessed in relation to quality management, examining how evolving approaches to technology enablers and off-site manufacturing contribute to performance enhancement and defect minimisation. The chapter applies international standards, as outlined in ISO 9001 and digital quality control tools to the task of understanding quality management, which must adapt to the inherent complexity of MMC. With the help of conceptual and theoretical analysis, the discussion emphasises that types of MMC technologies, such as prefabrication, modular construction and digital twins, require stringent quality assurance systems and methods that do not align with conventional construction processes. There was a suggestion that collaborative supply chain systems, an upskilled workforce and flexible regulatory structures would be necessary to secure quality in MMC operationalisation. The chapter aims to lay the groundwork for the reappraisal of quality assurance models by industry professionals and academics when considering the use of innovative and automated construction methods.

Keywords: modern methods of construction; quality management; prefabrication; lean construction; off-site manufacturing; ISO 9001

7.1. Introduction

The construction industry has a history of low productivity, cost overruns and variable quality results (Farmer, 2016). Modern methods of construction (MMC) have been developed as a solution to these age-old problems, with a focus on innovation and process reengineering.

MMC involves a variety of off-site and hybrid construction methods aimed at increasing efficiency, minimising waste and improving the quality of the build. The increasing popularity of MMC internationally demonstrates a move towards industrialised construction in line with lean production observed in manufacturing (Goodier and Gibb, 2007).

Quality management is also a crucial issue in the construction realm when parts are prefabricated off site but assembled on site. Compliance with design requirements, regulatory standards and avoidance of expensive rework necessitate an organised process for quality assurance (QA) and quality control (QC). As with any new technology, the use of MMC has also introduced novel quality issues, particularly in terms of retracing the structure, managing tolerances and integrating multiple parts into a unified whole (Pan *et al.*, 2018).

In the era of MMC, the gulf between invention and reliable quality supply remains. This gap will be addressed by exploring the relations between MMC and quality management in this chapter. This chapter aims to discuss how a quality management system can be a solid instrument to support the implementation of MMC, thereby avoiding the pursuit of innovation that may compromise performance and durability. It covers quality assurance mechanisms, then discusses relevant quality management models, and how QA and QC applications can be tailor-made to MMC environments. Case studies and rising digital technologies are discussed to showcase best practices and quality improvement in the construction process. The final part of the chapter provides practical advice for practitioners and policymakers seeking to integrate quality into their MMC systems.

7.2. Perspectives and concepts

7.2.1 Quality management in construction

Quality management plays a vital role within the construction industry, tasked with ensuring that the built environment, as constructed, complies with the design, develops within its legal framework and meets, or even exceeds, clients' wants and needs. The development of quality management in construction has followed the general trend in industry, focusing on the systematic control of procedures, materials and end products. This section examines the fundamental concepts of quality assurance and quality control, the application of international standards and guidelines, and traditional and contemporary approaches to quality management within the construction context.

7.2.2 Principles of quality QA and QC

QA and QC are the two main components of quality management in construction. QA focuses on planned and systematic activities implemented within a quality system, ensuring that quality requirements for an item are fulfilled. It includes components such as method statements, material specifications, supplier accreditation criteria and training programs (Oakland, 2014).

On the other hand, QC is a reactive activity used to identify and fix defects during or after construction. QC encompasses inspection tests conducted at the workshop, material verification, and dimensional and tolerance checks. Whereas QA is process driven, QC is product driven. The combination of both ensures that the processes not only drive the desired results, but also that any deviation is identified and corrected as soon as possible.

QA and QC are interdependent. Effective quality control should be integrated with a feedback loop to input the QC findings, to correct the QA and prevent problems from occurring in the future. Some of the primary principles of QA and QC in construction include consistency, compliance, continual improvement and customer satisfaction (Bravi *et al.*, 2019).

7.3. Methodological approach

Given the conceptual nature of this study, a qualitative approach rooted in secondary research has been adopted. The chapter synthesises insights from academic journals, industry reports and policy documents to examine the convergence of MMC and quality management practices. The use of a desk-based review is appropriate given the aim of evaluating evolving theoretical frameworks and best practices in MMC quality management.

The methodology follows a structured literature review process involving the identification, selection and critical analysis of relevant literature. Sources were drawn from databases including Scopus, Web of Science and Google Scholar, using search terms such as 'modern methods of construction', 'quality management', 'lean construction', 'prefabrication' and 'ISO 9001 in construction'. Selection criteria focused on relevance to the chapter's themes, recency (post-2000 literature), and the inclusion of peer-reviewed and high-impact publications.

Thematic analysis was applied to organise the findings into conceptual categories including: *(a)* quality frameworks applicable to MMC *(b)* integration of digital tools in quality assurance, and *(c)* challenges of quality control in MMC. This approach enables the chapter to derive insights from existing knowledge while drawing out new implications for the evolving construction landscape. To ensure practical relevance, the chapter includes case illustrations from recent MMC projects in the UK and globally that highlight how quality risks were managed and what tools or strategies were employed.

7.4. Results and discussion

7.4.1 ISO standards and regulatory frameworks

The International Organisation for Standardisation (ISO) has established frameworks on quality management that are widely adopted, including ISO 9001:2015, which specifies the requirements of a quality management system (QMS). ISO 9001 is founded on seven quality management principles: customer focus, leadership, people involvement, process approach, improvement, decision-making based on evidence and relationship management (Bravi *et al.*, 2019). ISO 9001 certification supports specific quality declarations for construction companies, which must demonstrate their ability to continuously supply products and services that meet customer and applicable regulatory requirements. The standard is process driven and requires organisations to identify, measure and improve their processes.

Development quality is, however, also governed by regional and national standards, in addition to the ISO codes. In the UK, for instance, the Building Regulations 2010 set up minimum requirements for design, construction and alterations. The National House Building Council (NHBC) sets technical standards for house building and offers warranties for quality inspections (NHBC, 2016). Even though no formal regulations exist, certification mechanisms, such as the Considerate Constructors Scheme, BRE's BREEAM and the Setto Buildoffsite Property Assurance Scheme (BOPAS), also complement formal regulations by incentivising best practices and offering market differentiation for quality-seeking companies. Furthermore, occupational health and safety rules (e.g. the CDM Regulations 2015 in the UK) overlap with quality requirements, as many defects related to the built environment are due to poor site management and a failure to observe safe practices (Gould and Joyce, 2009).

7.4.2 Traditional compared with modern quality management approaches

For centuries, construction quality was determined mainly by the skill of the tradesman and less formal supervision could ensure it. The more traditional approach would be to focus on end-of-process inspections and to depend, often considerably, on site managers and engineers to police quality after the event (Harris *et al.*, 2020). This 'reactionary' business model frequently resulted in rework, missed deadlines and cost overruns. Moreover, there were fragmented construction supply chains at the time, along with adversarial contractual arrangements that hindered collaborative quality improvement (Egan, 1998).

By comparison, today's quality management models are proactive, process-oriented and are applied throughout all stages of the construction process. They are broadly consistent with the principles of total quality management (TQM), lean construction and Six Sigma. TQM is directed towards creating a culture of quality at every level of the organisation, founded on continuous improvement, teamwork and customer satisfaction (Oakland, 2014). Lean Construction focuses on eliminating waste and maximising value using tools such as the Last Planner System and just-in-time delivery, which contribute to increased predictability and stability (Koskela, 2000).

Recent decades have seen the development of digital tools that have redefined contemporary quality management. Building Information Modelling (BIM) tools can help in early clash detection of design, scheduling and quantity take-offs. The ongoing comparison of construction today to historical measurements is made possible by drones, laser scanners and real-time sensors that provide a quality check. In addition, integrated project delivery (IPD) models promote early stakeholder involvement, shared risks and rewards that motivate joint accountability over quality (Kent and Becerik-Gerber, 2010). Contemporary approaches even analyse projects to identify trends for guiding future QA strategies based on project analytics. This decision to transition from reactive to predictive quality control processes signifies a paradigm shift in the industry's approach to and management of construction quality.

In summary, management in construction flows from reactive inspection to proactive process-based methods supported by global standards, digital tools and collaboration. Through the application of these contemporary paradigms, the construction industry can not only meet but also exceed the requirements for client satisfaction, performance and project success.

7.4.3 Interfacing MMC with quality management

The fusion between MMC and quality delivery systems has opened the doors for new paradigms of excellence in construction delivery. In contrast to conventional construction, MMC relies on prefabrication, modularisation and efficient site installation, which in turn necessitate stronger control over production and coordination. In this section, we will describe how

QA and QC procedures overlap at off-site and on-site stages, the necessity of traceability in prefabrication and the link between digital technologies, notably BIM, the Internet of Things (IoT) and quality monitoring.

7.4.3.1 Integrating QA/QC in off-site and onsite stages

The site-based nature of construction is at odds with the concept of prefabrication and assembly of manufactured components offered by MMC. This provides valuable opportunities to improve quality, as factory conditions are likely to be far less variable and more amenable to repeat high-precision work (Pan *et al.*, 2012). During the off-site phase, QA measures include standard operating procedures (SOPs), material qualification procedures, equipment calibration and trained personnel. Suppliers tend to implement ISO 9001-compliant systems for prefabricated components, uniformity and traceability – for example, volumetric modular construction can benefit from in-line inspection at various stages of fabrication, with 'DFT (design for testability) and off-site 'mill' realisation having the potential to eliminate defects and remediation in factory-based operations largely' (NHBC Foundation, 2016).

On the site level, QC involves the process of installing, aligning and attaching modules or panels. Tolerance in jointing systems, sealing performance, and mechanical and electrical interfaces needs to be carefully supervised. Here, the relationship between elements produced and manufactured in the factory and those assembled on-site reduces density or increases air permeability, potentially leading to misalignment between the elements, which can damage structural capacity and thermal performance (Lawson *et al.*, 2012).

Adoption of QA/QC throughout both off-site and on-site phases will require a comprehensive project quality plan that clarifies lines of accountability, inspection points, communication lines and resolution mechanisms. This is vital for DfMA methods where cross-discipline coordination must be accurate.

7.4.3.2 Traceability and documentation in prefabrication

Traceability implies the ability to trace the history, location and application of products, images and processes throughout the value chain. In the structure, where several precast elements are joined to form the final assembly, reliable history tracking is crucial from the perspectives of quality control, regulatory requirements and warranty security (Gould and Joyce, 2009). Each base component should be associated with a digital document that stores material sources, date produced, workshop, group, batch number and test result, among others. This information is necessary for conducting root cause analysis to investigate the causes of failures and for providing feedback in quality improvement.

The application of barcoding, QR codes and radio frequency identification (RFID) technology has ushered in a new era of MMC traceability. Parts can then be inspected at multiple stages and the resulting information can be input into an integrated quality database. This provides a real-time view of production status and alerts for exceptions.

It is also the case that digital records facilitate compliance with industry standards and/or requirements, such as the BOPAS certification, which requires the existence of traceable evidence of the durability and performance of modular buildings over a 60-year design life (Dara and

Hachem-Vermette, 2019). Also, standardised documentation templates for inspection and testing plans (ITPs), non-conformance reports (NCRs) and handover checklists promote uniformity throughout a project. Solutions hosted in the cloud also enable collaborative review and approval, removing barriers between manufacturers, general contractors and quality assurance auditors.

7.4.4 Digital tools and their roles in quality tracking

In MMC, digital technologies have become essential tools for quality monitoring. Chief among these is BIM, through which coordinated, data-rich digital models are developed to serve as a central reference point for design, fabrication and construction. BIM improves quality in several ways (see Figure 7.1). For example, clash detection software helps identify and resolve design conflicts before production, thereby minimising rework. According to Eastman et al. (2011), BIM also enables the consistent updating of project models as designs evolve, while supporting the visualisation of installation sequences and the capture of as-built information required for facility management.

Quality monitoring can be enhanced with IoT additional technologies, allowing for real-time sensor data collection from embedded components and site infrastructure. Temperature, humidity, vibration and structural loadings can be monitored by sensors, which can provide advanced alerts of out-of-specification deviations (Lu *et al.*, 2017). The development will be an iterative process, where designers work with the information they have today, and add further information that is required or can be obtained tomorrow. Some examples of how sensor technology can be added to improve project execution include the use of sensors that monitor the curing of concrete and notify a project team when it can start carrying loads, as well as the application of GPS technology to track modules and enhance the delivery plan.

Quality inspection and training in an augmented reality (AR) environment, combined with digital twins, can create a more intuitive experience for assessing quality and training users to simulate and evaluate a construction scene before performing the real process.

Figure 7.1 Application of BIM in quality management

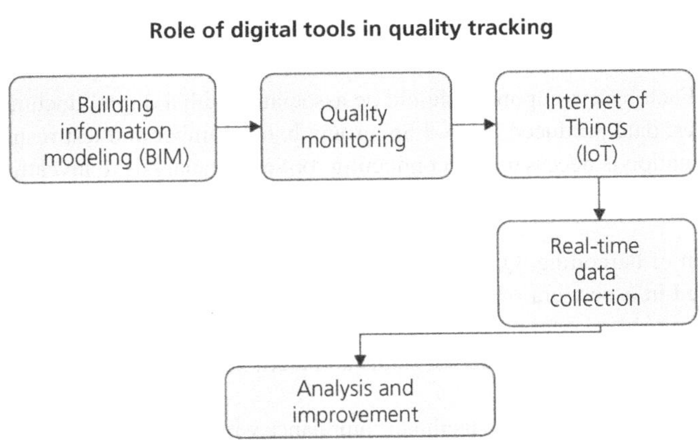

Source: Author's creation

Digital dashboards collect this information into quality performance indicators (QPIs) that identify areas of concern and enable continuous improvement. Systems such as these align well with Lean Six Sigma. They will aid in predictive analytics, representing a paradigm shift in quality control that is far removed from traditional historical quality inspections. For the true power of digital tools to be unlocked, integration between all software is key. Interoperable standards like Industry Foundation Classes (IFC) ensure data can move between BIM, scheduling, cost estimating and quality management systems without any loss of fidelity.

In summary, interfacing MMC and quality management implies an overarching approach that encompasses both pushing through processes and leveraging digital systems. From the factory to the job site, the integration of QA/QC guidelines, traceability programs and digital technologies is working to ensure that new building methods deliver high-performing, defect-free and long-lasting results.

7.4.4.1 Case studies and best practices

Robust quality systems as part of MMC have already been piloted on a few projects worldwide. This section presents several case studies to illustrate the implementation of QMS in MMC and to extract performance indicators from these cases.

EXAMPLES OF SUCCESSFUL MMC PROJECTS WITH ROBUST QUALITY SYSTEMS

BERKELEY MODULAR (UK)

As the flagship example of a vertically integrated MMC solution, Berkeley Group's off-site manufacturing factory in Northfleet, Kent, provides a fantastic case study. The factory manufactures volumetric modules with full mechanical, electrical and plumbing (MEP) installed, claddings and internal finishes. A strong quality management system supports every part of the production. The facility complies with the ISO 9001 standard and uses digital twins to model the assembly and quality aspects of a module. Each module is barcoded and associated with a digital quality file that documents inspections, compliance reports and transportation records (Berkeley Group, 2020). This ensures accountability, facilitates early defect identification and promotes data-driven improvement.

APEX HOUSE, WEMBLEY (UK)

This 29-storey student housing scheme incorporates volumetric modular construction from Tide Construction and Vision Modular Systems. Quality control, as over 679 modules were manufactured off-site before being built on-site, was also an essential factor that helped meet structural, acoustic and thermal performance criteria. Quality control procedures also encompassed pre-production samples, factory visits, and ITPs. BIM was used to resolve design clashes and coordinate MEP routing at an early stage. The project consequently obtained BOPAS accreditation and was delivered 30% quicker than a traditional build (McKinsey and Company, 2017).

LAING O'ROURKE'S EXPLORE INDUSTRIAL PARK (UK)

Laing O'Rourke's Explore Industrial Park is an off-site manufacturing facility for precast concrete elements serving extensive infrastructure and healthcare projects. Lean processes are utilised for continuous improvement, with quality control integrated into the design and delivery process. Six Sigma methods are employed at the factory to monitor and improve part precision, and real-time sensors capture quality data that is displayed by way of digital dashboards. The system monitors over 20 quality measures, ranging from curing conditions to installation tolerances (O'Rourke, 2019).

CLEMENT CANOPY, SINGAPORE

The 40-storey twin tower United Venture Development's residential development is the world's tallest concrete modular building. It was developed using the Prefabricated Prefinished Volumetric Construction (PPVC) system and over 1800 modules were produced off-site in Malaysia. The Building and Construction Authority (BCA) of Singapore has stipulated a stringent quality framework, including required pre-delivery inspections, third-party certification and Construction Quality Assessment System (CONQUAS) compliance. The project's performance was high in terms of dimensional tolerances and finish quality, with defects accounting for less than 30% of those in traditional construction (BCA, 2024).

7.4.4.2 Lessons learned and quality performance indicators

LESSONS LEARNED

The examination of these projects reveals several fundamental lessons essential to the successful introduction of quality into MMC.

(a) *Early quality planning integration*. Effective projects incorporate quality planning throughout the entire design-to-construction process. This means that inspection criteria, tolerances and the responsibilities of quality assurance teams at each juncture are outlined.
(b) *Use of digital technology*. BIM or other digital methods track coordination and detect design conflicts early on, providing more efficient documentation. The digital connectivity between design, construction and manufacturing is characteristic of high-performing projects.
(c) *Traceability focused*. Successful quality projects demonstrate the progression from the originating issue to its resolution, clearly identifying the components affected. This is achieved through the use of barcoding, RFID tags and cloud-based quality logs.
(d) *Collaborative quality culture*. Quality is not something assigned by general contractors as a need, but rather a responsibility shared among designers, fabricators, installers and owners. Frequent quality briefings, root cause analysis workshops and cross-functional quality reviews are typical.
(e) *Standardisation and repeatability*. Repetition in design and manufacturing enables the use of statistical process control, fostering the development and refinement of assembly methods that minimise variability and ultimately lead to consistency.
(f) *Training and competence assessment*. Ongoing training of staff in quality and digital protocols is key to success. The programs that invest in worker training are likely to have lower defects and higher productivity.

QUALITY PERFORMANCE INDICATORS (QPIS)

The most common QPIs are

- first-time pass rates
- rework percentage
- number of defects per unit/module

- inspection compliance rates
- customer satisfaction scores
- time and cost performance relative to benchmarks.

The use of these metrics enables project teams to gauge progress, benchmark against performance to date or other projects, and drive process improvement and organisational learning. Lastly, the most successful MMC projects are those that connect 'old-school' manufacturing discipline with construction's new ways of working, underpinned by a robust and transparent quality management system. These examples demonstrate that 'building quality in' rather than treating quality as a compliance activity can enhance delivery time, introduce better cost management and drive enhanced levels of client satisfaction.

7.4.5 Emerging technologies enhancing quality in MMC

The construction industry is undergoing a significant transformation with the integration of emerging technologies aimed at improving quality, efficiency and reliability. In MMC, which focuses primarily on industrialised processes and off-site production, technologies such as robotics, AI and digital twins open new doors to assurance and quality. In this section, we explore how automation in manufacturing, AI-driven analytics for real-time quality monitoring and digital twin technology contribute to predictive quality control in MMC.

7.4.5.1 Automation and robotics in manufacturing

Increased automation is a feature of many MMC facilities, most notably high-volume prefabrication. Automatic systems ensure accuracy, reproducibility and a reduction in human error, which can be crucial for maintaining product quality. In many applications, robots are now used to perform welding, cutting, handling of parts and even assembling parts. These can operate continuously, maintaining the consistency of product batches. Robots, such as those used in welding steel frames or installing wall panels, can achieve millimetre-level accuracy greater than that of manual work. Quality is also improved by automation in that it can be used to adjust production factors immediately. Computer numerical control (CNC) machines, for instance, maintain consistency with every cut and joint made according to the design. State-of-the-art vision systems mounted to robotic systems also provide dimension checks and anomaly detection during the manufacturing process.

There is even more to be gained from automation in MMC than accuracy. It also creates a healthier and safer working environment by minimising human exposure to drudgery and danger, thereby reducing the likelihood that human errors resulting from these will compromise quality (Rafa and Khalid, 2024). In addition, automated operations contribute to lean practices by minimising waste, shortening production cycle times and enabling just-in-time delivery. Such advantages, overall, enable a more controlled and superior end output in combination with a comprehensive QA/QC system.

7.4.5.2 Quality monitoring and AI analytics in real time

Real-time control involves continuously monitoring key quality variables throughout the manufacturing and assembly processes. With sensors, IoT devices and edge computing, MMC practitioners can receive live information on material properties, environmental conditions and operational performance. Moreover, parts with embedded sensors can monitor temperature, humidity, structural strain and vibration. This information is crucial for controlling the curing conditions of concrete elements, maintaining thermal insulation standards and detecting insensible displacements that may be related to a structural anomaly (Lu *et al.*, 2017).

AI analytics provide a whole new life to this data by rendering it actionable. When provided with historical defect data, machine learning models can predict the probability of failure in similar situations. Thus, it allows preventive QC measures, such as modifying environmental conditions or flagging equipment for recalibration. When producing bathroom pods, for instance, sensors can collect data that AI algorithms can process to identify variations in typical watertightness (down to the millimetre) of the pods before they even leave the factory. Likewise, AI image recognition can be used to detect surface defects on cladding panels or pre-cast concrete.

These digital dashboards aggregate sensor data and analytics into visual key performance indicators, informing decision-making. Such KPIs may include defect frequency, process efficiency and first-time pass rate. Quality managers can narrow their focus to a root cause investigation rather than a manual inspection with AI-powered anomaly detection. Furthermore, scheduling software AI integration enables dynamic resourcing, meaning you can reassign work or assets to mitigate quality risks you identify in real-time.

7.4.5.3 Digital twins and predictive quality control

A digital twin is a computational and dynamic model of a physical asset, process or system, used for prediction, analysis and control purposes. In MMC, the digital twin serves as a powerful tool for quality assurance, integrating design information, manufacturing data and operational feedback into a cohesive model. The simulation of stress loads, thermal performance and service life of components before they are manufactured opens the prospect of predictive quality control thanks to digital twins. Engineers, for example, could alter the type of structural support or insulation material to forestall any failure by modelling the real-world conditions for the transport and installation of a modular unit.

Digital twins can be synchronised with IoT-enabled devices to represent quality metrics in real time. Any variances between the design and as-built conditions, such as over-excavation, moisture content, or vibration tolerances, can be automatically identified and corrected before proceeding to the next stage. Digital twins also support life-cycle quality management. Manufacturing and assembly-phase collected data are used to predict maintenance and performance degradation, enabling pre-emptive asset care post-construction (Bolton *et al.*, 2018). Digital twin technology also simplifies compliance records. As-built records, inspection reports and compliance certificates are embedded within the digital twin, providing a single source of truth and enhancing transparency and accountability. From a project management perspective, digital twin technology facilitates collaboration in quality review by enabling stakeholders to interact with a standard virtual model. This capability provides a cross-disciplinary coordination facility, accelerates problem-solving and supports continuous improvement through feedback.

In conclusion, novel technologies are changing the paradigm of quality assurance in MMC. Automation and robotic systems also contribute to uniformity and minimising errors during production. Real-time tracking and AI analysis provide actionable data that can be used for immediate improvements in quality. Digital twins bridge the digital and physical worlds, offering an unprecedented level of control over the entire life cycle of building elements. As these technologies continue to develop, their incorporation into quality management systems will not only be beneficial but also necessary. Those construction companies that embrace these tools will be better positioned to build toward a sustainable, high-performing and increasingly client- and regulation-focused project.

7.4.6 Regulatory and compliance considerations

As MMC is being developed, so too must the speed of regulation and compliance. With the growing popularity of off-site and modular construction methods, it has become increasingly important to align these technologies with established building codes, standards and third-party certification programs. In this section, we examine how MMC code compliance is addressed, the role of third-party certifiers and the health, safety and environmental aspects of quality in industrialised construction.

7.4.6.1 Compliance with code and third-party certification

It establishes the minimum acceptable standards for construction quality, structural integrity, fire safety, energy efficiency, accessibility and other key aspects. For MMC, the complexity of enforcing these codes is higher due to the decentralisation of construction (from conventional sites to factories) and multi-jurisdictional involvement. One of the challenges is to infer intent in code that was generally already written with traditional discipline methodologies in mind – for instance, questions of fire resistance or acoustics may require validation by way of alternate means for volumetric modules or structural insulated panels (SIPs). Building control and standards practice are also being revised to suit MMC (MHCLG, 2019).

Third-party certification programs are used to close the regulatory loopholes. In the UK, BOPAS accredits MMC systems to a minimum 60-year expected design life for durability and maintainability. The BOPAS certification process encompasses an extensive certification program that includes manufacturer quality assurance systems, structural testing and site inspections. Additionally, NHBC Accepts is a certification scheme offered by the National House Building Council for assessing new MMC types for compliance with building control and warranty delivery. It gives confidence to developers, lenders and insurers in the system's reliability (NHBC 2020).

Globally, UL, Intertek and ICC Evaluation Service also offer third-party evaluation on compliance with local and global codes, including ISO. These are important certificates for marketplace considerations, for cross-border modular housing supply chains. Digitally facilitated verification processes, like 'checking against a BIM', are also becoming more popular. These systems enable the automatic code-based analysis of design models against building codes, thereby reducing the amount of manual code interpretation required and improving the safeguarding of regulatory analysis (Eastman *et al.*, 2011).

7.4.7 Health, safety and environmental quality

MMC promises to improve HSE by being delivered as part of a well-managed quality system. However, it also raises new challenges that need to be addressed to ensure quality. MMC also has several health and safety advantages. Workers in non-permanent structures are less exposed to inclement weather, working at heights and repetitive strain activities, all of which contribute to high levels of injury in conventional construction. Production environments in the factory are more controllable and a set of standardised safety procedures can be more easily introduced (Sadeghi *et al.*, 2025).

However, MMC presents transportation, craning and installation risks, especially for large volumetric modules. Parties involved want to avoid this high risk and high potential for liability, so quality systems will be expected to include robust lift plans, install procedures and safety inspections. Crucially, learning logistics, safety and assembly techniques are required to ensure the safety of working environments on MMC-enabled sites.

MMC also has some environmental benefits. It tends to be less wasteful, less resource-intensive and have a lower carbon footprint than building in situ, because off-site fabrication enables material usage and logistics to be better controlled. According to most accounts, high-performance, prefabricated elements can lead to improved thermal performance and reduced energy consumption throughout the building's lifespan (Pomponi and Moncaster, 2017).

Quality management systems should integrate environmental quality benchmarks, including

- material traceability and compliance with environmental standards (e.g. FSC-certified timber)
- waste reduction metrics and recycling rates
- energy performance verification (e.g. air tightness and u-value testing)
- life cycle assessment (LCA) documentation.

Incorporating environmental performance into QA/QC frameworks ensures not only code compliance but also alignment with broader sustainability goals such as those outlined in BREEAM or LEED certification standards.

Additionally, there is a growing reliance on post-occupancy evaluations (POEs) to ensure that MMC buildings meet their performance potential. POEs evaluate indoor air quality, noise levels, thermal comfort and user satisfaction, relating these to the quality of design, materials and building workmanship. HSE metrics and accountability are also a growing priority for regulatory agencies. Her Majesty's Government (HMG) has enshrined these principles into legislation with the UK Building Safety Act 2022 arising from the Grenfell Tower tragedy, which requires a 'golden thread' of building requirement throughout the building's lifetime such that developers and contractors must evidence how quality and safety have been achieved from design to occupancy (HM Government, 2022).

In summary, regulation and compliance issues are critical for the effective implementation of MMC. Negotiating code compliance, obtaining third-party certifications and integrating health, safety and environmental quality monitoring into quality management systems are necessary to ensure that industrialised construction methods deliver innovation and speed as well as reliability and long-term performance. As regulations continue to change, companies will be challenged to take more proactive approaches toward this by working closely with certifying bodies, adopting digital verification technologies and investing in comprehensive HSE QA procedures that protect the HSE of workers, users and the environment itself.

7.4.8 Training and competency in MMC quality systems

MMC has necessitated a shift in the skills and competencies required across the construction sector. Quality management in MMC is not just about embracing new technologies; it is also about bringing the workforce up to a decent level to deliver good quality. This section discusses the requirement for retraining workers for quality-centric MMC, elaborating on the roles of both academia and construction companies in developing the competence framework for quality assurance and quality control.

7.4.8.1 Upskilling the workforce for quality-focused MMC

MMC involves processes that are quite remote from what most people would regard as conventional construction but involve modular manufacturing, precision tower crane assembly, and harnessing digital technologies such as BIM, IoT and digital twins. Quality in such contexts requires technical and managerial skills.

The first domain of upskilling is digital literacy. Operators and managers need to know how to utilise digital tools for performing quality inspections, documentation and tracing – for example, BIM skills can allow people to visualise 3D models, check clash detection results and maintain design and build digital documents (Eastman *et al.*, 2011).

Secondly, effective production, especially under lean manufacturing systems, relies heavily on skills in precision measurement and quality control. The operators will need to be able to read standard operating instructions (SOI), take dimensional checks and adhere to clean room procedures of controlled environments.

Thirdly, training related to quality characteristics, such as root cause analysis, statistical process control (SPC), and the use of inspection and test plans (ITPs), is crucial. Such skills enable staff to recognise signs of quality deviation early and apply corrections efficiently (Oakland, 2014). Apprenticeships and modular training focused on MMC are growing trends. Other important qualities include communication, critical thinking and adaptability. Quality management efforts in MMC are frequently cross-cutting, challenging teams to collaborate across manufacturing, logistics and assembly on-site.

7.4.8.2 Institutional and contractor-level responsibilities

Competency in MMC quality systems needs to be developed and retained by both educational providers and constructors.

EDUCATIONAL BODIES

Universities, polytechnics and industry training organisations need to update training courses, so they are relevant to the MMC world. Traditional construction programs need to start offering modules on modular design, quality management frameworks and digital construction tools. Partnerships between academia and industry can support keeping training programs responsive and results focused.

They are supported by accreditors such as the Chartered Institute of Building (CIOB) and the Royal Institution of Chartered Surveyors (RICS), which are revising their competency frameworks to specify the knowledge required for non-traditional suppliers and updating the quality management standards for off-site construction. Mid-career individuals should have easier access to professional development programs and continuing education courses. This ensures that practising individuals aren't left out of the MMC shift and can lead on quality under new standards.

RESPONSIBILITY AT CONTRACTOR LEVEL

Construction companies are required to provide training programs, competencies and offer continuous professional development (CPD) for their employees. Large contractors with off-site manufacturing facilities may run their academies or collaborate with local colleges to offer tailor-made courses. Quality systems should have training matrices, competence assessment procedures and refresher training.

Contractors must also verify that their subcontractors and supply chain partners share the same position on quality. This includes reviewing subcontractors' training programs, induction sessions and disseminating of digital tools and standardised QA/QC protocols to the project team. Shadowing experienced staff during inspections in the factory, digital handovers and quality audits will help one gain practical skills. Mentoring and buddy systems also encourage skill transfer in teams. To institute these habits, some companies are incorporating quality metrics into performance evaluations and linking quality results to training outcomes. This not only promotes the importance of training but also creates a culture of quality that is shared by all. It is also government policy.

By applying a public procurement strategy that includes MMC experience and general quality provider certifications as prerequisites, companies are indirectly encouraged to invest in human capital development. Funding models supporting skills development, like the UK's Construction Skills Fund, could be adapted to provide subsidies for MMC-focused quality training.

Quality systems in MMC rely heavily on training and competency building. Equipping the workforce to navigate the technical, digital and procedural demands of industrialised construction is vital to ensuring that such innovations lead to excellent results. Institutions and contractors have separate but supporting roles to play in promoting a culture of competence and continuous improvement. As the industry progresses in its MMC maturity, early-stage investment in training will not only drive up quality outcomes but also increase efficiencies, going together with safety measures and a better work environment, to deliver lasting value for businesses and the built environment overall.

7.5. Summary and recommendations

MMC is transforming the construction industry on an international scale, introducing entirely new ways of working which combine speed, efficiency and ecological sustainability. Nevertheless, as shown in this chapter, the gains promised by MMC can be achieved only if supported by robust, coherent quality management systems. This final section reflects on the main findings of the chapter and provides specific lessons for industry players, policymakers and educators.

7.5.1 Summary of key insights

One of the key considerations in executing MMC projects is the successful integration of QA and QC methods into both off-site and on-site production. The controlled conditions of a factory enable enhanced accuracy and standardisation, but maintaining those standards throughout the logistics-to-site assembly process requires effective planning, a highly skilled workforce and the support of cutting-edge technology.

The significance of documents and traceability in pre-fabrication chains was demonstrated to be a fundamental aspect of the quality of MMC. Systems such as barcode scans and RFID tracking enable detailed, auditable quality trails, ensuring greater accountability and continuous improvement.

MMC quality is increasingly enabled by digital technologies. Software, such as BIM, IoT sensors and digital twins, provides a real-time view of performance, enabling predictive quality control and life-cycle optimisation. These technologies can enable the transition from reactive quality control to successful, proactive and predictive quality management. Real-life cases, such as Berkeley Modular, Apex House, Explore Industrial Park, or Clement Canopy, demonstrated best practices and highlighted the importance of early quality planning, digital integration and cooperation among all participants.

Legitimate access to regulatory and compliance systems, as well as third-party certifications such as BOPAS and NHBC, is essential for establishing MMC systems as credible. Additionally, HSE considerations will need to be integrated into quality frameworks, mainly since MMC will introduce new hazards and operational situations.

Learning and competency building also became an important pillar of quality. As MMC moves away from the traditional approach to building, workers at every level will need to acquire new skills. This combines technical skills in both digital and manufacturing methods and institutional obligation to modify curricular and workforce strategies.

7.6. Study implications

The key implications from the study are as follows.

(a) *Practice life cycle-oriented quality planning*. Quality in MMC should not be considered a one-time task, but rather a permanent responsibility throughout the product's lifetime, from design to operation. Parties involved in the project will need to implement life-cycle-based QA/QC plans that are compatible with performance-based goals throughout the asset's life cycle, utilising digital tools to monitor performance in real time.
(b) *Standardise quality benchmarks for MMC*. Specific MMC benchmarks can help establish industry acceptance if professionals and regulatory authorities collaborate in creating and implementing them. These would cover tolerances, performance indicators, inspection presets and traceability specifications specific to modularisation and off-site fabrication.
(c) *Digital quality tools throughout all phases of construction*. Digital solutions must be integrated into the entire MMC chain. Government and industry initiatives

should encourage investment in interoperable BIM, IoT and digital twin platforms, particularly among SMEs, which may lack the capabilities required for a full-scale digital transformation.

(*d*) *Reinforce third party certification and transparency*. There is a need for certification schemes to be as dynamic and progressive as the robotic manufacturing integration (RMI) invention. BOPAS, NHBC Accepts and equivalent international entities should have stringent, transparent and publicly accessible assessment criteria. Quality personnel and digital competency certification should also be considered.

(*e*) *Implement training and upskilling in procurement rules*. Public procurement rules should be designed to incorporate MMC-specific training requirements as a mandatory factor in tendering. Contractors should be required not only to demonstrate previous experience but also to provide evidence of workforce training and digital readiness as part of quality assurance submissions.

(*f*) *Boost collaboration and information sharing*. Fragmentation remains a significant obstacle to MMC quality. Forums, online platforms of knowledge and cross-sector alliances can help in increasing transparency and the sharing of information. Efforts such as the Construction Innovation Hub need to be adopted globally if we are to work towards international collaboration on MMC quality.

(*g*) *Policy alignment with performance-based outcomes*. Policy approaches should transition from exacting rules to performance-based benchmarks, which recognise the range of MMC systems' capacities. Standards should be based on outcomes, such as energy efficiency, durability and user satisfaction, rather than on fixed methodological requirements.

(*h*) *Encourage the use of post-occupancy evaluation (POE) and feedback loops*. POE and feedback loops should become standard practice for MMC project delivery. Collecting feedback on thermal comfort, air quality, maintenance problems and user satisfaction can be vital to optimising quality processes and guiding future designs.

(*i*) *Develop a 'quality maturity model' for MMC organisations*. Construction companies should evaluate and benchmark their quality management capabilities using a maturity model. This framework can be used to pinpoint deficiencies, bolster capabilities and inform clients and regulators of the readiness of quality.

(*j*) *Invest in research and technology*. Continued investment is also required in MMC quality research (empirical as well as theoretical). Funders and industry should support studies assessing quality performance, researching new risks and exploring new techniques, such as integrating AI, robotics and other novel interventions.

7.7. Conclusions

As MMC continues to expand, the quality of MMC becomes more critical. Building quality into every part of the process, underpinned by a competent workforce, digital systems and progressive policies, will be the factor that decides whether MMC can deliver on its promise of a smarter, swifter and greener built environment. Convergence is the future of quality in construction, including methods and management, technology and training, policy and practice. By embracing this convergence, stakeholders can create a new paradigm for construction that meets the evolving needs of society with integrity, innovation and excellence.

REFERENCES

BCA (2024) *Prefabricated Prefinished Volumetric Construction (PPVC)*. Building and Construction Authority, Singapore.

Berkeley Group (2020) Building for the Future: Modular Housing Report. chrome-extension://efaidnbmnnnibpcajpcglclefindmkaj/https://www.annualreports.com/HostedData/AnnualReportArchive/B/LSE_BKG_2020.pdf

Bolton A, Enzer M and Schooling J (2018) *The Gemini Principles: Guiding Values for the National Digital Twin and Information Management Framework*. Centre for Digital Built Britain and Digital Framework Task Group, Cambridge, UK.

Bravi L, Murmura F and Santos G (2019) The ISO 9001: 2015 quality management system standard: companies' drivers, benefits and barriers to its implementation. *Quality Innovation Prosperity* **23(2)**: 64–82.

Dara C and Hachem-Vermette C (2019) Evaluation of low-impact modular housing using energy optimization and life cycle analysis. *Energy, Ecology and Environment* **4(6)**: 286–299.

Eastman C, Teicholz P, Sacks R and Liston K (2011) *BIM Handbook: A Guide to Building Information Modeling for Owners, Managers, Designers, Engineers and Contractors*. Wiley, NJ, USA.

Egan J (1998) *Rethinking Construction: Report from the Construction Task Force*. Department of the Environment, Transport and Regions, UK.

Farmer M (2016) *The Farmer Review of the UK Construction Labour Model*. Construction Leadership Council. http://www.constructionleadershipcouncil.co.uk/wp-content/uploads/2016/10/Farmer-Review.pdf

Goodier C and Gibb A (2007) Future opportunities for offsite in the UK. *Construction Management and Economics* **25(6)**: 585–595.

Gould FE and Joyce N E (2009) *Construction Project Management*. Pearson Education, Upper Saddle River, NJ, USA.

Harris F, McCaffer R, Baldwin A and Edum-Fotwe F (2020) *Modern Construction Management*. John Wiley and Sons, NJ, USA.

HM Government (2022) Building Safety Act 2022. UK Parliament. https://www.legislation.gov.uk/ukpga/2022/30/contents

Kent DC and Becerik-Gerber B (2010) Understanding construction industry experience and attitudes toward integrated project delivery. *Journal of Construction Engineering and Management* **136(8)**: 815–825.

Koskela L (2000) *An Exploration Towards a Production Theory and Its Application to Construction*. VTT Technical Research Centre of Finland.

Lawson RM, Ogden RG and Bergin R (2012) Application of modular construction in high-rise buildings. *Journal of Architectural Engineering* **18(2)**: 148–154.

Lu Y, Wu Z, Chang R and Li Y (2017) Building Information Modeling (BIM) for green buildings: a critical review and future directions. *Automation in Construction* **83**: 134–148.

McKinsey and Company (2017) *Reinventing Construction: A Route to Higher Productivity*. McKinsey Global Institute, Washington, USA.

MHCLG (Ministry of Housing, Communities and Local Government) (2019) Modern Methods of Construction. House of Commons, London, UK. https://publications.parliament.uk/pa/cm201719/cmselect/cmcomloc/1831/1831.pdf (accessed 24/04/2026).

NHBC Foundation (2016) *Modern Methods of Construction: Views from the Industry*. IHS BRE Press, NHBC Foundation.

NHBC (2020) NHBC Accepts: Promoting Confidence in Modern Methods of Construction. https://www.nhbc.co.uk/technical/accepts

O'Rourke L (2025) Explore Manufacturing: Delivering Better Projects. https://www.laingorourke.com/company/our-businesses/explore-manufacturing/

Oakland JS (2014) *Total Quality Management and Operational Excellence: Text with Cases*. Routledge, London, UK.

Pan W, Gibb AGF and Dainty ARJ (2012) Strategies for integrating the use of off-site production technologies in house building. *Journal of Construction Engineering and Management*, **144(4)**: 04018014.

Pan W, Hu R, Linner T and Bock T (2018) A methodological approach to implement on-site construction robotics and automation: a case of Hong Kong. *ISARC. Proceedings of the International Symposium on Automation and Robotics in Construction, IAARC Publications* **35**: 1–8.

Pomponi F and Moncaster A (2017) Circular economy for the built environment: a research framework. *Journal of Cleaner Production* **143**: 710–718.

Rafa N and Khalid R (2024) *Modern Methods of Construction (MMC) for Net Zero Housing*. Energy Research Centre, London, UK.

Sadeghi H, Cheung C M., Yunusa-Kaltungo A and Manu P (2025) A systematic review of occupational safety and health in modular integrated construction. *Safety Science* **189**: 106897, 10.1016/j.ssci.2025.106897.

Emmanuel I Daniel and Fidelis A Emuze
ISBN 978-1-83708-437-1
https://doi.org/10.1108/978-1-83708-436-420261009

Chapter 8
Cost considerations and optimisation in off-site construction

Lovelin I. Obi[1] and Olayinka G. Omoboye[2]

[1]*Construction and Built Environment, School of Business, Leeds Trinity University United Kingdom* 0000-0003-0325-0091
[2]*Westminster Business School, University of Westminster, United Kingdom*
0000-0002-3565-7745

Abstract

Cost uncertainties are seen as a significant factor challenging the adoption and implementation of offsite construction (OSC) in housing delivery in many countries. Existing cost management frameworks often overlook the life-cycle orientation and front-loaded delivery structure of OSC. This chapter conceptualises a Lifecycle Cost Visibility and Optimisation (LCVO) framework to address this gap by examining the cost structure of OSC projects, including cost factors, categories, drivers and enabling conditions. Through a desk study methodology, it synthesises insights from literature and a comparative cost analysis of two UK-based case studies to conceptualise the framework. The LCVO framework integrates activity-based costing (ABC) and life-cycle costing (LCC) principles to provide a structured lens for understanding how costs arise and how they can be optimised across the entire OSC life cycle. It also incorporates circularity costs, recognising opportunities for value recovery at the end of life. Although conceptual, the LCVO framework provides practitioners and researchers with a tool for enhancing cost transparency, strengthening life cycle decision-making and promoting whole-life value. By extending analysis beyond capital expenditure to include operation, maintenance and disassembly, the framework reframes OSC cost management as a holistic, value-led process aligned with sustainability and circular economy goals.

Keywords: activity-based costing; circularity costs; cost optimisation; construction cost management; life cycle costing; offsite construction

8.1. Introduction

The pressing need to balance cost efficiency with sustainability and circular economy objectives is driving an ongoing shift in the construction industry to adopt non-traditional methods, such as off-site construction (OSC), in housing project delivery (Obi *et al.*, 2025). OSC, also known as off-site manufacturing, modern methods of construction, or industrialised construction, involves the design and fabrication of building elements and components (modules, panels and pods) in a factory, which are subsequently transported and installed on-site (Ofori-Kuragu *et al.*, 2022; Sutrisna *et al.*, 2019).

The growing OSC's uptake is evident in its increasing application across diverse project types and spread across countries (Ginigaddara *et al.*, 2022; Obi *et al.*, 2023; Ofori-Kuragu *et al.*, 2022; Zhang *et al.*, 2020), impacting on its market size, which is forecast to rise significantly, reaching USD 1.9 trillion in 2025 compared with USD 820 billion in 2020 (AGCS, 2021) – for instance, in the UK, it has been voiced as a viable approach to housing demand, labour shortages and sustainability imperatives. The UK OSC sector represents a market worth around £2–3 billion per year with about 16-17% of new homes integrating OSC techniques (KPMG, 2016; Obi *et al.*, 2023; UK Commission for Employment and Skills (UKCES), 2015). This growing appeal may be linked to its perceived benefits, such as controlled factory-based production, which fosters greater precision, standardisation and repeatability, enabling the kind of resource efficiency and waste reduction (Fulford and Standing, 2014; Jiang *et al.*, 2018; Kamali *et al.*, 2018). However, despite the benefits of OSC, its adoption remains a key barrier due to continued cost uncertainties (Agapiou, 2021; Ofori-Kuragu *et al.*, 2022; Sutrisna *et al.*, 2019).

The OSC process comprises several interrelated phases, typically including planning, design and procurement, fabrication, transportation, on-site assembly, use and disassembly (Obi *et al.*, 2023). OSC fundamentally alters the typical construction cost management approach. Unlike traditional methods, where expenditure is distributed across the project timeline, OSC has front-loaded costs at the project's early stages. This often limits flexibility for late-stage changes and, as a result, effective cost management becomes critical in earlier project phases, making accurate forecasting under conditions of high uncertainty critical (Razkenari *et al.*, 2020; Wen and Zhang, 2021). This accelerated commitment directly reflects the manufacturing-led nature of OSC, where substantial financial obligations need to be secured and invested well before physical on-site activities begin. Additionally, the interface between off-site and on-site activities often introduces hidden costs and coordination challenges (Agapiou, 2021).

Yet current cost management practices in OSC remain heavily conventional, with a focus on capital cost management, often neglecting later stages, such as in use and end-of-life. This fragmented view makes it difficult to fully understand cost implications across the OSC building life cycle, especially in relation to value generation, modular reuse and system optimisation. This has led to an increasing demand for a more integrated cost framework that aligns with OSC's manufacturing-led nature and to support informed decision making through the building life cycle. Such a framework must extend beyond capital costs to encompass long-term performance and the potential for circularity. Surprisingly, life cycle cost management models for OSC projects are rare, which limits both academic and professional efforts to manage and optimise costs effectively.

This chapter addresses the gap, integrating activity-based costing (ABC) and life-cycle costing (LCC) principles to conceptualise an LCVO framework for understanding how costs arise and how they can be optimised across the entire OSC life cycle. In doing so, the chapter advances OSC cost management by offering industry practitioners and researchers a structured lens for enhancing cost visibility and optimisation on OSC projects while supporting broader objectives of sustainability, efficiency and circularity. The chapter proceeds as follows: the next section reviews the literature and outlines the theoretical framework; the third outlines the methodology. This is followed by the findings and discussion and the chapter concludes with reflections and directions for future research.

8.2. Perspectives and concepts

8.2.1 Cost considerations in off-site construction projects

Cost plays a significant role in any construction activity and remains a concern in the discourse on OSC, where perceptions of higher initial capital requirements often cover its potential for long-term economic, environmental and operational benefits. Cost is defined as expenses made to achieve a specific outcome (Obi *et al.*, 2017). In the project context, Schram (2012) describes project cost as encompassing all direct and indirect financial commitments made by the client throughout the design and construction phases. Project cost is described as the total financial outlay by the client on cost factors/functions across the entire life cycle of the project (Ashworth and Perera, 2015; Obi *et al.*, 2025). In the OSC context, this would encompass costs associated with planning, design coordination, fabrication, transport and installation logistics, operational maintenance and end-of-life considerations.

Project costs can be classified along several dimensions, including by behaviour (fixed or variable), traceability (direct or indirect), function (e.g. manufacturing, marketing, design) and resource type (labour or materials) (Shihab and Prasad, 2017). Within OSC literature, cost discussions have primarily centred on classification by function, traceability and resources. Resource-based estimates typically aggregate labour, materials, plant, overhead and profit (Cartlidge, 2013).

Cost traceability classifies costs by direct and indirect costs. The former refers to costs traceable to production activities, such as labour, materials and specialised plants. Indirect costs are traced to project-supporting expenditures such as site preliminaries and attributed to project-supporting expenditures, including site preliminaries, essential office overheads, professional fees, contractor profit and risk allowances (Serugga *et al.*, 2023). A function-based cost classification links costs to distinct life-cycle phases, such as design, manufacturing, transportation, assembly, use, maintenance and disassembly.

Understanding cost categories provides a structure to capture and analyse expenditures across the project life-cycle. It enhances visibility on how expenses are incurred over the project timeline, which can be useful for accurate forecasting of total life-cycle costs. By explicitly linking cost categories with life-cycle stages, OSC clients and project managers can achieve a more holistic understanding of cost implications, supporting better optimisation, value delivery and alignment with sustainability goals.

While classification is great, it may not be sufficient. This is because OSC cost performance is also shaped by certain optimisation factors and enabling conditions that determine how these costs are incurred and managed. Scholars identify a wide array of influential factors that bear upon OSC costs (Liu *et al.*, 2023; Xue *et al.*, 2017; Zou *et al.*, 2024). These factors and conditions include the number of inspections, the size of moulds, contractors' capabilities, procurement methodologies, estimator expertise, prevailing market conditions, technological maturity, project design complexity, management capacity and specific certification or performance standards. While these conditions exert both direct and indirect influence on OSC project cost efficiency, they need to be integrated into a unified cost framework for effective OSC project cost management.

However, traditional cost categorisations remain heavily focused on capital expenditure, routinely overlooking both life-cycle-oriented classifications and these necessary enabling conditions. Addressing this limitation will therefore require a dual focus: understanding what costs are incurred (classification) and how these costs are generated and influenced (optimisation and enabling conditions). A project life-cycle cost approach enables a more informed evaluation of cost efficiency, allowing decision-makers to account for the long-term economic benefits of improved build quality, energy efficiency and reduced maintenance and replacement requirements inherent in OSC delivery models (Ashworth and Perera, 2015; Obi *et al.*, 2025).

LCC and ABC emerge as structured approaches to effectively conceptualise the OSC project cost structure. While LCC traces costs across all stages of the building life cycle, ABC highlights the activities and conditions that drive those costs. Together, they may offer a pathway towards a more integrated and effective approach to cost optimisation in OSC.

8.2.2 LCC and ABC in construction

LCC emerged from US military and government procurement during the 1950s and 1960s and expands cost analysis beyond capital costs to encompass the building life cycle. LCC is vital for fostering more sustainable, cost-effective outcomes because it embeds life-cycle thinking into the earliest stages of project development. Rather than prioritising the lowest upfront cost, LCC places emphasis on long-term value by evaluating all projected costs, from acquisition and operation to maintenance and end-of-life expenditure. At its core, LCC encourages early, integrated collaboration among project stakeholders to align design decisions with future performance and cost implications.

In OSC, LCC would enable data-informed strategies that can help clients and project managers balance initial investment with operational value. According to the RICS (2016) LCC report, the core components of LCC cost accounting and commitment structure are divided into the following.

- *Capital costs*. These encompass all expenditures incurred during the initial delivery phases of a construction project, specifically covering design, procurement, manufacturing, transportation and on-site construction.
- *Facility management costs*. These cover expenses associated with operating and maintaining the constructed facility throughout its lifespan.
- *End-of-life costs*. Such costs are related to deconstruction, demolition and disposal management.

While valuable, conventional LCC models appear deficient in capturing critical, OSC-specific variables, such as the prefabrication rate, complex transport logistics and the building's reuse or disassembly potential. The challenge, therefore, extends beyond simply quantifying expenditure over time, but also integrating those costs with the specific processes by which the building is manufactured, delivered and ultimately utilised. This is precisely why relying on LCC alone may prove insufficient in the OSC context.

ABC offers a cost management strategy that provides a framework to allocate and trace costs to products or projects to the specific activities that consume resources (Kaplan and Cooper, 1998). Rather than assigning costs broadly like the LCC, ABC traces costs to the actual activities that generate them, such as procurement, logistics or module installation, making it particularly well-suited to environments with diverse, interdependent activities, like in OSC projects. This capability can help project teams identify high-cost processes, inefficiencies, or pinpoint opportunities for process streamlining. Szczerbak and Wikarczyk (2023) highlighted that the core components of the ABC structure are divided into the following.

- *Cost resources*. These are inputs, such as labour, materials, equipment and energy that are consumed in the performance of specific project activities, such as excavating foundations, constructing walls and installing building services.
- *Cost drivers*. These are what causes costs to increase – for example, the frequency, duration and complexity of activities.
- *Cost categories*. Groupings of related costs by function or stage.
- *Cost factors*. Aggregate cost categories aligned with life-cycle stages.

While LCC effectively highlights when and why costs materialise throughout a building's life, ABC identifies how and where those expenditures are incurred. LCC provides a macro view of costs across the life cycle to support early strategic decisions, while ABC provides a micro-level insight for cost control during production and execution. When integrated, both could offer a multi-layered insight, enhancing cost visibility and supporting effective financial control across the project life cycle. This dual-lens methodology is particularly well-suited to OSC, which demands not only predictability but also adaptability in response to evolving project requirements and stakeholder needs.

8.3. Methodological approach

This chapter adopts an exploratory research approach to examine cost considerations and optimisation in OSC projects. A desk study methodology was adopted, combining a literature review and comparative case study analysis of two UK-based residential projects, one delivered using conventional methods and the other using off-site construction techniques.

The literature review was undertaken to explore and identify existing knowledge on OSC cost factors, categories, drivers and enabling conditions with the aim of informing strategies for optimising cost across the OSC project cycle. Literature reviews are crucial in academic research as they connect past and current studies, effectively building the foundation for future advancements in a discipline (Obi *et al.*, 2023). To ensure relevance, the review focused on peer-reviewed journal articles and reports published between 2020 and January 2025 – a period marked by significant advancements in OSC practices, especially in relation to sustainability and circular economy (CE) application in LCC.

Relevant documents were retrieved from the Scopus database using combined search terms structured as follows: ('off-site construction' OR 'modular buildings' OR 'prefabrication') AND ('life-cycle costing' OR 'construction cost' OR 'cost optimisation' OR 'activity-based costing') AND 'circular economy'. The RICS reports, specifically ICMS and the life-cycle costing guidance note, were also reviewed. A total of 20 documents were selected based on their relevance to cost management in off-site construction (OSC). Each document was manually screened and thematically analysed, enabling a systematic examination of textual data to identify attitudes, values and behaviours pertinent to the study's focus (Rimmel and Cordazzo, 2021).

The comparative case study aimed to understand how OSC diverges from traditional cost structures and the implications of these differences. The documents reviewed from the cases and the literature all provided insights into conceptualising the LCVO framework.

8.4. Results and discussion

OSC has been perceived by some as more expensive, stemming from limited cost transparency and a lack of standardised frameworks to capture the full scope of OSC's value, such as modularity, repeatability and circular performance. This section critically synthesises the key findings derived from the desk study. Initial findings establish the comparative value of OSC versus traditional methods, demonstrating that construction outlines how OSC differs from conventional cost structures. Findings from the review and case studies were synthesised to conceptualise the LCVO framework (Figure 8.1), which integrates cost categories, drivers and optimisation factors across three life-cycle cost factors: capital costs, facility management costs and end-of-life costs.

8.4.1 Comparative value of OSC compared to traditional construction

A comparative value of OSC relative to traditional construction was examined. A life-cycle perspective is required for effective cost management, as the findings show it is important to look beyond just the upfront capital costs. It requires assessing value across the full project life cycle, from early design through to end-of-life. As Fraser *et al.* (2015) note, meaningful comparisons must factor in broader indicators, such as programme certainty, operational savings and risk reduction. Table 8.1 summarises these comparisons, showing how OSC compares against traditional methods in terms of value across the project life cycle.

The comparative analysis of the two projects highlighted how OSC differs from conventional construction cost structure. A detailed understanding of the specific cost structure was essential to conceptualise the framework.

Project A was sourced from cost benchmarks from an earlier study conducted by Tatiya *et al.* (2018), who estimated deconstruction at £89.23/m^2 and demolition at £68.25/m^2 (2014 rates). This has been adjusted to 2020 using BCIS and the updated values of £128.24/m^2 for deconstruction and £84.30/m^2 for demolition are applied. Project B was sourced from an OSC project completed in the UK. For facility management and end-of-life costs, an economic life of 50 years was used to analyse both OSC and conventional buildings. The analysis of the project A and B cost structure is presented in Figure 8.2.

Figure 8.1 LCVO framework

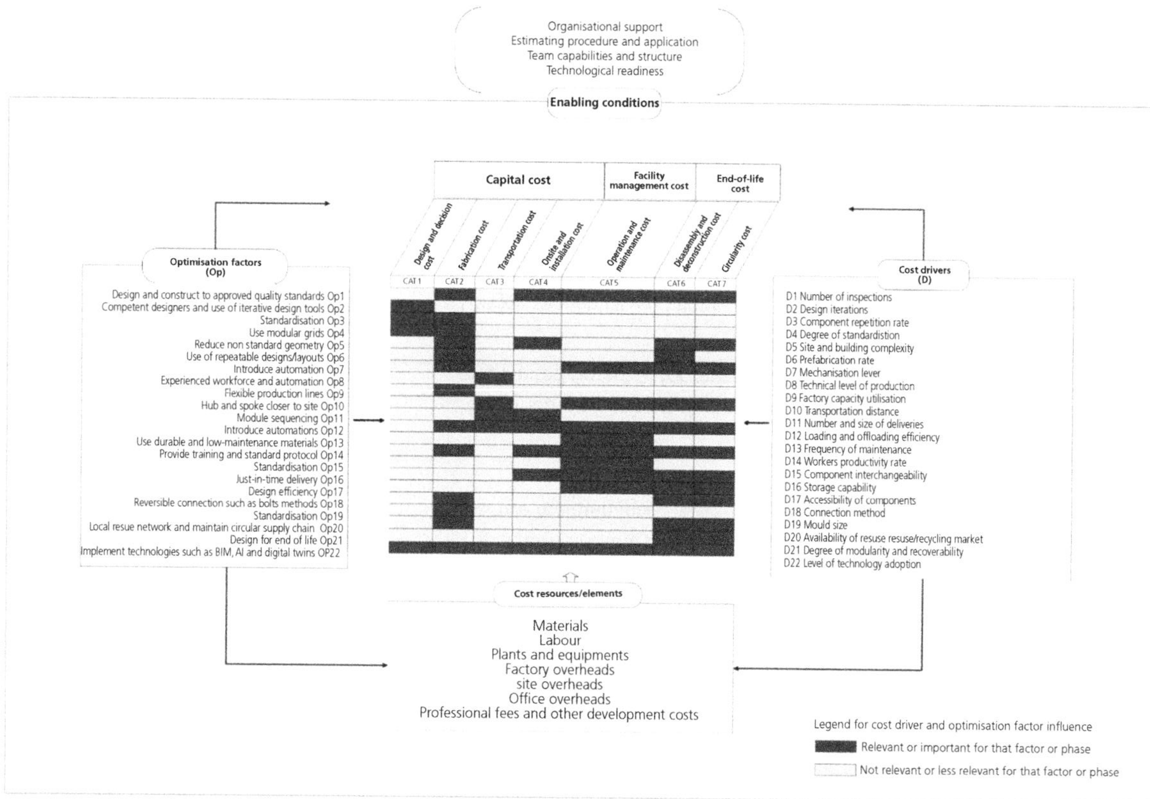

Source: Authors' creation

Table 8.1 Analysis of OSC compared to conventional construction

Metric	Off-site Construction	Conventional Construction	Sources
Capital cost	Higher upfront due to design freeze, factory mobilisation and transport logistics, but gains from efficient factory labour and reduced site trades.	Low initial costs but susceptible to variation orders and rework.	Agapiou, 2021; Ashworth and Perera, 2015; Hong *et al.*, 2018; Jansen *et al.*, 2020; Mao *et al.*, 2016; Ofori-Kuragu *et al.* 2022; Serugga *et al.*, 2023; Sutrisna *et al.*, 2019; Xue *et al.*, 2017; Zhou *et al.*, 2024
Facility management cost	Lower overall due to better building envelope, standardised systems and reduced reactive maintenance; also benefits from data-led facilities management (FM) planning.	Higher due to inconsistent quality, inspections and the burden of reactive maintenance.	Mao *et al.*, 2016; Muse *et al.*, 2021; Ofori-Kuragu *et al.* 2022; RICS, 2016, 2020; Zhang *et al.*, 2020
End-of-life cost	Lower due to opportunity for disassembly, potential for reuse and reduced demolition waste.	Higher labour-intensive demolition and limited potential for reuse.	Mao *et al.*, 2016; Muse *et al.*, 2021; Obi *et al.*, 2023; Sutrisna and Goulding, 2019
Time	Shorter, parallel production and accelerated delivery.	Longer, sequential workflows, delays vulnerable weather conditions.	Kamali *et al.*, 2018; Mao *et al.*, 2016; Ofori-Kuragu *et al.* 2022; Salama *et al.*, 2020
Risk profile	Front-loaded design and logistics risks, but manageable with planning.	Diffuse and harder to control such as site conditions, weather and subcontractor delays.	Agapiou, 2021; Kamali *et al.*, 2016; Ofori-Kuragu *et al.* 2022; Shen *et al.*, 2019;
Defects/quality	Better factory-based quality assurance and control processes, reducing rework and better alignment with design intent.	More variable higher defect rates due to on-site complexities.	Kamali *et al.*, 2016; Navaratnam *et al.*, 2019; Ofori-Kuragu *et al.*, 2022; Sutrisna and Goulding, 2019
Environmental impact	Potential for lower embodied and operational carbon,	Higher emissions, waste and transport impact due to site-built methods.	Cao *et al.*, 2015; Kamali *et al.*, 2016; Kyrö *et al.*, 2019; Ofori-Kuragu *et al.* 2022; Shen *et al.*, 2019

Source: Authors' creation

Figure 8.2 Cost analysis of OSC and conventional building (continued on next page)

CONVENTIONAL METHOD

Maintenance requirements:
Redecoration at 10 year intervals @0.6%
Replacement every 20 years @ 5.8%
Annual cleaning @ 1%
Repair @ 0.6%
Energy @ 1.9%

Bank of England Discount rate of 4.25%

Given
Discount rate 4.25%
CONVENTIONAL METHOD
Initial cost — £63,400.00 ($62m^2$ of 2 bedroom house Prototype)
£1,022.58 $/m^2$

Maintenance requirements:
Redecoration at 10 year intervals @0.6%
Annual cleaning @ 1%

Cost Type	Years	Discount Factor	Cost	Adjusted Cost	Present Value
Project/design fees and other development costs			£5,779.00		£5,779.00
Measured work (including prelimaries and overheads			£57,621.00		£57,621.00
Operation and Maintenance Cost					
	Redecoration				
	10	0.6595	£380.40		
	20	0.4350	£380.40		
	30	0.2869	£380.40		
	40	0.1892	£380.40		
	50	0.1248	£380.40		
		1.6954	£380.40	£645.00	
	Annual Cleaning				
	Year 1-50	20.5931	£634.00	£13,056.00	
	Replacement every 20 years				
	20	0.4350	£3,677.20		
	40	0.1892	£3,677.20		
		0.6242	£3,677.20	£2,295.00	
	Repairs				
	Year 1-50	20.5931	£380.40	£7,834.00	
	Energy				
	Year 1-50	20.5931	£1,204.60	£24,806.00	
					£48,636.00
Demolition Cost			£5,226.60		£5,226.60
Disposal					

Summary	
Project/design fees and other development costs	£5,779.00
Measured work (including prelimaries and overheads and profit)	£57,621.00
Operation and Maintenance Cost	48,636
Demolition Cost	£5,227
Disposal	
	£117,262.60

Source: Authors' creation

Figure 8.2 Continued

OFF-SITE CONSTRUCTION METHOD

Maintenance requirements:
Redecoration at 10 year intervals @0.53%
Replacement everyat 35 years @ 6.4%
Annual cleaning @ 0.9%
Repair @ 0.2%
Energy @ 0.5%

Bank of England Discount rate of 4.25%

Given
Discount rate 4.25%
OFF-SITE CONSTRUCTION METHOD
Initial cost — £79,255.76 (62m² of 2 bedroom house Prototype) £1,278.32 /m²

Maintenance requirements:
Redecoration at 10 year intervals @0.53%
Annual cleaning @ 0.9%

Cost Type	Years	Discount Factor	Cost	Adjusted Cost	Present Value
Design and Decision Cost (Project/design fees and other			£5,343.98		£5,343.98
Fabrication and Transportation Cost (including overheads and proft)			£46,323.09		£46,323.09
Onsite and Installation Cost (including overheads and proft)			£27,588.68		£27,588.68
Operation and Maintenance Cost					
	Redecoration				
	10	0.6595	£420.06		
	20	0.4350	£420.06		
	30	0.2869	£420.06		
	40	0.1892	£420.06		
	50	0.1248	£420.06		
		1.6954	£420.06	£712.00	
	Annual Cleaning Year 1-50	20.5931	£713.30	£14,689.00	
	Replacement @ Year 35				
	Year 35	0.2330	£5,072.37	£1,182.00	
	Repairs Year 1-50	20.5931	£158.51	£3,264.00	
	Energy Year 1-50	20.5931	£396.28	£8,161.00	£28,008.00
Disassembly/deconstruction cost			£7,950.76		£7,950.76
Circularity Cost					

Summary	
Design and Decision Cost (Project/design fees and other development costs)	£5,343.98
Fabrication and Transportation Cost (including overheads and proft)	£46,323.09
Onsite and Installation Cost (including overheads and proft)	27,589
Operation and Maintenance Cost	28,008
Disassembly/deconstruction cost	£7951
Circularity Cost	
	£115,214.52

Findings revealed that the initial capital cost for the OSC was much higher than that of a conventional building. This may have been due to a higher upfront capital cost required for factory-based production and module transportation. However, considering operation and maintenance, OSC yielded major savings of about 43% savings which could be due to lower replacement, reduced repair and lower energy consumption costs. This finding corroborates previous studies' findings that OSC buildings can reduce total costs by up to 10% through energy savings, reduced maintenance and increased asset durability (Zhang *et al.*, 2020).

From the case analysis, the demolition cost was cheaper than the deconstruction cost. While the deconstruction cost may initially appear costlier, it significantly reduces landfill waste and could be recovered through material reuse and salvage value from reclaimed components like panels and pods, also enhancing OSC circular economy outcomes.

Insights from these UK-based case studies affirmed a long-term value proposition. While OSC necessitates a higher capital investment early on, driven by accelerated design finalisation, factory setup and complex logistics, it frequently yields superior returns over the project life cycle. Specifically, lower maintenance requirements, improved energy efficiency and enhanced overall environmental performance collectively render OSC a more cost-effective choice over time. These empirical results are fully consistent with existing literature, which consistently highlights OSC's key benefits, including accelerated project delivery, enhanced quality control and waste reduction.

8.4.2 Conceptual LCVO framework

The findings of the desk study are synthesised into a conceptual LCVO framework (Figure 8.2).

The framework integrates insights from the comparative case analysis and the literature review. It brings together four main elements. Cost structure (cost classifications), cost drivers (activities that drive costs), optimisation factors (adjustment to improve cost efficiency) and the enabling conditions (organisational and capacity support). It highlights that OSC costs are structured into three life-cycle categories (capital, facility management and end-of-life) where cost resources are expended, and also how enabling conditions and optimisation factors can improve cost performance across the life cycle. These framework elements are further discussed below.

8.4.2.1 Cost resources

The analysis shows that OSC projects, like traditional construction, are underpinned by cost resources. However, the distribution of these resources (such as overheads) differs considerably from conventional delivery methods, reflecting the manufacturing-led nature of material costs. Material costs cover the direct costs for procuring raw and prefabricated components, which are procured in bulk and used in factory-based construction. Cost resource considerations include the following.

- *Labour costs*. These include concentrated factory-based activities such as component fabrication, assembly line operations and quality control, with a smaller but still significant share allocated to on-site installation crews and lifting operations.
- *Plant costs*. These include expenses for operating machinery, tools and production facilities, essential for OSC's mechanised and standardised production.
- *Site overheads*. These costs are tied to establishing and operating the temporary construction site enabling the contractor to deliver the works safely and efficiently.
- *Office overheads*. Project office administrative costs not directly linked to physical production or site activities, such as project management salaries, IT systems, accounting services, insurance, utilities and general office administration.
- *Factory overheads*. These are the indirect costs associated with maintaining the off-site factory environment. These include the cost of factory setup, security, staff welfare, office utilities and supporting infrastructure such as storage containers and logistics

- *Professional fees and other development costs*. Fees paid to architects, structural and MEP engineers, quantity surveyors, BIM coordinators, sustainability consultants and other specialist advisors for design, planning and compliance.
- *Profit and risk allowances*. These cover contractor margins and contingencies against uncertainties such as design freeze points.

These cost resources provide the foundation for OSC project expenditure and are OSC consistent with findings from previous studies (Muse *et al.*, 2021; Serugga *et al.*, 2023).

8.4.2.2 Cost factors, categories and drivers

Findings reveal three cost factors that are central to OSC cost management: capital costs, facility management costs and end-of-life costs, consistent with previous literature findings (Mao *et al.*, 2016). However, findings from the comparative analysis of OSC and conventional building cost structures, illustrated in Figure 8.1, show OSC projects were characterised by distinctive categories that diverge from those of traditional construction, including fabrication, installation, transportation, disassembly/deconstruction and circularity costs. This categorisation is directly consistent with the previous studies (Jansen *et al.*, 2020; Obi *et al.*, 2025; Serugga *et al.*, 2023). This distinctive distribution differs from traditional construction, underscoring the need for early cost visibility in OSC projects.

A range of cost drivers were associated with each cost category, explaining variations in OSC costs across the life cycle. These drivers represent the activities or variables that cause costs to rise or fall and are often stage specific – for instance, design complexity, procurement methods and technological maturity have been shown to significantly affect capital costs (Liu *et al.*, 2023; Xue *et al.*, 2017). The subsections unpack categories and drivers as highlighted in the LCVO framework.

CAPITAL COSTS

From Figure 8.1, the capital costs represent the most significant and front-loaded component of OSC expenditure. Capital costs encompass decision and design costs, fabrication, transportation, onsite preparation and installation.

DECISION AND DESIGN COSTS

Decision and design costs are expenditures incurred from professional and development inputs such as consultant and project team fees, research and development, feasibility studies, design development procurement strategy and stakeholder workshops. These design expenditures are important because early design finalisation is a prerequisite for achieving the necessary standardisation and precision inherent in factory production (Liu *et al.*, 2023; Sutrisna and Goulding, 2019). Within the LCVO framework, this finding highlights the importance of cost visibility in the early phase of the life cycle.

The literature findings reveal that OSC incurs higher upfront design costs than conventional methods. This is attributed to the need for early prototyping, coordination and digital integration (Razkenari *et al.*, 2020; Wen and Zhang, 2021). However, the case analysis revealed slightly lower design costs for OSC projects compared with conventional projects. This finding diverges from literature, which reports higher upfront OSC costs due to early standardisation and coordination requirements (Sutrisna and Goulding, 2019; Wen and Zhang, 2021). This difference could be based on the effective use of modular layouts and sophisticated digital tools within the OSC case study, which may have streamlined the entire design process and minimised costly rework.

While design costs remained a pivotal stage in shaping all downstream project efficiency, the empirical evidence strongly suggests that OSC design expenditure can be successfully contained where standardisation and digital integration are applied rigorously and effectively. Key cost drivers of this cost category include design complexity, the number of iterations, the extent of standardisation, stakeholder involvement and the incorporation of sustainability and circularity strategies, all of which increase specialist input and coordination effort (Xue *et al.*, 2017).

FABRICATION AND TRANSPORTATION COSTS

Fabrication and transportation together represent major capital expenditures in OSC, reflecting its manufacturing-led delivery model. Fabrication costs refer to capital expenditure incurred during the factory-based production of panels, pods or complete volumetric modules. Expenditures cover costs of labour, materials, plant and associated overheads, including factory supervision, rigorous quality assurance and utilities consumption.

The typical costs within this category include factory operatives and specialised technicians; raw materials (including necessary recycled and sustainable inputs); machinery operation and maintenance; specialised packaging and handling for transportation; and compliance processes implemented to guarantee modules meet all requisite structural, fire and acoustic standards.

Findings from the case analysis reveal fabrication as the largest cost category (£46 323) in OSC projects. Compared with the conventional project's measured works, this category shows a clear front-loading of expenditure into the manufacturing stage. However, this investment delivers downstream benefits through reduced site labour demand, fewer errors and higher quality control. These findings align with existing studies which emphasise fabrication as both a cost-intensive stage and a critical opportunity for cost optimisation in OSC (Mao *et al.*, 2016; Sutrisna and Goulding, 2019).

The literature review findings highlight fabrication costs are driven by certain factors. These include the degree of standardisation, component repetition, batch size and factory utilisation rates. Critically, high levels of standardisation and repetition facilitate economies of scale and significant waste reduction. Conversely, bespoke designs and small production runs inherently increase setup costs, lead to greater material waste and necessitate higher labour intensity (Liu *et al.*, 2023; Xue *et al.*, 2017). Inspections and managing materials take-back schemes also contribute significantly to this cost category (Hong *et al.*, 2018).

Costs incurred for transporting prefabricated modules, panels, or components from the factory to the construction site typically encompass haulage and vehicle hire, specialist handling, permits, skilled labour for loading and offloading, specialist equipment like cranes and forklifts, protective packaging, insurance and intelligent route optimisation systems

(Liu and Luo, 2025; Mao *et al.*, 2016; Obi *et al*, 2025; Serugga *et al.*, 2023). The case analysis did not highlight transportation as a separate cost category, suggesting that this cost is embedded in overheads. This lack of transparency reflects a broader challenge in OSC cost reporting, where transportation is often treated as an indirect cost despite its critical role in project delivery. However, the literature highlights transportation as a critical link in the OSC value chain and the costs are commonly driven by distance, volume and size of deliveries, site access limitations and load efficiency.

This finding highlights the importance of cost visibility; without explicit recognition of transport as a distinct category, opportunities for optimisation through logistics planning, load efficiency, or sustainable fleet integration are easily overlooked. The literature consistently shows that while OSC projects may carry proportionally higher transportation costs than conventional methods, these investments can generate life-cycle benefits through faster installation, reduced site disruption and improved overall efficiency when properly managed.

ONSITE AND INSTALLATION COSTS

On-site and installation costs cover the placement and integration of prefabricated components delivered to the site. These include site preparation and foundations, labour for installers, crane operators and supervisors, lifting equipment and access platforms, temporary facilities, storage areas, and safety and compliance processes (Liu and Luo, 2025; Mao *et al.*, 2016). The case analysis revealed that installation accounted for a notable portion of OSC capital costs and required specialist handling to ensure alignment and integration. Compared with conventional construction, OSC introduces distinctive cost pressures due to the size, weight and precision required for handling prefabricated units.

Key cost drivers of this cost category identified from the literature include site complexity, foundation type, module size, sequencing, total number of installations, storage capacity, the level of mechanisation and the requisite inspection and productivity rates. Within the LCVO framework, installation costs illustrate the need for visibility at this stage to manage costs – for example, modular standardisation and precise factory production reduced installation time and minimised on-site rework. This demonstrates how investments in accuracy at earlier stages can optimise onsite costs, aligning with LCVO's emphasis on tracing how design and production choices cascade through the life-cycle to affect overall cost performance.

FACILITY MANAGEMENT COSTS

Facility management costs encompass expenses incurred for activities such as planned maintenance, supervision, performance monitoring, replacement of worn components, and the management of take-back or recycling schemes during the building's operational phase (Muse *et al.*, 2021). The case analysis revealed that OSC projects achieved a lower facility management cost corroborating literature finding that the strategic use of modular design and durable materials can be effective in reducing long-term maintenance requirements. This efficiency could be associated with lower energy consumption and maintenance expenditure, resulting from improved design precision and the inherent quality of prefabricated components (Kamali *et al.*, 2018).

The literature revealed that operation and maintenance costs were strongly driven by the frequency and type of maintenance, accessibility of components, building size and complexity, occupancy patterns, and the modularity or interchangeability of building components (Kamali *et al.*, 2018). This suggests that issues like poor accessibility or insufficient standardisation will inevitably increase operational costs by complicating subsequent repairs and component replacements.

Within the LCVO framework, by providing visibility into cost drivers such as durability, accessibility and system modularity, the framework enables long-term cost optimisation. These drivers can be managed effectively to reduce overall life-cycle costs, while enhancing the asset performance of the building, which directly aligns with sustainability and circular economy objectives.

END-OF-LIFE COSTS

This relates to the costs associated with the disassembly and deconstruction of the building. End-of-life costs arise during the disassembly or deconstruction of prefabricated modules and panels, enabling reuse, recycling, or recovery.

DISASSEMBLY AND DECONSTRUCTION COSTS

Unlike conventional demolition, OSC promotes selective dismantling aligned with circular economy principles (Jansen *et al.*, 2020; Obi *et al.*, 2025). Typical components of this cost category include specialised labour and supervision, equipment and tools for disassembly, temporary storage and handling of recovered elements, safety and quality assurance, and waste management procedures such as sorting, recycling or redistribution of prefabricated components.

The case analysis reveals that higher costs are incurred for disassembly/deconstruction than for conventional demolition. This may be associated with salvage costs, but it may be offset by the salvage value from material recovery and reduced landfill charges. The literature highlights key factors driving deconstruction/disassembly costs. These variables include the degree of design for disassembly, the volume of elements to be disassembled, component connection methods, the level of skilled workforce, the mechanisation level, on-site storage facilities, safety protocols, the number of inspections and the nature of compliance required (Jansen *et al.*, 2020; Tatiya *et al.*, 2018).

Within the LCVO framework, end-of-life costs exemplify the importance of extending cost visibility to the final stage of the life cycle. By recognising disassembly costs upfront, stakeholders can make informed decisions during design and fabrication that reduce eventual disposal expenses and maximise salvage value opportunities.

CIRCULARITY COSTS

The increasing emphasis on circularity has highlighted the need to incorporate circularity costs into OSC cost frameworks. These circularity costs are linked to the 'R-imperatives' of reduce, reuse, recycle and recover, which help capture the financial implications of integrating circular practices in OSC (Janson *et al.*, 2020; Obi *et al.*, 2025). Their explicit inclusion is crucial for accurately assessing the necessary trade-offs between environmental benefits and financial implications.

Findings reveal that circularity costs are significant at the end of a building's life cycle, where reuse, recycle and recovery are often actioned. The costs associated with this category include inspection and testing of reclaimed modules, reverse logistics, cleaning and sorting, certification, energy-efficient processing, circular supply-chain schemes, and temporary storage and protection of reclaimed items (Obi *et al.*, 2025).

The case analysis provided no details for the circularity costs. However, literature findings indicate that OSC projects may face higher initial costs in establishing circularity protocols but could be offset by salvage value from component reuse, reduced landfill charges and potential circular supply-chain integration (Jansen *et al.*, 2020; Obi *et al*, 2025). The literature also highlights some key cost drivers influencing circularity costs. These include the degree of component modularity and recoverability, the availability of local reuse and recycling infrastructure, the presence of secondary markets for reclaimed materials, the complexity inherent in reverse logistics, and sorting, inspecting and grouping components for reuse or recycling (Jansen *et al*, 2020).

Within LCVO framework, cost visibility in this area is particularly important as circularity costs, although new, are underreported within conventional and OSC cost structures. Recognising and planning for these costs not only enables optimisation but also creates opportunities for long-term value generation, aligning OSC with broader goals of sustainability, efficiency and resource conservation.

8.4.3 Optimisation factors and enabling conditions

This section discusses the optimisation factors and enabling conditions within the framework. Optimisation factors are specific variables that can be used to optimise the cost drivers to improve OSC life-cycle cost outcomes, while enabling conditions are the support that make these changes possible. Together, these dimensions provide a foundation for improving cost efficiency in OSC projects.

8.4.3.1 Optimisation factors

Within the LCVO framework, these factors are embedded within the structure rather than being exclusive. The findings revealed a range of optimisation factors that could influence cost drivers across the OSC project life cycle. These factors act as enablers for cost reduction, value creation and life-cycle efficiency. These drivers directly link to the core cost categories. These are highlighted in Figure 8.2.

CAPITAL COST OPTIMISATION FACTORS

Capital costs are effectively optimised by focusing on design efficiency, standardisation, automation and precise logistics planning. Literature highlights common factors in achieving this optimisation, including the rigorous use of approved quality standards, iterative digital design tools, repeatable layouts, automated production lines, flexible factory configurations, and highly efficient logistics such as just-in-time (JIT) delivery (Sutrisna and Goulding, 2019; Zhang *et al.*, 2020). Collectively, these integrated strategies demonstrably reduce rework, minimise material waste and curtail site overheads, while simultaneously accelerating both the factory production and on-site installation timelines.

FACILITY MANAGEMENT COST OPTIMISATION FACTORS

Facility management optimisation is enabled by durable materials, maintainable systems, modular replacement strategies, digital monitoring technologies and workforce training

(Kamali *et al.*, 2018; Muse *et al.*, 2021). These factors reduce maintenance frequency, improve energy efficiency and extend asset life, ensuring operational costs are minimised while maintaining performance.

END-OF-LIFE AND CIRCULARITY COST OPTIMISATION FACTORS

Optimisation at the end-of-life stage focuses on designing for disassembly, using modular sizing, integrating reversible connections and establishing reuse/recycling networks (Jansen *et al.*, 2020; Obi *et al.*, 2025). Digital traceability tools further support the recovery and reintegration of components into new projects. These measures reduce disposal costs and landfill charges while creating opportunities for material recovery and secondary markets.

8.4.3.2 Enabling conditions

Literature highlights that OSC cost performance is strongly shaped by both project-specific factors and systemic conditions (Liu *et al.*, 2023; Liu and Lou, 2025; Mao *et al.*, 2016; Zou *et al.*, 2024). The latter creates the environment and capacity for actioning optimisation factors in the projects, leading to improved cost efficiencies in OSC. Findings from the literature review suggest the following enabling factors essential for supporting the management and optimisation of OSC life-cycle costs.

- *Organisational support.* This refers to top management commitment, supportive culture, communication structures and policies that create an environment for effective cost management of OSC project implementation (Liu *et al.*, 2023; Sutrisna *et al.*, 2019; Xue *et al.*, 2017; Zou *et al.*, 2024).
- *Team capabilities and structure.* This refers to early contractor engagement, experienced, skilled and collaborative team, knowledge of data requirements necessary to implement cost management and optimisation effectively (Fulford and Standing, 2014; Razkenari *et al.*, 2020).
- *Estimating procedure and application.* This involves accurate identification and application of cost drivers and access to quality cost data that informs cost decisions (Liu and Luo, 2025).
- *Technology readiness.* This refers to the adoption of appropriate digital tools and systems such as BIM, AI and digital twins and their compatibility with existing systems to support cost visibility and optimisation (Wen and Zhang, 2021; Zhang *et al.*, 2020)

Without these conditions, the benefits of optimisation factors cannot be fully realised – for example, technology readiness ensures that digital cost data can inform design and facility management decisions, while organisational and team support provide the collaborative structures necessary for implementing strategies such as design for disassembly.

8.5. Implications

The framework has revealed how enabling conditions can create the needed environment in where the optimisation factors can be applied to strategically drive cost down of the project. It suggests that optimisation factors can help streamline processes and improve cost efficiency and outcomes. These interactions shape the OSC cost structure across the building life cycle. Hence there sits a dynamic relationship between enabling conditions, optimisation factors and the cost structure as together they determine the life-cycle performance of ODSC projects. These findings have important implications for both theory and practice.

8.5.1 Theoretical implications

Many OSC frameworks are derived from traditional cost management, failing to capture the nuances of sequencing, integration and logistics in OSC. This lack of reliable data contributes to stakeholder hesitancy, especially among public and affordable housing providers who operate under tight budget constraints (Agapiou, 2021). While the literature identifies cost factors, categories and drivers, they are often fragmented and not cohesively modelled.

Traditional LCC approaches underrepresent operational and circular aspects, whereas ABC tends to focus narrowly on capital expenditure (Mao *et al.*, 2016; Serugga *et al.*, 2023). Existing LCCs are limited in accounting for operational and circular aspects that OSC can offer, while ABC tends to focus narrowly on capital costs. By conceptualising the LCVO framework, this chapter contributes theoretically to this discourse by integrating the OSC cost structure with optimisation strategies and enabling conditions into a single framework.

LCVO demonstrates that cost drivers can be optimised across the project life cycle. Importantly, the framework highlights the mediating role of enabling conditions such as organisational support, team expertise and technology readiness (Liu and Luo, 2025). The chapter, therefore, fundamentally reframes OSC cost management into a dynamic life-cycle system where both cost visibility and optimisation are actively extended beyond predominantly capital-focused exercises.

8.5.2 Practical implications and recommendations

The findings highlight that prioritising capital expenditure often leads to the underestimation of long-term OSC project life-cycle costs, such as facility management and end-of-life activities. The LCVO framework directly addresses this deficiency by providing a structure for visualising costs across the project life cycle. This capability enables practitioners to identify not only where costs materialise but, critically, why they occur and how they can be effectively optimised.

The study confirmed that while OSC involves higher upfront costs in design, factory setup and logistics, these can be offset by downstream benefits such as reduced maintenance, fewer defects and improved energy performance (Wen and Zhang, 2021; Zhang *et al.*, 2020). Based on the findings and the LCVO framework, the following recommendations are proposed for industry practitioners and policymakers.

- *Adopt life-cycle-oriented cost reporting.* The LCVO framework highlights that costs extend beyond capital expenditure into operation, maintenance and end-of-life. Systematically reporting operation, maintenance and end-of-life costs, including circularity, will improve visibility and allow stakeholders to assess true value and make more informed investment decisions on OSC.
- *Invest in digital tools and integration.* Digital readiness is an enabling factor in the framework. Broadening the use of BIM, digital twins and traceability systems can improve design coordination, procurement planning and long-term cost monitoring across the life cycle.
- *Promote modular standardisation.* Optimisation factors identified in the framework show that repeatable layouts, standardised connections and optimised sizing reduce

fabrication and installation costs. Encourage repeatable layouts, standardised connections and optimised component sizing to reduce waste, improve productivity and support more predictable cost outcomes.
- *Embed circularity principles*. The framework reframes end-of-life costs as opportunities for recovery. Design-for-disassembly, reuse networks and secondary markets can lower disposal costs and unlock long-term value. Incorporate design-for-disassembly, reuse networks and secondary markets for reclaimed components into project planning. This transforms end-of-life from a cost burden into an opportunity for value recovery.
- *Build skills and organisational capacity*. The enabling factors embedded within the framework underscore the necessity of robust organisational support and refined team capability. Investment in specialised training, covering life-cycle costing, digital modelling and circular economy practices is therefore essential to fully realise the optimisation potential of OSC.

Collectively, these recommendations illustrate a core tenet: OSC cost optimisation is never achieved through technical efficiencies alone. Instead, it is intrinsically linked to life-cycle thinking that must be rigorously embedded across project design, holistic management and procurement strategies.

8.6. Conclusions

This chapter successfully introduced the conceptual LCVO framework for enhancing cost management in OSC. It leverages the principles of LCC and ABC to capture OSC cost structure, highlighting the cost factors, categories, drivers and enabling conditions. This could help practitioners and researchers see how costs are incurred and implement necessary actions to improve cost efficiency.

The study demonstrated that while OSC inherently entails higher upfront capital expenditures, they are offset through reduced maintenance, improved energy performance and robust opportunities for reuse and recycling. Whereas traditional cost models treat end-of-life primarily as a simple demolition expense, LCVO fundamentally reframes this as a clear economic recovery opportunity through structured reuse, recycling and remanufacturing. Within this context, disassembly and material recovery are correctly positioned as cost-generating, yet ultimately value-enhancing, activities. The expenditures incurred in detailed inspection, reverse logistics and storage are actively balanced against the salvage value derived from reclaimed components, such as structural pods and panels. This approach effectively aligns OSC with core principles of the circular economy and underscores its substantial long-term sustainability potential. Moreover, the LCVO framework crucially brings transportation and circularity costs into clear focus.

This chapter contributes to knowledge on OSC cost management studies by conceptualising the LCVO framework that extends cost management beyond capital expenditure to include operational, end-of-life and circularity costs, while positioning enabling conditions as central to achieving optimisation.

Practically, it offers a decision-support lens for the client and the project team to understand the cost structures and the cost drivers, which helps to prioritise actions that can reduce waste and streamline activities and inspections. In addition, contractors can see how optimisation factors could lower production and site-related costs, improving cost efficiency on the projects. It also offers insights into the enabling conditions needed to support OSC adoption and implementation, which could support policymaking.

As a conceptual framework, it does have its limitations, and its effectiveness will depend on how it is operationalised. It was developed based on literature findings and case study analysis of two projects in the UK. The context, market maturity, regulation and industry structure may vary within regions and across different contexts. The scope of the LVCO framework lies in providing a structural perspective to understand relationships between elements (cost structure, cost drivers, optimisation factors and enabling conditions) that can guide analysis, especially at the strategic planning level, where decision makers require a holistic view of how costs are shaped and can be optimised.

Future research can test and refine the framework through case studies, cross-national comparisons, digital simulation and action-based industry projects. With further refinement, LCVO can evolve into a robust tool for guiding the industry toward cost-efficient, sustainable and circular construction systems, supporting the long-term viability and competitiveness of OSC.

REFERENCES

Agapiou A (2021) An exploration of the best value perceptions of small housebuilding developers towards offsite construction. *Sustainability* **13(7)**: 4054.

Allianz Global Corporate and Specialty (2021) Managing the New Age of Construction Risk: 10 Trends to Watch as the Sector Builds Back Better. chrome-extension://efaidnbmnnnibpcajpcglclefindmkaj/https://commercial.allianz.com/content/dam/onemarketing/commercial/commercial/reports/agcs-10-trends-construction-industry.pdf (accessed 1/3/2023).

Ashworth A and Perera S (2015) *Cost Studies of Buildings*. Routledge, London, UK.

Cao X, Li X, Zhu Y and Zhang Z (2015) A comparative study of environmental performance between prefabricated and traditional residential buildings in China. *Journal of Cleaner Production*, **109**: 131–143.

Cartlidge D (2013) *Estimator's Pocket Book*. Routledge.

Fraser N, Race GL, Kelly R, Winstanley A and Hancock P (2015) An offsite guide for the building and engineering services Sector. *BuildOffsite, Apr*.

Fulford R and Standing C (2014) Construction industry productivity and the potential for collaborative practice. *International Journal of Project Management* **32**: 315–326.

Ginigaddara B, Perera S, Feng Y and Rahnamayiezekavat P (2022) Development of an offsite construction typology: a Delphi Study. *Buildings*, **12(1)**: 20.

Hong J, Shen GQ, Li Z, Zhang B and Zhang W (2018) Barriers to promoting prefabricated construction in China: a cost–benefit analysis. *Journal of Cleaner Production* **172**: 649-660.

Jansen BW, van Stijn A, Gruis V and van Bortel G (2020) A circular economy life-cycle costing model (CE-LCC) for building components. *Resources, Conservation and Recycling* **161**: 104857.

Jiang R, Mao C, Hou L, Wu C and Tan J (2018) A SWOT analysis for promoting off-site construction under the backdrop of China's new urbanisation. *Journal of Cleaner Production* **173**: 225–234.

Kamali M, Hewage K and Milani AS (2018) Life cycle sustainability performance assessment framework for residential modular buildings: aggregated sustainability indices. *Building and Environment* **138**: 21–41.

Kaplan RS and Cooper R (1998) *Cost & Effect: Using Integrated Cost Systems to Drive Profitability and Performance*. Harvard Business Press.

Kyrö R, Jylhä T and Peltokorpi A (2019) Embodying circularity through usable relocatable modular buildings. *Facilities* **37(1/2)**: 75–90.

Liu M and Luo M (2025) Cost estimation model of prefabricated construction for general contractors based on system dynamics. *Engineering, Construction and Architectural Management*, **32(1)**: 621-638.

Liu Y, Chang R-D, Zuo J, Xiong F and Dong N (2023) What leads to the high capital cost of prefabricated construction in China: perspectives of stakeholders. *Engineering, Construction and Architectural Management* **30(2)**: 805–832.

Mao C, Xie F, Hou L, *et al.* (2016) Cost analysis for sustainable off-site construction based on a multiple-case study in China. *Habitat International* **57**: 215–222.

Muse A, Horner M, O'Sullivan G, *et al.* (2021) ICMS: Global consistency in presenting construction life-cycle costs and carbon emissions. International Construction Measurement Standards Coalition (ICMSC).

Navaratnam S, Ngo T, Gunawardena T and Henderson D (2019) Performance review of prefabricated building systems and future research in Australia. *Buildings* **9(2)**: 38.

Obi LI, Arif M and Kulonda DJ (2017) Prioritizing cost management system considerations for Nigerian housing projects. *Journal of Financial Management of Property and Construction* **22(2)**: 135–153.

Obi LI, Arif M, Lamb SM and Ashayeri I (2023) Establishing interrelationships and dependencies of critical success factors for implementing offsite construction in the UK. *Smart and Sustainable Built Environment*, ahead-of-print.

Obi LI, Awuzie B, Asare O, Lamb S and Thurairajah N (2025) A circularity cost framework for offsite housing projects. *Construction Innovation*, ahead-of-print.

Ofori-Kuragu J, Robert O-K and Wanigarathna N (2022) Offsite construction methods – what we learned from the UK Housing Sector. *Infrastructures* **7**: 164.

Razkenari M, Fenner A, Shojaei A, Hakim H and Kibert C (2020) Perceptions of offsite construction in the United States: an investigation of current practices. *Journal of Building Engineering* **29**: 101138.

RICS (Royal Institution of Chartered Surveyors) (2016) *Life Cycle Costing*, 1st ed. London, UK.

Rimmel G and Cordazzo M (2021) Deductive versus inductive content analysis: a methodological research note to disclosures studies in intellectual capital research. In *Research Handbook on Intellectual Capital and Business*. Edward Elgar Publishing, pp. 109–124.

Salama T, Figgess G, Elsharawy M and El-Sokkary H (2020) Financial modeling for modular and offsite construction. *Proceedings of the 37th International Symposium on Automation and Robotics in Construction (ISARC 2020), Kitakyushu, Japan*, 27–28.

Schram JFJ (2012) *Real Estate Appraisal*. Rockwell Publishing Company.

Serugga J, Kagioglou M, Edwards D, Talebi S and Johns N (2023) Key influences to cost modelling and analysis in off-site construction: research trends and GAP analysis. *Architectural Engineering and Design Management*: 1–22.

Shen K, Cheng C, Li X and Zhang Z (2019) Environmental cost-benefit analysis of prefabricated public housing in Beijing. *Sustainability* **11(1)**: 207.

Shihab WK and Prasad S (2017) Activity based costing system. *International Journal of Scientific Research* **8(7)**: 18288–18306.

Sutrisna M, Cooper-Cooke B, Goulding J and Ezcan V (2019) Investigating the cost of offsite construction housing in Western Australia. *International Journal of Housing Markets and Analysis* **12(1)**: 5–24.

Szczerbak M and Wikarczyk A (2023) The usefulness of management accounting tools in reducing waste. *Economics and Environment* **86(3)**: 186–201.

Tatiya A, Zhao D, Syal M, Berghorn GH and LaMore R (2018) Cost prediction model for building deconstruction in urban areas. *Journal of Cleaner Production* **195**: 1572–1580.

UKCES (UK Commission for Employment and Skills) (2015) *A 10 Year Change Programme - Developing Offsite Construction Skills for the Challenges Ahead.* chrome-extension://efaidnbmnnnibpcajpcglclefindmkaj/https://assets.publishing.service.gov.uk/media/5a74e33040f0b65c0e8454a6/151021_OSCS_brochure_for_web_reduced.pdf#:~:text=Page%202.%20Recent%20estimates%20suggest%20that%20the,the%20build%20process%20and%20cleaner%20working%20conditions (accessed 16/3/2024).

Wen Z and Zhang H (2021) Research on cost management of prefabricated construction based on BIM. *IOP Conference Series: Earth and Environmental Science* **768(1)**: 012110.

Xue H, Zhang S, Su Y and Wu Z (2017) Factors affecting the capital cost of prefabrication—a case study of China. *Sustainability* **9(9)**: 1512.

Zhang Y, Lei Z, Han S, Bouferguene A and Al-Hussein M (2020) Process-oriented framework to improve modular and offsite construction manufacturing performance. *Journal of Construction Engineering and Management* **146(9)**: 04020116.

Zhou J, Tam VW, Qin Y and Illankoon C (2024) A critical review and comparative analysis of cost management on prefabricated construction research (2000–2022). *International Journal of Construction Management*, **24(9)**: 997–1006.

Zou C, Zhu J, Cao J *et al.* (2024) Critical factors influencing the cost in engineering–procurement–construction (EPC) projects: a case study of prefabricated buildings (PBs) in China. *Journal of Intelligent & Fuzzy Systems*: JIFS-240206.

Emmanuel I Daniel and Fidelis A Emuze
ISBN 978-1-83708-437-1
https://doi.org/10.1108/978-1-83708-436-420261010

Chapter 9
Lean and collaborative costing in the UK construction industry: challenges, benefits and optimisation to MMC

Sa'id Ahmed Namadi[1], Mark Danso-Amoako[2] and Abba Tahir Mahmud[2]

[1]*School of Engineering and the Environment, Kingston University, London*
0000-0002-3815-0683
[2]*School of Civil Engineering and Surveying, University of Portsmouth, UK*
0000-0001-8972-7481; 0000-0002-9854-1651

Abstract

The adoption of modern methods of construction (MMC) is transforming the construction industry by enhancing efficiency, sustainability and productivity. However, the UK construction industry faces ongoing challenges in cost management, with inefficiencies often leading to budget overruns and project delays. Lean and collaborative costing (CC) approaches have emerged as potential solutions, particularly in the context of modern methods of construction (MMC). Using secondary data, this study explores the key challenges of implementing lean and CC practices within UK construction projects, including cultural resistance, integration complexities and fragmented supply chain coordination. The study also examines the benefits, including cost reductions, waste minimisation and enhanced project efficiency. Furthermore, the research investigates optimisation strategies tailored for MMC, leveraging digital tools, early stakeholder engagement and process standardisation. Reviewing existing literature and industry case studies, this study provides insights into how lean and CC can enhance cost predictability and efficiency in MMC projects. The findings will offer practical recommendations for industry stakeholders aiming to improve cost management through lean principles and collaborative practices.

Keywords: lean construction; CC; MMC; cost optimisation; supply chain integration

9.1. Introduction

The UK construction industry has long struggled with inefficiencies such as cost overruns, project delays and fragmented delivery systems (Afana *et al.*, 2024; Ahmed and El-Sayegh, 2021).

Decades of reports, including Rethinking Construction (Egan, 1998), have highlighted that up to 30% of construction costs may be attributed to waste or non-value-adding activities. These inefficiencies undermine productivity, profitability and client satisfaction, while adversarial relationships among stakeholders frequently lead to disputes and claims that further escalate costs. Research indicates that construction projects have experienced an average cost overrun of approximately 28% over the past 70 years (Afana *et al.*, 2024), illustrating the systemic nature of the problem.

A central contributor to these inefficiencies lies in traditional cost estimation and management practices. Conventional costing methods typically occur after the design phase, which creates a disconnect between design intent and budgetary constraints (Ahmed and El-Sayegh, 2021). When estimates exceed available budgets, projects often require redesign, commonly referred to as 'value engineering', leading to time delays, additional costs and compromised quality. This reactive approach prevents cost optimisation during the critical early stages of project development and perpetuates the cycle of inefficiency across the construction supply chain.

In response to these persistent challenges, MMC have emerged as innovative approaches aimed at enhancing efficiency, sustainability and cost-effectiveness in the built environment (NHBC Foundation, 2016). MMC encompass off-site manufacturing, prefabrication, modularisation and other industrialised techniques that reduce on-site labour, minimise waste and accelerate delivery timelines. However, despite their potential, MMC projects also introduce new complexities in cost planning and financial control. Their reliance on early design finalisation, integration of digital tools and coordination across multiple supply chain partners requires new, collaborative approaches to cost management.

Against this backdrop, lean and CC frameworks have gained attention as potential solutions to the cost management challenges in MMC. By promoting transparency, early stakeholder involvement and shared decision-making, these approaches aim to reduce waste, improve cost predictability and enhance overall project performance. This study explores the challenges, benefits and optimisation strategies of implementing CC within MMC projects in the UK construction industry.

Research Aim

This study aims to examine the challenges and benefits of implementing CC within MMC projects in the UK and to identify strategies for optimising cost management through lean and collaborative approaches.

9.2. Perspectives and concepts

9.2.1 MMC and CC

MMC are formally defined by the UK Government as 'a wide range of innovative construction processes and techniques that improve productivity, quality and sustainability outcomes across the built environment' (HM Government, 2020). Ministry of Housing, Communities and Local Government (2019) further categorises MMC into seven categories, encompassing off-site manufacturing, 3D volumetric and panelised systems, sub-assemblies, component-based construction and on-site process innovations. These definitions provide a structured understanding of MMC as both a technological and procedural transformation within the UK construction industry.

While often associated with recent advances in digitalisation and automation, the concept of MMC has historical roots in post-war reconstruction, when prefabricated housing was deployed to meet urgent housing needs. Those early applications laid the groundwork for the industrialised building systems of today. In contemporary practice, MMC represents a strategic response to the UK's productivity challenges, skills shortages and sustainability targets (Ministry of Housing, Communities and Local Government, 2019; Pan and Sidwell, 2011).

By transferring significant portions of construction activity from site to controlled manufacturing environments, MMC offers substantial potential to improve efficiency, quality control and waste reduction. Digital technologies, particularly building information modelling (BIM), further enhance MMC by enabling design precision, data integration and process standardisation across the supply chain. Together, these innovations contribute to faster delivery, reduced rework and improved life-cycle performance (NHBC Foundation, 2016).

However, these benefits can only be fully realised through effective cost management and coordination frameworks. MMC introduces unique financial and logistical challenges, including higher upfront capital investment, complex supply chain interfaces and the need for early design finalisation. Traditional cost management approaches often struggle to accommodate these factors, as they are designed for sequential, site-based processes rather than integrated, manufacturing-led workflows (Pan and Sidwell, 2011).

Consequently, there is a growing need for collaborative and transparent costing practices that align financial decision-making with design, production and delivery. Approaches such as CC and lean construction provide the structural and cultural mechanisms to achieve this integration, facilitating early stakeholder involvement, open-book cost sharing and joint accountability. These principles are explored in the following section, which examines the foundations and mechanisms of CC as a key enabler of MMC efficiency.

Collaborative costing is a proactive cost management approach that involves all stakeholders (clients, designers, contractors and suppliers) from the early stages of a project in the budgeting and decision-making process. It emphasises transparency, shared goals and integrated decision-making to optimise value and control costs (Namadi *et al.*, 2017). Key principles include early involvement, open communication and accountability. Collaborative costing aligns with lean construction by reducing waste and improving efficiency through joint planning and real-time cost feedback. It also supports delivery models like target value design (Namadi *et al.*, 2017) and integrated project delivery, which prioritise client-defined value over lowest-cost bidding. This approach fosters trust, reduces rework and enhances project outcomes.

CC is defined as an integrated approach where stakeholders (clients, contractors, designers and suppliers) jointly contribute to the cost estimation and control process. This approach often uses open-book accounting, target costing and BIM to foster transparency (Namadi *et al.*, 2017). Ahmed *et al.* (2019) opined that CC is an approach that engaged stakeholders (upstream and downstream players) around a wider scheme budget – creating a sense of ownership, driving positive behaviours to achieve desired cost outcomes.

9.2.2 Traditional costing in the UK construction industry

Traditional costing practices in the UK construction industry have long been characterised by fragmented procurement methods, adversarial relationships and a linear approach to project execution (Humphreys *et al.*, 2003; Yang *et al.*, 2023). Under this model, project stakeholders, including clients, designers, contractors and suppliers, often work in silos. Each party focuses primarily on their contractual obligations, with limited integration across disciplines. This segmentation typically results in the late involvement of key supply chain members, such as subcontractors and manufacturers, who are only brought in after the design has been finalised. Consequently, their opportunity to contribute cost saving ideas or constructability input during the early, more influential design stages is minimal.

Moreover, traditional costing relies heavily on post design quantity surveying and cost estimation methods, which are reactive rather than proactive. Cost overruns are often identified too late, resulting in redesigns, delays and budget escalations. Value engineering exercises, intended to reduce costs, are frequently conducted late in the process, often leading to compromises in quality or function. These inefficiencies are exacerbated by a lack of transparency in cost data and limited communication between project participants.

Given these recurring issues of lack of integration, poor cost predictability, adversarial relationships and reactive cost control, there is a growing consensus that traditional costing models are no longer fit for purpose in today's complex construction environment. These shortcomings necessitate a new, more collaborative approach to project costing that aligns with modern construction philosophies such as lean construction.

9.2.3 Demand for CC models

CC offers a paradigm shift from traditional cost management practices by promoting early stakeholder engagement, transparent cost data sharing and integrated decision-making processes (Mayouf *et al.*, 2019). Rooted in lean construction principles, emphasising value creation, waste reduction and continuous improvement, it aligns the goals of all project participants, from clients and designers to contractors and suppliers (Ballard and Howell, 2003).

Key advantages of CC include the early involvement of contractors and supply chain partners, leading to more accurate and realistic budgeting, better design decisions and fewer surprises during construction. Cost planning becomes dynamic and iterative, with continuous feedback enabling refinement of both design and budget.

Transparency is central to CC. By openly sharing cost data, trust is fostered among stakeholders and disputes are reduced. This openness supports informed decision making and a shared understanding of cost drivers and trade-offs.

Furthermore, CC facilitates interdisciplinary integration. By breaking down organisational silos and encouraging multidisciplinary collaboration, project teams more readily identify and eliminate waste, streamline workflows and enhance constructability (Mayouf *et al.*, 2019) which ultimately leads to improved project performance in both cost and delivery.

9.2.4 The advent of 5D BIM

BIM has transformed how construction projects are designed, costed and delivered. In particular, 5D BIM integrates cost information as the fifth dimension into 3D models, enabling real time cost estimation and cost control as design evolves (Mayouf *et al.*, 2019). Unlike traditional quantity surveying, 5D BIM allows teams to instantly visualise how design modifications affect costs, supporting decisions made early in the project life cycle.

5D BIM aligns closely with lean construction and CC, enabling early contractor involvement through a centralised, common data environment (CDE) (British Standards Institution, 2018) providing a single source of truth for project information, accessible to all project participants, As outlined in the ISO 19650-1 standard, this shared CDE enhances collaboration, reduces miscommunication and aligns design and cost objectives from the outset. The ability to simulate design and material alternatives supports optimisation of value while reducing redesign costs.

Transparency is improved as cost data become visible and accessible to all stakeholders, enabling real time budget monitoring, deviation identification and corrective action. Integration across disciplines, architects, engineers, cost consultants and contractors supports a shared understanding of scope and budget, reducing the likelihood of disputes.

In essence, 5D BIM is a cornerstone of CC strategies. It enables proactive cost management, transparency and efficiency across the project life cycle (Moses *et al.*, 2020, Das *et al.*, 2025). Its adoption in the UK construction industry, particularly in public sector and infrastructure projects, is growing rapidly.

9.2.5 CC models in use in the UK

In recent years, a growing number of CC models, including target value design, open-book costing and alliance contracting, have gained traction within the UK construction industry, reflecting a broader shift toward lean construction and integrated project delivery (IPD) philosophies (Ballard and Howell, 2003; Lahdenperä, 2012; Mosey, 2019; RICS, 2021). These frameworks promote transparency, shared risk and early stakeholder engagement, aligning cost management with value creation across the project life cycle (Davies and Mackenzie, 2014; BEIS, 2013).

9.2.6 The use of CC in MMC projects

MMC, including off site manufacturing and prefabrication, require early design finalisation and precision. Design changes late in the process can be extremely costly due to prefabrication timelines. Unlike traditional builds, MMC projects often have minimal flexibility during construction. Collaborative costing models such as 5D BIM, TVD and IPD are essential in addressing this challenge. These models integrate cost data with design and engage stakeholders early, ensuring financial implications are understood well before fabrication begins – for example, 5D BIM enables real time visualisation of design cost impacts, reducing the risks and expenses associated with late changes (Mayouf *et al.*, 2019; Moses *et al.*, 2020).

Additionally, collaborative environments foster collective responsibility and support value engineering from the start. In MMC contexts, this alignment ensures coherence between design intent, budget constraints and manufacturing processes. As the adoption of MMC grows in the UK, CC becomes critical for delivering projects on time, on budget and to specification.

9.3. Methodological approach

9.3.1 Research design

This study adopts a qualitative, multiple case study design underpinned by secondary data analysis. The aim is to explore how lean and collaborative costing principles are applied in MMC within the UK construction industry and to identify both challenges and optimisation opportunities. The qualitative case study approach is particularly suited to this research because it enables a detailed understanding of complex, context-dependent processes that are difficult to quantify. Collaborative costing and MMC practices are influenced by organisational culture, procurement models, digital maturity and stakeholder dynamics, all of which are best examined through interpretive, case-based inquiry.

This research design is exploratory and descriptive in nature. It does not seek to test hypotheses but to identify patterns, themes and insights from documented real-world projects. The approach allows for cross-case comparison to uncover common success factors and barriers associated with CC implementation in MMC projects.

9.3.2 Rationale and justification

This study employs a qualitative case study design based exclusively on secondary data to investigate the application of lean construction and CC within MMC in the UK. This approach is justified by the evolving and context-specific nature of MMC practices, which are influenced by procurement structures, digital capabilities and stakeholder dynamics. Case-based analysis enables in-depth examination of these interrelated factors and is consistent with prior research on integrated and collaborative construction models (Ahmed *et al.*, 2019; Davies *et al.*, 2009; Monyane *et al.*, 2020; Pishdad and Onungwa, 2024).

The reliance on secondary data is further justified by the extensive availability of credible academic and industry sources on MMC, BIM and CC. Given the proprietary nature of cost data and the restricted access to live project information, secondary data provide an ethical, practical and sufficiently rich basis for analysis. This approach aligns with established construction management studies that have effectively utilised secondary data to examine innovation in costing and procurement (Namadi, 2019).

Overall, the chosen methodology offers a rigorous and contextually grounded means of exploring CC practices within MMC, ensuring analytical depth and research validity.

9.3.3 Research steps

The research followed a structured, multi-stage process designed to ensure rigour, validity and clarity of analysis.

9.3.3.1 Literature review and theoretical framing

- A comprehensive review of academic literature, industry reports and government publications was conducted to establish the conceptual foundation of lean construction, CC and MMC.
- Key frameworks such as TVD, IPD and 5D BIM were examined to inform the analytical lens for the case study review.

9.3.3.2 Identification of case studies

- Using predefined selection criteria (see below), representative construction projects that have demonstrably implemented lean or CC practices were identified.
- The case studies were drawn from both UK and international contexts to enable comparative insight and broader applicability of findings.
- Data collection and synthesis.
- Secondary data were collected from peer-reviewed journal articles, official project reports, technical publications and government reviews (e.g. National Audit Office reports, BRE documentation).
- Each source was screened for credibility, relevance and depth of information.
- Extracted data focused on implementation processes, cost management approaches, collaboration mechanisms, outcomes and lessons learned.

9.3.3.3 Cross-case analysis

- The selected case studies were comparatively analysed to identify recurring themes, patterns and divergences.
- Factors such as stakeholder engagement, cost transparency, use of digital tools and procurement models were coded and interpreted.
- This step aimed to determine which aspects of CC are most effective in supporting MMC delivery.

9.3.3.4 Synthesis and interpretation

- Insights from the case studies were synthesised with findings from the literature review to build an integrated understanding of challenges, benefits and optimisation strategies for lean and CC in MMC.
- The findings were then contextualised within the UK construction sector to ensure practical relevance.

9.3.4 Case study selection criteria

To ensure representativeness and analytical richness, the case studies were selected based on the following criteria.

(a) *Relevance to MMC or lean construction*. The project must involve significant use of MMC or LC practices.
(b) *Demonstrated collaborative costing application*. Projects must provide documented evidence of collaborative or integrated cost management approaches (e.g. open-book costing, target costing, 5D BIM, or IPD).

(c) *Availability of reliable secondary data*. The case must have credible and detailed documentation from peer-reviewed publications, government reports, or recognised industry sources.
(d) *Scale and complexity*. Medium to large-scale projects were prioritised to ensure sufficient complexity and stakeholder diversity for meaningful analysis.
(e) *Geographical and sectoral representation*. Preference was given to projects within the UK construction context (e.g. Heathrow Terminal 5, Crossrail, London 2012 Olympics), while including international examples (e.g. Sutter Health Castro Valley Medical Centre, USA) for comparative benchmarking.

9.3.5 Data source

The study relies entirely on secondary data drawn from multiple verified sources to ensure validity and triangulation.

(a) *Academic sources*. Peer-reviewed journals, conference proceedings, doctoral theses and scholarly publications relating to lean construction, CC and MMC (Ahmed *et al.*, 2019; Mayouf *et al.*, 2019; Namadi, 2019).
(b) *Industry and government reports*. Publications by the National Audit Office (2012), Constructing Excellence (2020) and the NHBC Foundation (2016), HM Government (2020).
(c) *Technical and digital sources*. Documentation on BIM (ISO 19650-1, 2018) and case-specific project delivery reports.
(d) *Case-based publications*. Studies and project analyses detailing implementation of collaborative or lean practices in real-world projects (e.g. Bryde *et al.*, 2013; Davies *et al.*, 2009; Kent and Becerik-Gerber, 2010).

The use of secondary data ensures breadth and depth of insight across multiple contexts and mitigates the challenges associated with accessing proprietary or confidential cost data in live construction projects.

9.3.6 Summary

In summary, this research employs a qualitative, secondary-data-based multiple case study design to examine how lean and CC are being implemented within MMC projects. Through systematic literature analysis, case selection and cross-case synthesis, the study provides a robust understanding of challenges, benefits and optimisation strategies for achieving cost efficiency and collaborative success in the UK construction industry.

9.4. Results and discussion

9.4.1 Overview

The findings draw on the synthesis of secondary data and case-based evidence from projects such as Heathrow Terminal 5, Crossrail, London 2012 Olympics and the Sutter Health Castro Valley Medical Centre. These projects provide insights into how lean and CC principles are practically implemented within complex construction environments. The analysis identifies five dominant themes that underpin the effective integration of CC into MMC, as itemised in Table 9.1,

Table 9.1 Summary of key findings

Theme	Key Findings	Implications for MMC
Early cost alignment	Early engagement and target costing improve cost certainty and minimise rework.	Embed collaborative budgeting at concept stage.
Digital integration	BIM and CDE enhance transparency and coordination.	Invest in data interoperability and digital skills.
Risk and trust	Shared risk/reward promotes cooperation and reduces disputes.	Encourage trust-based contracts (e.g. IPD, NEC Option C).
Supply chain coordination	Early supplier involvement enhances accuracy and reduces waste.	Build capacity among off-site and manufacturing partners.
Value optimisation	Collaborative approaches deliver better life-cycle value and cost efficiency.	Shift focus from lowest price to best value outcomes.

Source: Authors' creation

(*a*) early cost alignment and front-end integration
(*b*) digital integration and data transparency
(*c*) risk allocation, trust and collaborative culture
(*d*) supply chain coordination and stakeholder engagement
(*e*) performance outcomes and value optimisation.

Each theme is explored below, illustrating both the opportunities and constraints identified through the analysis.

9.4.1.1 Early cost alignment and front-end integration

A consistent finding across all reviewed cases is the critical importance of establishing cost alignment at the earliest design stages. Projects such as Heathrow Terminal 5 and the London 2012 Olympics demonstrated that early cost involvement of all key actors, clients, designers, contractors and suppliers, enabled more accurate budgeting and reduced the frequency of redesigns or 'value engineering' later in the process.

In both projects, the adoption of target cost models and joint budgeting created shared ownership of financial objectives. This early alignment enabled continuous value optimisation and informed trade-offs between cost, performance and design flexibility. Similarly, the Sutter Health project in the USA evidenced how IPD models allowed concurrent design and cost development, reducing delivery time by nearly one-third compared to conventional methods (Kent and Becerik-Gerber, 2010).

From a lean perspective, early integration eliminates waste associated with rework and late design alterations. In MMC contexts, where design finalisation and prefabrication occur early, this alignment is particularly crucial. The evidence supports the notion that cost certainty and value management must begin at project inception to avoid the high-cost implications of design revisions once manufacturing has commenced.

However, the data also highlight a recurring challenge: public sector procurement frameworks often delay contractor involvement due to rigid tendering requirements. This restricts early collaboration and undermines the very benefits CC seeks to achieve. Therefore, policy reform and procurement restructuring remain essential enablers for front-end integration in MMC projects.

9.4.1.2 Digital integration and data transparency

The integration of digital technologies, particularly BIM and 5D cost modelling, was identified as a cornerstone of successful CC. Crossrail, as a flagship UK project, adopted a 5D BIM framework and CDE, enabling real-time cost and progress tracking across more than 60 contractors and consultants (Bryde *et al.*, 2013). This digital integration provided transparency and traceability of cost data, supporting lean decision-making and early identification of deviations. Projects that implemented CDEs demonstrated significantly reduced design errors, faster approvals and enhanced coordination among geographically dispersed teams, all of which are critical in MMC's modular supply chains.

The analysis suggests that data interoperability and standardisation are key success factors. While BIM enhances collaboration, inconsistent data formats and software incompatibility remain barriers to seamless cost integration. Moreover, many construction organisations lack the digital maturity required to fully exploit the analytical potential of BIM-based costing. The findings indicate that digital tools are not merely technical enablers but cultural catalysts, they reinforce openness and accountability, fostering trust between stakeholders through shared data visibility. However, to achieve full benefits, training and standardisation across the supply chain are required to ensure data quality and uniform interpretation.

9.4.1.3 Risk allocation, trust and collaborative culture

Another prominent theme concerns the way risk and reward mechanisms shape collaboration. Projects employing gain/pain share agreements (e.g. Heathrow T5) demonstrated that transparent cost structures and shared incentives promote cooperative behaviour and reduce adversarial disputes. When risks and benefits are collectively owned, stakeholders are more likely to prioritise project success over individual gain.

This cultural shift from contractual protectionism to mutual accountability reflects the core philosophy of lean construction, continuous improvement through teamwork and transparency. However, the findings also reveal that such trust-based relationships take time to develop and are often undermined by traditional procurement and hierarchical management structures (Lahdenperä, 2012, p.63). In MMC, where off-site production requires precision, any breakdown in trust or data sharing can lead to significant cost and schedule consequences. Thus, establishing a collaborative culture supported by clear governance frameworks is essential.

The evidence from Sutter Health and the Olympics projects indicates that leadership commitment, transparent communication channels and aligned incentives are vital in embedding collaboration at all project levels. Nevertheless, several UK projects reviewed (particularly in public infrastructure) continue to exhibit partial adoption, embracing open-book costing but maintaining risk-averse contracting. This partial implementation limits the transformative potential of CC.

9.4.1.4 Supply chain coordination and stakeholder engagement

Effective cost collaboration depends heavily on how well the supply chain is integrated. MMC projects often involve multiple tiers of manufacturers, logistics providers and subcontractors. The findings show that early supplier engagement significantly improves cost accuracy, reduces procurement delays and enhances constructability – for example, Crossrail's early collaboration with suppliers enabled accurate prefabrication costing and reduced material waste, and, similarly, Heathrow T5's integrated team structure facilitated direct coordination between designers and fabricators, ensuring that cost data informed production planning.

The analysis highlights that lean and collaborative practices strengthen communication across the value chain, transforming traditionally fragmented relationships into strategic partnerships. However, the findings also reveal barriers: smaller suppliers often lack digital capacity and contractual flexibility to fully engage in collaborative frameworks (Blismas and Wakefield, 2009). Thus, sustainable MMC delivery requires capacity-building initiatives that extend collaboration beyond Tier 1 contractors to the broader supplier system. Training, digital literacy and contractual inclusivity are recurring prerequisites identified across all case studies.

9.4.1.5 Performance outcomes and value optimisation

Across the analysed cases, projects implementing CC consistently reported improvements in cost predictability, schedule performance and waste reduction. The London 2012 Olympics, for instance, achieved on-budget delivery by combining performance-based incentives with integrated cost planning and target value design.

Similarly, Heathrow T5's open-book framework enabled dynamic adjustment of costs and risks during execution, leading to enhanced transparency and dispute avoidance. In both cases, the integration of lean and CC principles resulted in measurable performance improvements, including reductions in cost overruns and improved delivery certainty (Eastman *et al.*, 2008; Namadi *et al.*, 2017)

The analysis also indicates a shift from focusing on lowest cost to best value. Collaborative environments promote life-cycle thinking, ensuring that design, cost and performance decisions optimise long-term outcomes rather than short-term savings. This is particularly significant in MMC, where manufacturing precision and standardisation directly influence both cost efficiency and sustainability. Nonetheless, while the evidence supports the efficiency benefits of CC, quantifying its financial impact remains challenging due to differences in project scope, reporting standards and data confidentiality. Future research should therefore focus on developing standardised metrics for evaluating CC performance in MMC contexts.

The findings shows that CC provides a viable pathway toward achieving lean and sustainable outcomes in MMC. Its success, however, depends on systemic integration of culture, technology and procurement reform. Projects that embedded CC principles holistically, combining digital tools, open-book contracting and early engagement, consistently achieved superior performance. The evidence reinforces that collaboration must be institutionalised, not merely procedural. Without supportive digital infrastructure, consistent leadership and procurement innovation, CC remains a fragmented practice rather than a transformative paradigm.

In conclusion, while significant progress has been made in integrating lean and CC within MMC, achieving sector-wide adoption will require a shift from compliance-driven to value-driven project delivery – one that places collaboration, transparency and shared accountability at its core.

9.4.2 CC application in the construction sector

CC has increasingly been recognised in the construction sector as a strategic approach to improve cost management, stakeholder alignment and project delivery outcomes – particularly in projects utilising MMC. A few studies and real-world case examples have demonstrated the impact of CC in reducing cost overruns, enhancing early-stage certainty and fostering a more cooperative project culture.

Namadi (2019) proposed a CC framework tailored to the UK construction industry, advocating for the alignment of both upstream (clients, designers) and downstream (contractors, suppliers) stakeholders around a shared budget. The study critiqued the traditionally linear RIBA Plan of Work, arguing that its fragmented stages contribute to siloed operations and inefficiencies. Instead, Namadi (2019) identified core elements such as target costing, optioneering and integrated value engineering as mechanisms to support more predictable and performance-driven project outcomes. The research called for a systemic shift toward life-cycle-oriented costing models that embed collaboration throughout the project.

Sheamar *et al.* (2024) extended this discussion by investigating how innovative procurement models can alleviate cost overruns. Focusing on contractor perspectives, the study identified early contractor involvement (ECI), open-book costing, integrated project insurance (IPI) and shared risk/reward mechanisms as critical enablers of CC. These procurement innovations were found to improve transparency and stakeholder engagement, key ingredients for collaborative success.

9.4.2.1 Case examples of CC in practice

Real-world projects provide compelling evidence for the application of CC within construction (see Table 9.2 below). These principles have been increasingly applied in complex construction projects globally, demonstrating tangible benefits in cost predictability, stakeholder integration and schedule performance. The following cases – Heathrow Terminal 5 (UK), London 2012 Olympics, Sutter Health Castro Valley Medical Centre (USA) and Crossrail (UK) – illustrate how lean and CC frameworks have been implemented at scale, providing valuable lessons for MMC adoption.

HEATHROW TERMINAL 5 (UK)

Heathrow Terminal 5 represents one of the earliest large-scale UK projects to integrate collaborative delivery and open-book cost management. Delivered under BAA's innovative 'Integrated Team Working' model, the £4.3 billion programme replaced traditional contracting with partnering agreements and gain/pain share mechanisms, where contractors and consultants jointly managed risks and rewards.

A target cost framework allowed budgets to be continually refined through early engagement and transparent financial tracking. Stakeholders, including BAA, design consultants and tier one contractors, collaborated through shared data systems and co-located project teams. This model eliminated adversarial cost disputes and encouraged proactive problem-solving. As documented by Davies *et al.* (2009), this model facilitated early engagement, reduced adversarial relationships and supported cost predictability, an early example of lean principles in UK infrastructure.

Table 9.2 Summary of projects that utilised CC and impacts on MMC in construction

Project / Example	Approach	CC Features	Benefits	Impact on Timeliness	Impact on Communication
Sutter Health Castro Valley Medical Centre (USA)	IPD with shared budget	Open-book agreements, early contractor involvement, real-time cost tracking. Use of 3D modelling tools.	30% reduction in time, budget adherence, fewer change orders.	Delivered 30% faster than comparable builds due to concurrent design and construction.	Daily coordination meetings and digital collaboration tools (e.g. BIM) enhanced decision-making.
London 2012 Olympic Venues (UK)	NEC3 Target Cost Contracts	Gain/pain share, risk register, cost transparency.	£430m savings, stronger collaboration, low dispute rate.	Projects delivered on schedule to meet the Olympic deadline. Early target costing streamlined designs.	Cross-organisational KPI tracking and regular reviews enhanced coordination among stakeholders.
Heathrow Terminal 5 (UK)	Project partnering	Joint cost planning, collaborative design, no fixed-price contract.	No major disputes, innovation, high satisfaction.	Reduced legal disputes and adversarial delays. However, initial phases faced approval-related delays.	Co-location and real-time platforms (e.g. Primavera) improved communication and eliminated litigation.
Crossrail (UK)	BIM-enabled costing	5D BIM, real-time updates, digital integration. Over 60 firms connected through a digital cost environment.	Accurate forecast, better cost control, less rework.	Avoided delays through accurate forecasting and minimised rework.	Real-time cost and design updates improved cross-party communication and reduced silos.

Source: Authors' creation

LONDON 2012 OLYMPICS (UK)

The London Olympics programme, managed by the Olympic Delivery Authority (ODA), demonstrated the value of performance-based incentives and integrated cost management on a national scale. Using the NEC3 contract framework and target value design (TVD) principles, the ODA implemented collaborative budgeting, open-book accounting and shared risk/reward systems across its £9 billion portfolio.

Cost management was digitally supported by project dashboards and data integration between design, cost and procurement teams. Despite the scale and complexity, the project was completed on time and within the public spending envelope, a remarkable achievement credited to transparent cost planning, joint forecasting and value-driven decision-making. The model established new benchmarks for collaborative public-sector delivery and provided a template for MMC projects requiring multidisciplinary coordination and cost certainty. According to the National Audit Office (2012), the project benefited from cost-linked milestones, integrated cost planning and a culture of collaboration across an extensive supply chain.

SUTTER HEALTH CASTRO VALLEY MEDICAL CENTRE (USA)

The Sutter Health project in California adopted an IPD framework – one of the most advanced applications of CC globally. A multi-party agreement between the client, designers and contractors aligned all stakeholders to shared cost and performance targets.

A 5D BIM system integrated design and cost information in real-time, enabling teams to assess the financial impact of design choices instantly. This digital integration reduced rework, improved constructability and shortened delivery by around 30% compared to traditional models (Kent and Becerik-Gerber, 2010). The IPD model fostered trust, joint problem-solving and transparency through open-book collaboration, demonstrating the efficiency and predictability benefits possible when technology and culture align. The project's outcomes are directly applicable to MMC environments that depend on digital coordination and early cost certainty.

CROSSRAIL (UK)

Crossrail, one of Europe's largest infrastructure projects, exemplified the power of digital-enabled collaboration. The £18 billion programme utilised 5D BIM and a CDE accessible to over 60 contractors and consultants (Bryde *et al.*, 2013). Collaborative cost management allowed real-time monitoring of progress, financial performance and supply chain data.

The integration of lean principles and digital workflows reduced duplication, scope creep and rework, major causes of cost overruns in traditional projects. This approach demonstrated how digital transparency, early supplier involvement and integrated design-to-cost strategies can deliver efficiency gains even in megaprojects with extensive stakeholder networks. Crossrail's framework directly informs MMC adoption by showing how digitally integrated CC can coordinate multiple off-site manufacturing and delivery partners.

9.4.3 Cross-case insights and implications for MMC

The cases reviewed collectively reinforce the strategic benefits of CC, particularly for MMC projects that demand precision, prefabrication and integration across geographically dispersed stakeholders. See Table 9.3 below.

Table 9.3 Comparative summary of CC case studies in construction

Project	Scale and Sector	Collaborative Mechanism	Digital / Lean Tools Used	Key Outcomes	Relevance to MMC
Heathrow Terminal 5 (UK)	£4.3bn / Aviation Infrastructure	Integrated team working, gain/pain share, target costing	Shared data systems, co-located teams	Enhanced cost predictability, reduced disputes	Demonstrates early UK adoption of lean cost integration
London 2012 Olympics (UK)	£9bn / Public Infrastructure	NEC3 contracts, target value design, performance incentives	Digital dashboards, collaborative forecasting	On-time, on-budget delivery, improved transparency	Model for collaborative public-sector MMC programmes
Sutter Health (USA)	$320m / Healthcare	IPD, multi-party agreement	5D BIM, real-time cost simulation	30% faster delivery, cost reduction, high trust	Closely mirrors MMC manufacturing integration principles
Crossrail (UK)	£18bn / Transport	Collaborative cost planning, supply chain integration	5D BIM, CDE	Reduced rework, efficient procurement, improved data flow	Strong evidence of digital-enabled CC for modular, multi-stakeholder projects

Source: Authors' creation

9.4.3.1 Early stakeholder integration enables cost predictability

All four cases confirm that cost reliability improves when key actors are involved early in design development. TVD and early contractor involvement foster shared cost ownership, critical for MMC where design freeze occurs earlier due to off-site manufacturing. In contrast, projects that delayed stakeholder integration faced higher risk of rework and design misalignment. This reinforces the need for MMC frameworks to institutionalise front-end cost collaboration through open-book budgeting and early supply chain consultation.

9.4.3.2 Digitalisation is a catalyst for collaboration

The widespread use of BIM and digital data environments in these cases demonstrates that technology enhances collaboration, but only when paired with clear governance and skills capacity. 5D BIM, as employed in Crossrail and Sutter Health, provides real-time visibility of design and cost data, aligning decisions across disciplines. For MMC, where precision manufacturing and modular sequencing are critical, digital integration is not optional – it is foundational for achieving lean, cost-efficient delivery.

9.4.3.3 Trust and shared risk-reward mechanisms drive cultural transformation

Projects using gain/pain share or multi-party agreements consistently reported stronger collaboration and fewer disputes. Heathrow T5 and Sutter Health show that transparent risk allocation builds accountability and reduces adversarial behaviour. For MMC projects involving multiple contractors and manufacturers, such models ensure equitable participation and shared commitment to value optimisation rather than lowest-cost competition.

9.4.3.4 Supply chain collaboration strengthens MMC delivery

Cross-case analysis shows that early and continuous engagement with suppliers improved cost accuracy, reduced material waste and enhanced constructability. This finding is particularly relevant to MMC supply chains, where manufacturers must commit to specifications before production begins. Collaborative procurement frameworks (e.g. NEC Option C, IPD) can therefore bridge the design-manufacture gap and enhance production efficiency.

9.4.3.5 Value focus replaces cost minimisation

All projects demonstrate a shift from 'lowest cost' to 'best value.' CC supports life-cycle thinking, considering long-term operational savings, quality and sustainability. In MMC, this perspective aligns with lean principles of eliminating waste and optimising value across the life cycle, rather than simply reducing upfront costs.

9.4.4 Summary

CC holds transformative potential for improving cost outcomes and operational efficiency in MMC projects. Its success lies in promoting early engagement, transparency and integrated decision-making, all of which align closely with the principles of modern construction delivery.

However, to scale CC across the sector, changes are required in procurement practices, regulatory frameworks and professional capability development. Despite these benefits, effective CC implementation requires significant organisational maturity, robust communication systems and a supportive procurement environment. When executed well, the initial setup complexities are outweighed by long-term gains in efficiency, cost predictability and relationship quality.

9.5. Implications and strategic pathways for implementing CC in MMC

9.5.1 Implications for practice and policy

The empirical and case-based findings reviewed in this study reveal several significant implications for the integration of CC within MMC. These insights are critical for industry stakeholders seeking to enhance cost predictability, value delivery and operational efficiency in the context of off-site and modular construction paradigms.

Firstly, the adoption of CC necessitates a fundamental cultural transformation within the construction industry. Prevailing adversarial practices and fragmented project delivery models stand in stark contrast to the collaborative ethos required for CC to be effective. Cultural inertia and resistance to change remain formidable barriers to implementation, suggesting that behavioural change initiatives must accompany technical solutions.

Secondly, digital infrastructure emerges as a key enabler of CC. The integration of BIM, 5D cost modelling and CDE facilitates real-time collaboration, allowing stakeholders to concurrently assess design, cost and scheduling decisions. However, disparities in digital maturity and interoperability issues across project actors may hinder seamless collaboration unless addressed through standardisation and capacity-building initiatives.

Thirdly, traditional procurement structures often act as inhibitors to CC, particularly those predicated on lowest-cost tendering. These models fail to support the early involvement and joint accountability mechanisms that underpin collaborative practices. Alternative procurement frameworks such as IPD and target cost contracts demonstrate greater alignment with CC principles and should be more widely adopted.

Fourthly, the importance of front-end planning is particularly pronounced in MMC projects due to their reliance on early design freeze and manufacturing lead times. CC supports the establishment of target costs and collaborative risk management at the conceptual stage, thereby reducing rework, cost escalation and supply chain disruption during implementation.

Lastly, the findings underscore the persistent skills and knowledge gaps that limit the uptake of CC in practice. Many professionals lack the requisite training in lean construction, integrated cost modelling and collaborative behaviours, highlighting a pressing need for targeted education and professional development.

9.5.2 Strategies for successful implementation of CC in MMC

To operationalise CC in MMC contexts, the following strategies are proposed. These are grounded in both the case study evidence and the theoretical constructs reviewed.

9.5.2.1 Early integration of CC principles

Embedding CC at the early stages of project development – particularly from concept and schematic design – enables stakeholders to align on a shared allowable cost and collaboratively manage risks from the outset. Techniques such as optioneering and integrated value engineering allow for informed trade-off decisions, reducing design rework and cost overruns

(Namadi, 2019; Sheamar *et al.*, 2024). Early-stage financial alignment is particularly critical in MMC, where design changes during manufacturing can lead to significant delays and cost escalation.

9.5.2.2 Collaborative procurement models

Traditional procurement approaches often foster fragmented decision-making and adversarial relationships, which are incompatible with the interdependent nature of MMC. Instead, frameworks such as NEC Option C (target cost), early contractor involvement (ECI) and IPD promote transparency, joint risk-sharing and aligned incentives (Ahmed *et al.*, 2019; Davies *et al.*, 2009). These models facilitate earlier collaboration and enable cost efficiency through shared accountability and trust-building mechanisms.

9.5.2.3 Digital integration and interoperability

Digital technologies are essential enablers of CC in MMC. The use of BIM, cost-integrated design tools and CDE supports real-time cost visibility and coordination across all stakeholders (Bryde *et al.*, 2013; Kent and Becerik-Gerber, 2010). These platforms enable concurrent design and cost analysis, which is crucial for modular construction where precision and sequencing are paramount.

9.5.2.4 Capability development and collaborative culture

The effectiveness of CC depends not only on systems but also on people. Structured training on CC methodologies, value-based decision-making and team integration techniques is required to develop a culture of collaboration. Assigning collaborative leadership roles, such as integration managers, supports alignment and enhances communication across diverse project actors (Ahmed *et al.*, 2019). Capacity building at all levels of the supply chain is essential for sustained collaborative practice.

9.5.2.5 Feedback mechanisms and continuous improvement

Post-project reviews, performance metrics and lessons-learned processes should be built into CC workflows to drive continual improvement. By institutionalising learning loops, project teams can refine their collaborative frameworks, enhance stakeholder engagement and evolve practices based on experience (Zimina *et al.*, 2012). These insights contribute to more mature and adaptive cost management across successive MMC initiatives.

9.5.2.6 Policy support and industry alignment

To scale CC effectively, supportive policies and institutional frameworks are required. Public sector clients can lead by example by embedding CC in their procurement and project delivery strategies. Regulatory alignment and standardised frameworks for collaborative cost planning – especially within public infrastructure programmes – will provide clarity and reduce adoption barriers (Latham, 1994; National Audit Office, 2012). Incentivising early contractor involvement, open-book costing and life-cycle-based budgeting can accelerate industry-wide adoption.

9.5.3 Summary

CC, when effectively applied, has the potential to transform the cost management landscape of MMC projects. Its emphasis on early engagement, joint decision-making and value optimisation aligns closely with the modular, integrated and fast-paced nature of MMC delivery. However, for CC to achieve mainstream adoption, a multi-faceted implementation strategy is required – encompassing procurement reform, digital integration, cultural change and capability development.

Future research should focus on the development of standardised MMC-specific cost frameworks, exploration of real-time digital costing platforms and longitudinal studies evaluating the long-term impact of CC on project performance.

9.6. Conclusions

MMC are revolutionising the UK construction sector through improved efficiency, sustainability and speed of delivery. These techniques, which include modular construction, off-site manufacturing and digital integration, demand an equally modern approach to cost management. Traditional costing methods, long embedded in the industry, are no longer sufficient to support the dynamic and integrated nature of MMC projects. Instead, the rise of CC – a lean, transparent and integrated approach – represents a necessary evolution.

Collaborative costing fosters early-stage engagement between stakeholders, aligning financial decision making with design and construction phases. Unlike traditional models that often lead to fragmented responsibilities and budget overruns, collaborative models promote shared accountability and trust. The increasing demand for such models is further driven by the need for greater cost certainty, risk mitigation and life-cycle value in a sector pressured by tight margins and high client expectations.

The advent of 5D BIM has acted as a technological enabler, facilitating real-time cost modelling, scenario planning and data-driven decision making. Within this context, UK construction firms are gradually embracing CC, with various models demonstrating success in aligning budgets with performance, especially in MMC projects. The cases studied indicate that these approaches can yield significant benefits, including reduced waste, better risk sharing, enhanced value engineering and improved project timelines.

Conversely, the study also found that implementing CC in MMC environments is not without challenges. Barriers such as cultural resistance, lack of standardisation, technological fragmentation and unclear ownership of data continue to hinder widespread adoption. Despite these obstacles, the study demonstrated benefits – from more efficient procurement to enhanced project outcomes – making a compelling case for strategic adoption.

Furthermore, insights from the case-studies reinforce the idea that CC, when embedded into MMC workflows, enhances project integration and fosters innovation. Conceptual developments suggest a growing maturity in models that integrate lean thinking, digital tools and collaborative governance structures.

Strategically, the path forward involves fostering industry-wide cultural change, investing in digital skills and developing clear implementation frameworks that integrate CC with MMC processes. Government incentives, client advocacy and cross-sector learning will be essential to driving this transformation. Ultimately, CC represents more than a technical adjustment – it is a foundational shift toward smarter, more sustainable construction in the UK.

9.7. Study contribution

This study contributes to existing knowledge by synthesising evidence on the application of lean and CC within MMC in the UK. Drawing on secondary data and case studies, it deepens understanding of the key challenges, benefits and optimisation strategies associated with CC adoption. This enhanced understanding will enable the construction industry to recognise how CC can improve efficiency, cost predictability and value delivery in MMC projects. The findings also provide practical insights for industry stakeholders seeking to strengthen collaboration, reduce waste and achieve more sustainable and integrated project outcomes through lean and collaborative practices.

9.8. Study limitations

This study is subject to several limitations that should be acknowledged. Firstly, it relies exclusively on secondary data sources, including published case studies and academic literature. While these sources provide valuable insights, they may not fully capture the most recent or unpublished industry practices and the researcher has limited control over data quality, consistency and contextual depth. Moreover, the absence of primary data, such as interviews or surveys, restricts the ability to validate findings through direct stakeholder perspectives.

Secondly, as a qualitative, document-based case synthesis, the findings are interpretive and context-specific. The results, therefore, cannot be generalised across all construction projects or regions without caution.

Future studies could address this limitation by integrating mixed methods or conducting empirical fieldwork to test and expand on the conclusions presented here. Despite these constraints, the study offers a comprehensive overview of existing practices and provides a solid foundation for future empirical research on CC and MMC implementation in the UK construction industry.

REFERENCES

Afana O, Al Zubaidi R, Abu Dabous S and Ibrahim F (2024) Categories and factors of cost overrun in construction projects: a systematic review. *Engineering, Technology & Applied Science Research* **14(6)**: 18330–18347.

Ahmed S and El-Sayegh S (2021) Critical review of the evolution of project delivery methods in the construction industry. *Buildings* **11(1)**: 11.

Ahmed SN, Pasquire C and Manu E (2019) Exploratory study of costing collaboratively in the UK construction industry. In, *Proceedings of the 27th Annual Conference of the International Group for Lean Construction* (Pasquire C and Hamzeh FR (eds)), Dublin, Ireland, pp. 1163–1174.

Ballard G and Howell G (2003) Lean project management. *Building Research and Information.* **31(2)**: 119–133.

Blismas N and Wakefield R (2009) Drivers, constraints and the future of off-site manufacture in Australia. *Construction Innovation* **9(1)**: 72–83.

Bryde D, Broquetas M and Volm JM (2013) The project benefits of Building Information Modelling (BIM). *International Journal of Project Management* **31(7)**: 971–980.

BSI (British Standards Institution). (2018) ISO 19650-1:2018: Organization and digitisation of information about buildings and civil engineering works, including building information modelling (BIM) – Information management using building information modelling – Part 1: Concepts and principles. BSI Standards Limited, London, UK.

Constructing Excellence (2020) The CE 2020: Report on Collaborative Construction. BRE Ltd., Watford, UK.

Das K, Khursheed S and Paul VK (2025) The impact of BIM on project time and cost: insights from case studies. *Discover Materials* **5**: 25. https://doi.org/10.1007/s43939-025-00200-2

Davies A, Gann D and Douglas T (2009) Innovation in megaprojects: systems integration at London Heathrow Terminal 5. *California Management Review* **51(2)**: 101–125.

Davies A and Mackenzie I (2014) Project complexity and systems integration: constructing the London 2012 Olympics and Paralympics Games. *International Journal of Project Management*, **32(5)**: 773–790.

Eastman C, Teicholz P, Sacks R and Liston K (2008) BIM Handbook. Wiley.

Egan J (1998) Rethinking Construction: Report of the UK Construction Task Force. Constructing Excellence, UK.

HM Government (2020) The Construction Playbook: Government Guidance on Sourcing and Contracting Public Works Projects and Programmes. https://www.gov.uk/government/publications/the-construction-playbook (accessed 10/10/2025).

Humphreys P, Matthews J and Kumaraswamy M (2003) Pre-construction project partnering: From adversarial to collaborative relationships. *Supply Chain Management* **8(2)**: 166–178.

Kent DC and Becerik-Gerber B (2010) Understanding construction industry experience and attitudes toward integrated project delivery. *Journal of Construction Engineering and Management* **136(8)**: 815–825.

Lahdenperä P (2012) Making sense of the multi-party contractual arrangements of project partnering, project alliancing and integrated project delivery. *Construction Management and Economics* **30(1)**:57–79.

Latham M (1994) *Constructing the Team (The Latham Report)*, HMSO, London, UK.

Mayouf M, Gerges M and Cox S (2019) 5D BIM: an investigation into the integration of quantity surveyors within the BIM process. *Journal of Engineering, Design and Technology* **17(3)**: 537–553.

Ministry of Housing, Communities and Local Government (2019) Modern Methods of Construction Working Group: Developing a Definition Framework. https://www.gov.uk/government/publications/modern-methods-of-construction-working-group-developing-a-definition-framework (accessed 9/10/2025).

Monyane G, Emuze F, Awuzie B and Crafford G (2020) Evaluating a collaborative cost management framework with Lean construction experts. In and *Proceedings of the 10th International Conference on Engineering, Project and Production Management* (Panuwatwanich K and Ko C-H (eds)). Springer, Singapore, pp. 383–393.

Moses T, Heesom D and Oloke D (2020) Implementing 5D BIM on construction projects: Contractor perspectives from the UK construction sector. *Journal of Engineering, Design and Technology* **18(6)**: 1867–1888.

Mosey D (2019) *Collaborative Construction Procurement and Improved Value.* 2nd edn Wiley-Blackwell, Chichester, UK.

Namadi (2019) A framework for CC in the UK construction industry. PhD, Nottingham Trent University. https://irep.ntu.ac.uk/id/eprint/40295.

Namadi S, Pasquire C and Manu E (2017) Discrete costing versus CC. *Proceedings of the 25th Annual Conference of the International Group for Lean Construction*, Heraklion, Greece, pp. 3–10.

National Audit Office (2012) *The London 2012 Olympic Games and Paralympic Games: Post-Games Review*.

NHBC Foundation (2016) Modern methods of construction: Views from the industry. https://www.nhbc.co.uk/insights-and-media/foundation/publications/modern-methods-of-construction-views-from-the-industry (accessed 3/2/2021).

Pan W and Sidwell R (2011) Demystifying the cost barriers to offsite construction in the UK. *Construction Management and Economics* **29(11)**: 1081–1099.

Pishdad P and Onungwa IO (2024) Analysis of 5D BIM for cost estimation, cost control and payments. *Journal of Information Technology in Construction* **29**: 525–548.

Sheamar S, Wedawatta G, Tennakoon M, Palliyaguru R and Antwi-Afari MF (2024) The potential of new models of construction procurement to counter cost overruns in construction projects: an exploratory study from a contractor's perspective. *Journal of Financial Management of Property and Construction* **29(2)**, 211–228.

Yang Y, Wei Z and Zhang Z (2023) Stakeholder relationship in construction projects: a mixed methods review. *Buildings* **13(12)**, 3122, 10.3390/buildings13123122

Zimina D, Ballard G and Pasquire C (2012) Target value design: using collaboration and a Lean approach to reduce construction costs. *Construction Management and Economics* **30(5)**: 383–398.

Emmanuel I Daniel and Fidelis A Emuze
ISBN 978-1-83708-437-1
https://doi.org/10.1108/978-1-83708-436-420261011

Chapter 10
Timber-based modern methods of construction for sustainable construction: challenges and opportunities

Steve Bertasso[1], Benjamin Shirley[1] and Robert Hairstans[1,2]

[1]*New Model Institute for Technology and Engineering, Hereford, United Kingdom*
0009-0000-5195-8743; 0009-0004-6860-9872; 0000-0003-3993-7944
[2]*Edinburgh Napier University, Edinburgh, United Kingdom*

Abstract

This chapter explores the integration of timber and modern methods of construction (MMC) as a response to the urgent environmental and efficiency challenges faced by the industry. Focusing on timber-based systems such as panelised construction, volumetric modular and mass timber hybrids, it examines how these methods offer low-carbon, high-performance alternatives to traditional steel and concrete construction. Through discussion of life-cycle thinking, embodied carbon reduction and circular economy principles, the chapter presents timber not only as a renewable material but as a platform for innovation in off-site fabrication, waste minimisation and design efficiency. The regulatory landscape is examined in detail, focusing on fire safety concerns and the evolving UK and EU policy frameworks – including the Timber in Construction Roadmap (2025) – that are reshaping the role of timber in practice. Supply chain risks, material sourcing and economic viability are also addressed, with evidence supporting the use of locally sourced timber to enhance resilience and reduce emissions. The chapter concludes with strategies for advancing industry adoption through education, policy and research, including proposals for living lab environments that enable real-time performance monitoring. By promoting integrated, data-driven construction processes, timber-based MMC can support a more sustainable and scalable built environment.

Keywords: timber construction; off-site manufacturing; mass timber; prefabricated systems; circular economy; fire safety in timber buildings

10.1. Introduction

Creating sustainable spaces efficiently is the key challenge facing the building industry in the developed world. Between embodied and operational carbon, the built environment remains one of the largest contributors to the climate crisis. As global climate targets become more urgent, the construction sector must transition to building methods and materials that not only reduce emissions but also deliver quality, performance and speed. Timber and other bio-based materials, when integrated with a range of MMC – including panelised systems, volumetric modular units and hybrid timber technologies – offer a viable, scalable pathway to addressing these challenges without sacrificing function, form, or aesthetic integrity (UK Government, 2025).

Today's construction processes are more regulated and scrutinised than ever. Planning, compliance, building safety and environmental standards are now embedded in every phase of development – from concept design through to completion and occupancy. While these changes are essential for producing better-performing and safer buildings, they also contribute to longer project timelines, higher upfront costs and an increasing reluctance to adopt innovative approaches. As a result, owners and developers often default to familiar construction methods, even when more sustainable, cost-effective options are available when viewed over a building's lifetime. This short-term thinking locks in inefficiencies and foregoes opportunities for long-term carbon and cost reductions.

What is often overlooked in traditional project assessments is the integrative potential of MMC, which compresses the construction timeline, dramatically raises quality standards and improves outcomes by streamlining multiple stages of a building's life cycle. Off-site systems – such as closed timber frame panels, cross-laminated timber (CLT), glued-laminated timber (GLT) and fully volumetric modular solutions – enable precise, factory-controlled production, reducing on-site labour and waste. When combined with digital design and manufacturing technologies, these methods enable buildings to be delivered more quickly and with greater consistency, while significantly reducing embodied carbon and enhancing operational performance (Churkina *et al.*, 2020).

This chapter examines how a diverse range of off-site timber-based systems is redefining the approach to sustainable construction. These include high-performance panelised wall systems with integrated insulation and services, volumetric modules designed for rapid urban infill, and mass timber systems capable of meeting commercial structural demands. Each system contributes uniquely to sustainability goals – whether through material efficiency, energy performance, or ease of disassembly and reuse – making MMC not only relevant but essential to a low-carbon future.

Additionally, this work examines the policy, regulatory and economic frameworks necessary to support wider adoption. It explores how life-cycle thinking, circular economy principles and procurement reform can foster adoption, and how shifts in education, training and data sharing can mitigate industry resistance to change. Barriers such as supply chain vulnerabilities, building regulations and perception of risk are explored, alongside examples of innovation from practice and policy that signal a new direction for the industry.

Finally, it identifies opportunities for future research and innovation, particularly in the development of digitisation and real-world monitoring, or 'living labs'. These environments enable buildings to be assessed based on performance data, rather than assumptions, providing a powerful feedback loop that can validate design intent, inform policy and support continuous improvement. This is particularly relevant when considering the potential impact of climate change and the corresponding environmental conditions. By transitioning from prescriptive codes to performance-based outcomes and from fragmented practices to integrated delivery models, off-site timber and MMC provide a vital foundation for constructing a more efficient, resilient and sustainable built environment.

10.2. Perspective and concepts

In recent years, timber has reemerged as a prominent material in the construction industry, effectively addressing both structural and environmental challenges through innovative applications and technologies. The advancement of modern construction techniques, such as enhanced panelisation systems (incorporating insulation, windows, mechanical, electrical and plumbing (MEP), and cladding), volumetric modular systems, structural insulated panels (SIPs), mass timber – including CLT and GLT – and hybrid systems (combining different approaches optimally) has revolutionised the utilisation of timber in buildings. These methodologies not only enhance precision, speed and efficiency in construction but also significantly reduce embodied carbon, thereby aligning with global sustainability objectives. The renewable nature of timber, coupled with advancements in prefabrication and engineered wood products, positions it as an adaptable and high-performance alternative to preferred materials such as steel and concrete (Ramage *et al.*, 2017).

10.2.1 Timber as a building material

Timber, particularly in its engineered forms, presents a compelling alternative to traditional structural materials, such as steel and concrete, offering comparable strength, stability and design flexibility while significantly reducing environmental impact. Products such as CLT, GLT and Laminated Veneer Lumber (LVL) have transformed timber from a conventional framing material into a high-performance structural component capable of supporting multi-storey and even high-rise construction. These engineered systems distribute loads efficiently and demonstrate favourable strength-to-weight ratios, minimising the need for heavy foundations and facilitating prefabricated solutions that enhance on-site efficiency.

In hybrid systems, timber is often strategically combined with minimal amounts of steel or concrete to leverage the best properties of each – utilising timber for vertical and lateral load-bearing components and steel or concrete for specific functions such as core stabilisation or fire protection in high-stress areas. This synergy reduces overall material consumption, enabling the creation of lighter and more adaptable structures – for example, hybrid mass timber buildings may utilise CLT floor panels alongside GLT beams with steel-braced frames and concrete cores, creating systems that are structurally robust, seismically resilient and considerably lower in embodied carbon (Ramage *et al.*, 2017).

Embodied carbon – the total greenhouse gas emissions associated with material extraction, processing, transportation and construction – remains a critical metric in the assessment of building materials. Steel and concrete are among the largest contributors to embodied carbon in construction, due to energy-intensive production processes and high transportation costs. In contrast, timber naturally sequesters carbon during its growth phase and retains that carbon when used in buildings. Studies have shown that switching from concrete or steel frames to timber-based or hybrid systems can reduce embodied carbon by up to 60% (Churkina *et al.*, 2020). When timber is sourced sustainably and used within circular design frameworks, it not only replaces carbon-intensive materials but also contributes to carbon storage throughout the building's life span and in future construction.

Thus, the evolution of timber from a traditional to a technologically advanced structural material signals a transformative shift in building practices – one that aligns strength, sustainability and innovation in the pursuit of climate-conscious construction.

10.2.2 Circular economy of timber

Timber's role in a circular economy is one of its most compelling attributes, reinforcing its environmental superiority over finite, high-carbon materials such as concrete and steel. Unlike many traditional construction materials that follow a linear life cycle – from extraction to use and eventual disposal – timber can be designed within regenerative loops that extend its usefulness far beyond its initial application. This circularity is rooted in timber's natural origin, ease of disassembly and the adaptability of its byproducts for reuse, recycling or upcycling into new construction materials (Pomponi and Moncaster, 2017).

At the core of timber's circular potential is its ability to be reused across multiple building life cycles. Components such as beams, panels and floor systems can be recovered from deconstructed buildings and repurposed for new projects with minimal reprocessing. Unlike concrete, which is often crushed and downcycled, or steel, which requires significant energy to melt and reform, timber elements can retain structural integrity through careful design and handling. Prefabricated timber systems, including modular and panelised units, are particularly well-suited for deconstruction, making reuse not only possible but also economically viable.

Beyond whole-member reuse, timber fibres and wood waste can be recycled and upcycled into a wide array of new products. Engineered wood products, such as particleboard, oriented strand board (OSB) and medium-density fibreboard (MDF), as well as newer technologies such as wood-fibre insulation and bio-based composites, are often manufactured using reclaimed wood waste. These materials extend the life of timber far beyond its initial structural use and reduce the demand for virgin timber resources. As circular economy principles gain traction, innovations in binding agents and fibre processing have further enhanced the structural and environmental performance of upcycled timber-based materials.

Waste streams from both manufacturing and on-site construction are also being redirected toward productive ends. Offcuts, sawdust and shavings – once considered waste byproducts – are now key inputs for bio-fuels, wood-plastic composites and acoustic or thermal insulation products. This closed-loop thinking transforms what was once landfill-bound waste into valuable industrial feedstock, contributing to carbon sequestration and resource efficiency (Ramage *et al.*, 2017).

Critically, the ability of timber to be part of a circular economy is amplified when it is sourced from responsibly managed forests and incorporated into buildings with future disassembly in mind. Design for deconstruction (DfD) strategies, combined with digital tools for material tracking (such as BIM and digital passports), are accelerating the integration of timber into circular supply chains. The result is a regenerative material flow in which timber not only stores carbon during its life but also continually re-enters the material economy, driving down emissions and resource depletion (Pomponi and Moncaster, 2017).

In summary, timber exemplifies the principles of circularity in the built environment: it is renewable, reusable, recyclable and regenerative. Its capacity to be transformed across life cycles – from structural applications to recycled composites – positions it as a cornerstone of low-carbon, resource-conscious construction now and into the future.

10.3. Methodological approach

This work is informed by a qualitative research methodology aligned with established research practices. The approach explored technical and historical methods before examining modern practice and proposing future interventions for the built environment.

A review of historical construction techniques, including traditional carpentry and timber framing, as well as masonry, steel and reinforced concrete systems, establishes a baseline for evaluation. Traditional carpentry is considered not only a craft-based practice but also an early example of material efficiency, structural logic and prefabrication. This historical overview provides a vital baseline for assessing modern construction methods, particularly with respect to structural performance, material optimisation, durability and environmental impact.

Building on this foundation, the research investigates the development and formalisation of MMC, with a focus on off-site and industrialised systems. Panelised timber construction, volumetric modular systems, mass timber assemblies and hybrid structural solutions are examined using peer-reviewed literature, technical standards, industry guidance and policy documentation. Emphasis is placed on performance criteria such as fire safety, quality assurance, embodied carbon and construction efficiency.

A case-study-based analytical framework is used to link theoretical principles to practical application. Selected contemporary projects are examined to assess how timber-based MMC is realised within real-world constraints, including regulatory compliance, supply chain logistics and programme delivery. Case studies help identify transferable technical lessons and performance outcomes rather than promoting individual projects.

Finally, the findings are synthesised to support a forward-looking analysis that considers implications for regulation, education and future research. Emphasis is placed on performance-based evaluation, post-occupancy monitoring and the role of living laboratory approaches in generating evidence to support outcome-focused design. This methodology reflects engineering research traditions that prioritise measured performance, repeatability and technical rigour, supporting informed advancement of timber and off-site construction practices.

10.4. Results and discussion

10.4.1 Case study: Skylon Campus, Centre for Advanced Timber Technology – a flagship for modern timber construction in education

Project overview

The Centre for Advanced Timber Technology (CATT), located at NMITE's Skylon Campus in Hereford, represents a pioneering example of MMC and mass timber technology applied to an educational environment. CATT serves as a regional hub for timber innovation and education, showcasing best practices in sustainable building design and construction. The 2500 m^2 facility features learning spaces, design studios and workshops, setting new standards for low-carbon, high-performance academic buildings in the UK (Constructing Excellence, n.d.).

10.4.1.1 Off-site timber construction methodology

The construction of Skylon Campus relied heavily on off-site fabrication and mass timber systems. The structural frame was constructed using a combination of CLT and GLT, resulting in a hybrid engineered timber system. These elements were precision-manufactured off-site and delivered to the site as pre-cut, ready-to-assemble components, enabling rapid, accurate construction with minimal waste.

In addition to the main structural elements, the external envelope utilised off-site-manufactured timber wall panels. These insulated panels were pre-fabricated and included integrated linings and vapour control layers, contributing to a high-performance building envelope. This approach ensured tight tolerances, improved on-site efficiency and reduced exposure to weather during construction.

10.4.1.2 Construction programme and efficiency

The use of mass timber and off-site panelisation significantly streamlined the construction timeline. Early coordination between the design team, suppliers and contractors ensured that services and junctions were resolved before fabrication began. This reduced the need for on-site modification and rework.

The pre-manufactured timber systems also contributed to a cleaner and safer site environment. With major components delivered in a just-in-time sequence and assembled with minimal equipment, disruption to neighbouring developments and construction logistics was minimised. These efficiencies align with the project's role as a demonstration of modern construction practices.

10.4.1.3 Sustainability and carbon performance

Skylon Campus was designed to meet ambitious sustainability targets, with a core focus on reducing embodied and operational carbon. The use of CLT and GLT not only enabled carbon sequestration but also eliminated the need for high-carbon materials such as concrete and steel in the superstructure. This resulted in significant reductions in embodied carbon, supporting NMITE's broader commitment to low-impact construction.

Table 10.1 Case Study 1 – Skylon Campus, CATT

MMC Feature	Description and Key Benefits
Off-site manufacturing	Efficient construction with high levels of precision and minimal site waste
Hybrid timber structure	Combined CLT and GLT for a robust, flexible and sustainable frame
Sustainability	Significant embodied carbon reduction, low-carbon timber used throughout.

Source: Authors' creation

The building fabric was designed to achieve high thermal performance, contributing to long-term operational energy efficiency. Insulated timber wall panels, low-air-leakage detailing and passive solar design strategies further enhanced the building's environmental performance.

Wherever possible, materials were sourced from UK suppliers to reduce transport emissions and support local supply chains, reflecting the principles explored in NMITE's broader research into regional timber utilisation.

10.4.1.4 Project significance and outcomes

The project significance and outcomes are summarised in Table 10.1.

Summary

The Skylon Campus of NMITE is a benchmark project for sustainable educational infrastructure. It illustrates how modern timber technologies and off-site manufacturing can deliver high-performance buildings that meet strict environmental, quality and efficiency targets. As a national exemplar of best practice in timber construction, CATT highlights the growing role of engineered timber in delivering resilient and low-carbon solutions across the built environment.

10.4.2 Case study: Moxy Hotel, Fountainbridge – a modular timber solution for urban hospitality

Project overview

The Moxy Hotel in Fountainbridge, Edinburgh, is a prominent example of modern construction using mass timber and off-site manufacturing methods to achieve sustainable and efficient outcomes in urban hospitality development. Completed in 2022, the project forms part of the broader New Fountainbridge regeneration scheme and includes 262 guest rooms, a rooftop bar and ground-floor offices. The development adopts a hybrid structural strategy integrating off-site timber systems with a steel-and-concrete podium base (BE-ST).

10.4.2.1 Off-site timber construction methodology

The hotel's primary superstructure comprises fully modular CLT bedroom pods. These units were manufactured off-site in Italy under factory-controlled conditions and delivered to the site as turnkey modules. Each pod was pre-fitted with floor slabs, wall panels, internal linings, service penetrations, insulation, windows and bathrooms (wet-boxes), significantly reducing onsite labour and construction duration. The use of CLT enabled high levels of dimensional accuracy, airtightness and thermal performance.

Supporting the CLT modules is a steel-framed podium structure with in-situ concrete slabs, which form the base level and support commercial space. This hybrid system enables the vertical repetition of timber modules while ensuring robustness at ground level, where higher loading and service integration are required. Pre-insulated and sealed CLT wall panels were also used, delivered with windows and membranes pre-installed to expedite the closure of the envelope.

10.4.2.2. Construction programme and efficiency

The integration of off-site manufacturing resulted in a rapid installation phase for the superstructure. The CLT pods were installed over 16 weeks, with the repetitive, modular nature of the rooms enabling efficient stacking and minimal onsite adjustment. Compared with traditional construction methods, the off-site approach reduced the project timeline and significantly limited disruption in the city centre.

Factory conditions also ensured improved consistency and finish quality, particularly in the integration of mechanical, MEP systems, which were coordinated during the manufacturing stage. The reduced requirement for wet trades and follow-on works onsite contributed to a cleaner, safer and more predictable construction process.

10.4.2.3 Sustainability and carbon performance

The use of mass timber for the superstructure delivered a substantial reduction in embodied carbon. Approximately 2000 m^3 of CLT was used in the project, with each cubic metre storing an estimated 676 kg of CO_2(BE-ST). This results in a total sequestration of over 1300 t of CO_2, equivalent to the annual emissions of hundreds of passenger vehicles. The material efficiency and low-waste nature of the off-site fabrication process enhance these carbon benefits.

In addition to carbon sequestration, the pre-manufactured systems contributed to improved building performance through enhanced airtightness, continuity of insulation and reduced thermal bridging. These attributes support long-term operational energy efficiency and align with broader sustainability goals for the development.

10.4.2.4 Project significance and outcomes

The project significance and outcomes are summarised in Table 10.2.

Summary

The Moxy Hotel demonstrates the potential of modular timber construction in large-scale hospitality projects. By combining prefabricated CLT modules with hybrid structural systems, the project delivers high performance in terms of sustainability, construction speed and build quality. Its success highlights the growing relevance of mass timber and modern methods of construction in complex urban developments, particularly where carbon performance and efficiency are key drivers.

Table 10.2 Case Study 2 – Moxy Hotel

MMC Feature	Description and Key Benefits
Off-site manufacturing	Accelerated construction with consistent quality and reduced site disruption
Hybrid timber structure	Combined the advantages of mass timber with robust steel and concrete podium
Sustainability	Over 1300 t CO_2 sequestered, low material waste and high thermal performance.

Source: Authors' creation

10.4.3 The role of off-site and mass timber in delivering efficient, sustainable construction

CATT and the Moxy Hotel in Edinburgh represent two distinct applications of modern timber construction. Yet both demonstrate how off-site fabrication and mass timber systems can transform conventional building practices into more sustainable, efficient alternatives.

In both projects, mass timber – specifically CLT and GLT – served not only as a structural material but also as a core component of each project's sustainability strategy. By replacing more carbon-intensive materials, such as concrete and steel, with engineered timber, both developments achieved substantial reductions in embodied carbon. The CATT project, rooted in a rural educational setting, emphasised the use of localised supply chains whenever possible to reduce transport emissions and support the circular economy. Meanwhile, the Moxy Hotel capitalised on the carbon-storage properties of CLT, with over 1300 t of CO_2 locked away in its modular timber pods.

Off-site construction was pivotal in both contexts, although deployed in different forms. At CATT, off-site manufacturing enabled the production of precision-cut structural elements and insulated wall panels that reduced construction time, minimised site waste and improved envelope performance. The Moxy Hotel took this one step further by adopting a volumetric modular approach, with entire rooms – including finishes and services – being fabricated off-site and delivered as plug-and-play units. This strategy dramatically accelerated the build process and ensured quality control in a city centre location where time, access and disruption were critical constraints.

What unites both case studies is the way in which off-site methodologies and engineered timber converged to support sustainability targets without compromising quality, programme, or performance. These projects demonstrate that modern timber systems are not merely alternatives to traditional construction – they are leading solutions for delivering buildings that are both high-performing and climate-conscious.

Together, they underscore a growing shift in the construction industry: one that embraces prefabrication, values carbon accounting and recognises timber as a primary material in the future of sustainable building.

10.4.4 Challenges in adopting timber MMC

While timber-based construction has gained significant momentum as a sustainable and structurally viable alternative to conventional materials, regulatory concerns – particularly those related to fire safety – continue to present challenges, especially within the UK's building landscape. Following high-profile fire incidents, most notably the 2017 Grenfell Tower tragedy, there has been increased scrutiny of the materials used in mid- and high-rise construction (MHCLG, 2018). As a result, regulations such as the UK's Building (Amendment) Regulations 2018 introduced restrictions on the use of combustible materials in the external walls of residential buildings over 18 metres, significantly impacting the adoption of mass timber in certain contexts.

10.4.4.1 Regulatory constraints and opportunities

Regulatory measures, while driven by legitimate safety concerns, have led to misconceptions that all timber systems inherently pose unacceptable fire risks. Advances in engineered timber products and fire-safety design have demonstrated that, when properly implemented, timber can meet and even exceed required safety performance standards. Mass timber elements, such as CLT and GLT, predictably char when exposed to fire, forming a protective layer of charred wood that slows combustion and maintains structural integrity. This behaviour is well-documented and can be modelled and accounted for in structural and fire safety engineering.

Modern timber systems are designed with passive fire resistance in mind, often incorporating layers of fire-resistant insulation, intumescent coatings and sacrificial layers that enhance their fire performance. Additionally, timber structures can be compartmentalised to prevent fire spread and where needed, integrated with non-combustible materials in hybrid designs to meet specific regulatory thresholds. Crucially, adherence to strict design codes, such as BS EN 1995-1-2 (Eurocode 5: Design of timber structures – Part 1-2: General – Structural fire design) and relevant guidance from the UK's National Fire Chiefs Council ensure that timber buildings are rigorously assessed for fire safety (STA, 2021).

Off-site construction plays a pivotal role in addressing these and other regulatory challenges by delivering a higher degree of control, consistency and compliance. Prefabricated timber systems, including volumetric and panelised solutions, are manufactured under factory-controlled conditions where quality assurance, moisture control and fire-resistant detailing can be precisely implemented and verified. This not only mitigates on-site construction errors but also enables better documentation and traceability – key components in demonstrating regulatory compliance.

Furthermore, off-site methodologies allow for early integration of fire safety strategies during the design phase. This results in more coordinated outcomes across disciplines, improving the alignment between architectural intent, structural capacity and fire performance. As digital tools like Building Information Modelling (BIM) become more integrated with regulatory pathways, they further support transparency and evidence-based approval processes for timber projects.

In summary, while fire safety concerns remain a key regulatory hurdle for timber construction in the UK, they are not insurmountable. With engineered solutions, robust fire modelling and the quality assurance benefits of off-site construction, timber-based systems can meet the stringent demands of modern building regulations. Unlocking this potential will require ongoing education, regulatory evolution and collaboration among industry, policymakers and fire safety experts to strike a balance between innovation and public safety.

10.4.4 2 Timber supply chain considerations

While timber-based off-site construction is increasingly recognised for its sustainability and efficiency, it remains exposed to the vulnerabilities of globalised supply chains. High-performance timber products, such as CLT, GLT and LVL, are often imported from Central Europe, North America and Scandinavia. Although these sources offer consistent quality, relying on international suppliers introduces considerable financial, environmental and logistical risks – risks that recent global events have highlighted with increased sharpness.

Disruptions, such as Covid, wars, supply shortages, port congestion and natural disasters, have delayed material deliveries and driven up costs, undermining the time and cost efficiencies that off-site construction is intended to deliver. Such events disrupt just-in-time manufacturing models, creating delays that ripple through project schedules and budgets. Additionally, reliance on long-distance transport increases the embodied carbon of construction materials, thereby diminishing the environmental benefits that timber construction would otherwise provide.

In contrast, sourcing timber locally offers a more resilient, environmentally responsible and economically regenerative approach. The Building from England's Woodlands project, led by CATT at the New Model Institute for Technology and Engineering (NMITE) in collaboration with Built Environment-Smarter Transformation, Edinburgh Napier University, dRMM Architects and Ecosystem Technologies, highlights the untapped potential of the UK's home-grown timber resources (CATT, 2023). Findings from the research demonstrate how local species such as oak, beech, birch, ash and sweet chestnut can inform building design, reduce environmental impacts and support rural economies. The study emphasises the importance of design-led engagement with local material availability, encouraging architects and engineers to consider underutilised resources in their region rather than defaulting to imported solutions.

Locally sourced timber provides several key advantages, including the following.

- *Supply chain resilience.* Utilising regional wood products mitigates the risk of global disruptions, resulting in greater reliability in scheduling and procurement.
- *Carbon reduction.* Shorter transportation distances significantly reduce emissions related to logistics, thereby contributing to lower embodied carbon throughout the project life cycle.
- *Economic localisation.* Supporting domestic forestry and processing industries stimulates job creation, fosters skill development and strengthens rural economies.
- *Circular resource use.* Local sourcing complements circular economy principles, especially when combined with design for deconstruction (DfD), reuse and biobased upcycling strategies.

However, barriers to widespread adoption remain. In the UK, for example, the timber sector currently lacks sufficient domestic infrastructure to produce large-scale engineered products at scale. Furthermore, industry standards and structural testing protocols have historically favoured commercial softwood species such as spruce, limiting market access for locally grown underutilised alternatives. Addressing these limitations will require coordinated investment in processing facilities, research into species performance and updates to standards and regulations that reflect the potential of the domestic timber resource base.

While global markets have driven much of the growth in engineered timber construction, future resilience and sustainability lie in creating and embracing local supply chains. The Building from England's Woodlands project highlights the potential to redefine construction practices by designing with regional resources, thereby reducing dependency on global systems while enhancing environmental performance and economic equity within local construction sectors.

10.5. Implications

A key challenge is to develop a broader understanding of the practices to encourage inclusion of newer building methods. Promoting timber and MMC across the built environment requires a coordinated approach that reaches beyond individual projects and into the broader systems that shape education, policy and industry practice. A growing body of evidence now demonstrates that engineered timber products, when deployed through off-site fabrication and MMC processes, offer a credible pathway to meeting the construction sector's net-zero goals, while also improving productivity, build quality and resource efficiency.

To capitalise on this potential, both academic and professional education must evolve to integrate the knowledge and skills required to design and deliver timber buildings effectively. This includes embedding material understanding of engineered timber, digital design processes, environmental assessment methods and off-site construction workflows into built environment curricula. As the construction sector moves toward performance-based standards and low-carbon delivery models, these competencies will become essential rather than optional.

In parallel, professional training and industry upskilling are vital. The construction workforce, from site operatives to senior consultants, needs greater familiarity with prefabricated timber systems and the distinct detailing, sequencing and logistics they require. Industry-led initiatives, manufacturer training schemes and collaborative research and development programmes have a significant role to play in building this capacity. Enhanced integration between supply chain actors – designers, manufacturers, contractors and clients – is equally essential to unlock the full benefits of MMC and reduce inefficiencies that often arise from fragmented project delivery.

Policy intervention has also begun to align with these ambitions. The UK's Timber in Construction Roadmap, adopted by Parliament in 2023 and reaffirmed in 2025, marks a significant shift in the policy environment. The roadmap sets out a national strategy to increase the use of timber in construction, driven by a need to decarbonise the built environment. Key actions include developing embodied carbon targets, supporting the domestic timber supply chain, reforming procurement practices to favour renewable materials and funding research into timber systems. This policy foundation offers clear support for MMC as a mainstream approach to low-carbon building (UK Government, 2025).

Regulatory progress is reinforcing this direction. The UK's evolving building regulations, including updated guidance on the use of structural timber and fire performance, are beginning to support the safe and widespread deployment of mass timber systems. Across Europe, similar shifts are evident. France's bio-based material targets for public buildings and the EU's recognition of timber under the Green Deal's circular economy goals signal a broader policy and regulatory alignment in favour of timber and MMC.

Taken together, these developments represent a critical shift in how the industry understands material performance, construction delivery and environmental impact. Timber and off-site construction are no longer niche or experimental – they are fast becoming central to future-proofing the built environment. To support this transition at scale, greater emphasis must be placed on education, regulation and industrial transformation to ensure that the knowledge, skills and systems are in place to enable a sustainable and resilient construction sector.

Building on the growing momentum behind timber and MMC, there are several strategic opportunities to further encourage adoption across the construction landscape. These strategies target the systemic barriers that still hinder wider adoption – such as a conservative risk culture, fragmented delivery models and economic uncertainties – and offer actionable pathways to integrate sustainable timber systems into mainstream practice.

(a) *Incentivise low-carbon construction through fiscal policy and procurement*. A major lever for change lies in the way buildings are financed and procured. By aligning fiscal policy with environmental performance, governments and local authorities can accelerate the shift toward timber and MMC. Financial incentives such as tax relief, enhanced capital allowances, or reduced planning levies for developments that meet embodied carbon thresholds could reward low-impact building practices. Public sector procurement, in particular, can lead by example – mandating or scoring highly for timber-based MMC in housing, schools and civic buildings sends strong market signals. The prioritisation of life cycle carbon assessment in procurement frameworks is especially powerful, ensuring that upfront carbon emissions – often a significant weakness of traditional construction – are addressed at the outset.

(b) *Establish a national data infrastructure for timber and MMC performance*. The lack of centralised, transparent data on the performance of timber and off-site systems remains a significant barrier to confidence and consistency. Creating a national open-access database of post-occupancy performance, embodied carbon metrics, cost comparisons and delivery timelines would allow industry stakeholders to make informed decisions based on real-world outcomes. Data from built projects – covering durability, energy performance, fire safety, acoustics and cost-efficiency – would reduce uncertainty and help overcome perceived risks. It would also support the development of more robust digital tools for specifying and designing timber-based buildings. A standardised evidence base would accelerate regulatory acceptance and simplify compliance pathways.

(c) *Support innovation hubs and demonstrator programmes across regions*. Scaling the adoption of timber and MMC requires both innovation and regional adaptability. Investment in regional innovation hubs – similar to the CATT – can provide spaces for prototyping, skills training and collaborative development between academia and industry. These hubs would serve as testing grounds for new materials, hybrid systems and off-site techniques tailored to local supply chains and climate contexts. Similarly, government-backed demonstrator projects can validate emerging practices in live construction scenarios. By documenting process improvements, cost implications and environmental benefits, these projects serve as replicable models that reduce the perceived risk associated with first-mover investment.

Together, these three strategies reinforce a construction culture that values material responsibility, rewards low-carbon innovation and builds trust in emerging systems. When paired with supportive policy and education, they provide a comprehensive foundation for accelerating the transition to a sustainable and resilient built environment.

10.6. Conclusions

The evolving landscape of timber construction and MMC offers a powerful opportunity to address some of the most pressing challenges in the built environment – namely, carbon reduction, resource efficiency and construction productivity. Engineered timber systems, when coupled with off-site manufacturing and digital design, are demonstrating their value not only through reduced embodied carbon but also through faster delivery, enhanced quality and reduced waste.

To move from promise to mainstream adoption, however, these methods must be embedded into the core of how we educate, regulate and build. As outlined, integration of timber and MMC principles into architectural, engineering and construction management curricula is essential. Upskilling the current workforce and supporting collaborative design and procurement models will be critical in enabling these innovations to scale.

Government intervention, through mechanisms such as the UK's Timber in Construction Roadmap, is beginning to provide the policy framework needed to accelerate the uptake. Regulatory reform in both the UK and across Europe is increasingly accommodating timber-based solutions, provided they are supported by robust performance data and delivered to exacting standards. This alignment of education, industry and regulation is necessary but not sufficient. The next step is evidence.

To support the continued development and uptake of timber and MMC, future research should prioritise performance-based evidence over prescriptive assumptions. This can be achieved through the creation of long-term monitoring projects and living laboratories, where the performance of timber buildings can be studied in real-world use. These living labs would collect data on energy use, internal comfort, moisture and durability, operational carbon, and maintenance patterns – helping to close the gap between predicted and actual building performance. Such research would not only enhance individual projects but also inform regulations, mitigate risks associated with innovative methods and refine procurement criteria.

Additional areas for future research should include the following.

- Fire performance in hybrid mass timber systems, to expand design confidence and regulatory acceptance.

- Material circularity and end-of-life pathways for timber elements, including reuse, recycling and integration into new product streams.
- Comparative life-cycle and whole-life carbon analysis across different building typologies and geographic contexts.
- Supply chain resilience and local sourcing models, particularly in light of global disruptions and the environmental impacts of long-distance material transport.
- Digitally integrated design-to-manufacture workflows, focusing on increasing automation, reducing waste and enhancing productivity in the off-site environment.

These research efforts must be interdisciplinary, bringing together material scientists, building physicists, architects, engineers, economists and sociologists to address the complex systems that affect construction choices. They must also be embedded into real projects – not abstracted into theoretical models – so that results can be grounded, credible and immediately actionable.

The future of the built environment will not be shaped by one material or one method alone. But timber, as a renewable, carbon-storing and high-performance material, is uniquely positioned to lead a shift toward regenerative, resource-efficient construction. By investing in research, embracing performance-led design and institutionalising new knowledge through education and policy, the construction industry can realise this opportunity, not only to build better, but to build responsibly for generations to come.

REFERENCES

CATT (Centre for Advanced Timber Technology) at NMITE (2023). Local resources inform design: mid-point results from the Building from England's Woodlands project. [online] https://bondbryan.co.uk/project/centre-for-advanced-timber-technology/ (accessed 28/05/2025).

Churkina G, Organschi A, Reyer CPO *et al.* (2020) Buildings as a global carbon sink. *Nature Sustainability* **3**: 269–276.

Constructing Excellence (n.d.) Skylon Campus. [online] https://constructingexcellence.org.uk/skylon-campus/ (accessed 28/05/2025).

Ministry of Housing, Communities and Local Government (MHCLG) (2018) Building (Amendment) Regulations 2018. [online] https://www.gov.uk/government/publications/building-amendment-regulations-2018 (accessed 28/05/2025).

Pomponi F and Moncaster A (2017) Circular economy for the built environment: a research framework. *Journal of Cleaner Production* **143**: 710–718.

Ramage MH, Burridge H, Busse-Wicher M *et al.* (2017) The wood from the trees: the use of timber in construction. *Renewable and Sustainable Energy Reviews* **68**: 333–359.

STA (Structural Timber Association) (2021) Fire Safety in Use Guidance. [online] https://www.structuraltimber.co.uk (accessed 28/05/2025).

UK Government (2025) Timber in Construction Roadmap 2025. [online] https://www.legislation.gov.uk/uksi/2018/1230/contents/made (accessed 28/05/2025).

Modern Methods of Construction: Foundations, Processes and Applications

Emmanuel I Daniel and Fidelis A Emuze
ISBN 978-1-83708-437-1
https://doi.org/10.1108/978-1-83708-436-420261012

Chapter 11
The benefits to mental health and well-being of off-site construction

David Emery

Supply Chain Sustainability School, London, United Kingdom 0009-0007-1553-379X

Abstract

Working on construction sites can be arduous and even dangerous. The industry takes the physical well-being of its workforce seriously: safety procedures and equipment are obvious, ubiquitous – and welcome. But besides physical safety, there is an equally important aspect of workers' health that the industry needs to protect and that is their mental health. The mental well-being of site workers is notably worse than in the general population, especially among supervisors. This can lead to harmful behaviours such as substance abuse and even suicidality. Besides diminishing the quality of life for workers and their families, mental ill-health results in productivity and profit losses for organisations, estimated at several £billions annually. The industry must find ways to reduce the stressors on the workforce to alleviate the harm they suffer and encourage new entrants into the industry. This qualitative research project sought to establish whether the greater project certainty that offsite construction (OSC) delivers, such as improved cost and programme certainty, higher quality and less hazardous procedures, reduces the stressors on the workforce and hence delivers consequent positive benefits for supervisors' and site managers' mental health and well-being. It took the form of a literature review, followed by semi-structured interviews with a small sample of site managers currently working on OSC projects, but who also had extensive experience of delivering traditional construction (TC) projects and hence could make informed comparisons. The literature review highlighted the poor mental health of the UK workforce in general and the construction workforce in particular and analysis of the respondents' opinions revealed unanimous positivity towards OSC in terms of reducing stressors. Hence, the author proposes that clients commissioning construction projects should consider the potential impact of the specified solution on the welbeing of their workforce.

Keywords: well-being; mental health; offsite; site manager; project delivery

11.1. Introduction

This chapter is based on original research which comprised a literature review, followed by the development of case studies derived from semi-structured interviews with industry practitioners, specifically site managers with experience in delivering complex projects by way of both TC methods and OSC using modern methods of construction (MMC). They were thus able to contribute experiences of both methods, in terms of project success, as well as any link between the success of their projects and their personal sense of mental well-being and that of their teams.

The construction industry represents one of the world's largest industrial sectors, employing approximately 7% of the global workforce (Sunindijo and Zou, 2012). Significantly, one of UK construction's most pressing problems is the decline in its workforce. Without the recruitment and retention of younger entrants, this decline will be inexorable, leading – as we shall see – to severe consequences (Farmer, 2016). Hence, the question arises about the value the industry – indeed, society – assigns to the workforce's welfare. Traditional measures of 'value', such as, time cost and quality, usually only consider project-specific factors; 'softer' issues such as health, safety, quality, human and environmental factors are almost totally excluded (Blismas *et al.*, 2006).

There are, though, other means by which project success might be measured, but Frefer noted 'a lack of agreement concerning the criteria by which success is judged' and there are, indeed, many other measurable key performance indicators (KPIs) (Frefer *et al.*, 2018). Unfortunately, 'softer' issues are either completely disregarded or, at best, are merely implicit within these comparison exercises (Blismas *et al.*, 2006). The potential improvements in these very issues are some of the most significant benefits of OSC, but until project solutions evaluations are more holistic and value based, the uptake of off-site production (OSP) in construction is likely to remain slow (Blismas *et al.*, 2006).

Mental health in construction appears to be poorly understood, and the industry would benefit from a reliable and valid evidence base to support mental health literacy initiatives and treatments (Hutton *et al.*, 2022). Nevertheless, Bryson and Duncan claim that compared to the general population, the construction industry suffers higher rates of mental ill-health and suicide risks (Bryson and Duncan, 2018). Chan *et al.* (2020) found that the foremost risk factors to mental illness of construction workers were those related to job demand and job control – for example, the lack of good communications or clarity in task arrangements could create potential pitfalls for workers' mental health (Li *et al.*, 2022), and supervisors experience higher psychological demands and need for recovery (Boschman *et al.*, 2013). If these psychological demands are a result of stressors in managers' work, then logic suggests that successful and efficient projects (if indeed, that is what OSC is found to deliver!) may be beneficial for workers' well-being.

The concept of mental health is poorly understood. 'Good' mental health is commonly thought of as merely the absence of any mental disorders, but this definition fails to consider a more positive measure which might be thought of as the importance of mental well-being (Keyes, 2005; Seligman, 2011; Westerhof and Keyes, 2010). Although work can be a cause of ill health, in most cases 'work is good for health, especially if it is good quality work' (Waddell and Burton, 2006). However, workers employed in male-dominated industries, like construction, report elevated levels of mental illness, including anxiety and depression.

Worse, they are particularly vulnerable to suicide, more so than the general population (Bryson and Duncan, 2018; Nwaogu *et al.*, 2022).

There is a clear need to address the mental health of these workers and the workplace provides an important but often overlooked setting to develop tailored strategies (Chapman *et al.*, 2020). This is especially important for construction supervisors, as they are responsible for site management, a job with particularly high mental demands (Hutton *et al.*, 2022).

Mental health problems can be observed everywhere, with more than 4% of the world's population suffering from depression alone. The World Health Organization (WHO) predicts that, by 2030, mental health problems (particularly depression) will be the leading cause of mortality and morbidity globally (Rees-Evans, 2020).

In the UK, approximately 15% of the workforce displays symptoms of an existing mental health condition. This can lead to disastrous consequences: 300000 of those experiencing long-term mental health problems lose their jobs each year (Rees-Evans, 2020). This is especially unfortunate since long-term worklessness can be further harmful, not only to mental health but also to physical health, including higher mortality, poorer general health, long standing illness and so on (Waddell and Burton, 2006). It should be noted from the same report, however, that re-employment can re-establish some degree of self-esteem, improve general and mental health and reduce psychological distress, so the consequences are somewhat reversible.

The overall aim of the research was to ascertain whether there was evidence that OSC delivered improved project certainty – by way of, for example, higher quality and reduced defects, fewer health and safety risks and other factors – and, if so, does this bring benefits to the site management workforce in terms of their mental health and well-being by reducing some of the stressors? If it transpires that OSC does indeed provide an opportunity to improve workforce well-being, then it will be recommended that procurement decisions should include this as a material consideration.

The objectives of the research were as follows.

(a) *Literature review.* To research the relevant thinking concerning mental health and well-being amongst UK construction workers, and whether this might be affected by the success or otherwise of the projects in which they are engaged, seeking comparisons between workers' mental health in TC compared to OSC.
(b) *OSC in use.* To draw on the knowledge of experienced site managers to understand the range and scope of OSC and MMC currently in use on residential projects in the UK, including any risks and benefits in delivering project success, and to draw on site managers' experiences and observations about their own well-being when compared to project delivery success to illustrate any correlation between project success and workers' well-being.
(c) *Value assessment.* To propose workers' well-being as a value measurement in assessing OSC projects (if the link between project success and well-being is established).

11.2. Perspectives and concepts

Far from being burdensome or harmful, work is beneficial to individual well-being and has wider societal benefits. In their research on whether work is good for human health and well-being, Wardell and Burton (2006) suggest that there are multiple reasons to consider work as being a way to improve the well-being of individuals, families and communities. They can be viewed in multiple ways, including economic, social and moral, albeit health benefits are dependent on the nature and quality of work. Such benefits are significant in the light of their assertion that 'health is fundamental to human well-being and work is an integral part of modern life'.

Work also meets important psychosocial needs. Work can be central to an individual's sense of identity, their social roles and their social status (Waddell and Burton, 2006). The Chartered Institute of Building (CIOB) reports that stable employment can be an important factor for maintaining good mental health to the extent that there are examples where being employed by a business has improved well-being, independence and social support, even for those who have severe mental illness (Rees-Evans, 2020). It might be assumed, therefore, that the sheer size of the construction industry would provide correspondingly numerous potential benefits, especially considering the wide range of jobs that exist in construction (Boschman *et al.*, 2013).

Before exploring whether these theoretical benefits to health and well-being are being realised for individuals, it would be instructive to investigate the 'health' of the industry itself; after all, if construction is in decline and/or is suffering from endemic problems, can it realistically deliver benefits to those who participate in it? Worse, might any 'ill health' in the industry – which would likely manifest itself as poor project performance – cause harm to some of its participants?

Previous reports – for example, Egan (1998) and Farmer (2016) – identify practices and processes in TC that have long been abandoned by other industry sectors. The epithet 'traditional' construction characterises an industry which has seen little evolution until recently and, compared to other sectors, it lags in adopting innovation (Razak *et al.*, 2022). The wider UK construction industry continues to underachieve relative to other sectors in terms of profitability (McCord *et al.*, 2015). Delays are inherent throughout the industry and these create major difficulties in terms of project performance (McCord *et al.*, 2015). This matter is especially relevant to this chapter, looking, as it is, for correlation between project performance and workers' well-being.

Further concern about TC can be found in 'The Farmer Review of the UK Construction Industry'. The extent of the industry's problems is illustrated by the report's subtitle: 'Modernise or Die'. Farmer singles out the declining workforce as potentially calamitous for the sector: 'The … ticking 'time bomb' is that of … workforce size and demographic … we could see a 20–25% decline in the available labour force within a decade'. Farmer predicts that such a decline '… would undermine the UK's ability to deliver critical social and physical infrastructure, homes and built assets required by other industries to perform their core functions' (Farmer, 2016).

The importance of achieving a strong and healthy construction industry workforce can, therefore, hardly be overstated. Farmer's report has a 'strong medical process analogy' in which he seeks to 'identify the symptoms, diagnose the root causes, provide a prognosis, establish a treatment plan for recovery and keep the industry under observation' (Farmer, 2016). This medical analogy is apt, as this chapter now turns to the well-being of the workforce.

The construction workforce is employed in more than 80 different job roles; about 55–70% is made up of workers with manual skills (such as bricklayers, carpenters, electricians and so on), with the remaining 30–45% in managerial roles (Boschman *et al.*, 2013).

There is conflicting research concerning the level of mental ill-health in the workforce. Stocks *et al.* (2010) found a lower incidence (4%) of mental illness among construction workers in the UK compared to all other employment sectors. They reported that compared to other diseases or conditions, mental illness had the lowest incidence among construction workers. This suggests that mental illness is the least important health issue in an industry whose workforce appears more mentally robust than those in other sectors.

In contrast, other sources regard workers' mental ill-health to be a statistically far larger problem. The Union of Construction, Allied Trades and Technicians (UCATT) reported that 35% of [their] respondents were suffering from – or had suffered from – a mental illness, with 44% having had time off due to mental ill-health issues; the two symptoms which caused most pain were depression and anxiety, with 75% having experienced depression and 58% anxiety (Sumner, 2016). The contrast between the two authors' data is stark, perhaps only partially explained by the fact that the definitions of 'mental illnesses' includes a broad range of conditions. According to research carried out for CIOB, in England, one in six people will experience a common mental health problem such as depression or anxiety in any given week, the most common mental health diagnoses being mixed anxiety and depression, affecting 7.8% of the population (Rees-Evans, 2020). Remember, too, that 'good' mental health is not merely the absence of any 'bad' mental illness disorders.

There are significant commercial consequences when the workforce cannot function to its full capability. The review for CIOB found that the cost of poor mental health to the UK economy stood at between £74 billion and £99 billion, and caused employers to lose £33–42 billion annually, because of employees taking time off work (Rees-Evans, 2020). Bevan declares that 'the UK is already facing the economic and social consequences of a 'wellness' crisis' (Bevan, 2010).

The Office for National Statistics (ONS) states that 12.4% of sickness absence days were attributed to mental health conditions – equivalent to 4.4 days per employee (Rees-Evans, 2020). The resulting social and economic burden is associated with an estimated 50–60% of absenteeism, costs related to lost productivity and increased risk of physical health problems and involvement in accidents (Chapman *et al.*, 2020).

The conclusion is that mental ill-health accounts for a significant proportion of all work absences, at huge cost to the economy. Is the incidence of mental illness in construction significantly higher than in other sectors, though? The research suggests that this is indeed the case.

There are, however, potentially more serious consequences than simply financial, as evidenced by the propensity of construction workers towards suicide. Nwaogu *et al.* (2022) found that mental ill-health symptoms and suicide (suicidal ideation) are prevalent in the construction industry (Chan *et al.*, 2020). The nature of the work and the culture in construction may be contributing factors to these phenomena. Workers employed in male-dominated industries, such as construction, report elevated levels of mental illness, with suicide rates 80% higher in male low-skilled construction workers than in the general population, but this group is less likely to seek access to treatment than the general population (Chapman *et al.*, 2020). This is unfortunate as mental ill-health, when left unattended, is a precursor to suicide (Kamardeen and Hasan, 2023; Sunindijo and Zou, 2012). The research did not reveal any comparisons between the levels of mental health experienced by workers on TC compared to workers on OSC projects. Indeed, none of the companies whose employees contributed to the research interviews currently records employee data which differentiates between TC and OSC teams.

What are the types of mental ill-health suffered by construction workers, what are the causes and what are the consequences for the sufferers and, indeed, the organisations? To reiterate, mental health is poorly understood and is commonly considered to be merely the absence of a mental illness disorder, without considering mental well-being (Keyes, 2002, 2005; Seligman, 2011; Westerhof and Keyes, 2010). Mental health is ‘a state of well-being in which the individual … can cope with the normal stresses of life’.

WHO asserts that mental health is a basic human right (World Health Organisation, 2023). The most common mental health diagnoses are mixed anxiety and depression, anxiety being a persistent feeling of worry, fear or nervousness (Rees-Evans, 2020). It was found that among building construction practitioners the prevalence rate for anxiety and depression was a huge 87% and 70%, respectively (Nwaogu *et al.*, 2022).

Stress is a state of mental or emotional strain or tension resulting from adverse or demanding circumstances. Typically, 82% of people feel stressed at least some of the time and 8% feel stressed all the time. Some 26% of respondents said this was due to problems at work (Rees-Evans, 2020). The HSE defines stress as ‘the adverse reaction people have to excessive pressures or other types of demands placed on them’ (HSE, 2023). According to research by Boschman, among building construction supervisors there was an 20% prevalence rate of post-traumatic stress disorder (PTSD) symptoms, compared with 18% for depression, perhaps because accidents are a major hazard in the construction industry and can affect the mental health of other workers (Boschman *et al.*, 2013).

When left unattended, mental ill-health is a precursor to suicide (Kamardeen and Sunindijo, 2017). How people respond to stress varies and can be both a positive and a negative experience. On average, men – more than women – respond to stress by taking risks such as misusing alcohol and/or drug abuse, which can contribute to the likelihood of suicide (Rees-Evans, 2020). Male construction workers are at higher risk of mental health problems, but they are less likely to seek access to treatment than is the general population (Chapman *et al.*, 2020).

'Stressors' that can arise from working on construction sites vary across the workforce and depend on several factors, such as age, gender, occupation or the type of work, the ability to cope with stressors being influenced by workers' individual resilience and their genetics (Jiang *et al.*, 2021). Younger and older workers respond to different stressors; the former experience more stress due to job demands and their occupational climate, the latter are stressed where there is poor job control (Lim *et al.*, 2017). Furthermore, younger workers (25–44) appear to experience a prevalence of high psychological distress due to reduced social and emotional support systems (Chapman *et al.*, 2020), and there is a higher mental health risk factor in females compare to men (Jiang *et al.*, 2021).

According to Li *et al.* (2022), construction workers perceive excessive workload as one of the most significant job demands that can lead to stress, burnout, lower satisfaction and other negative psychological well-being (PWB) outcomes (Li *et al.*, 2022). Labour-led capacity shortages – implying excessive workload for those left working in the industry – are likewise identified as a source of stress in TC (Farmer, 2016).

Other negative factors on workers' mental health include lack of good communications or clarity in task arrangements, role conflict and task stressors are important causes of negative PWB constructs. These might be explained by the complexity of large-scale construction projects, requiring extensive collaboration between multiple stakeholders, which could be further compounded by the inherent challenges arising from matters such as inaccurate project estimates, poor site management and unforeseen risks and so on (Li *et al.*, 2022).

Construction projects are one-off ventures and operations are conducted by temporary organisations and structures, and construction professionals' work is typically performed in an uncertain, dynamic and erratic environment (Asquin *et al.*, 2010). These factors increases the likelihood that workers would suffer under stressful conditions (Lingard and Francis, 2004). Such elevated levels of job demands have been linked to cynicism and emotional exhaustion, both of which are attributes of burnout (Pinto *et al.*, 2014).

Farmer reflects on an industry typified by poor health and safety, physicality, austere working conditions, embedded prejudices and (a lack of) job security. Farmer (2016) states that a 'combination of an ageing workforce, low levels of new entrants and ... accelerated shrinkage, now threatens the very sustainability of the industry'. These issues can offer nothing positive to the workers in terms of their sense of well-being, especially as employees crave stability at work rather than constant change or uncertainty (Bevan, 2010). The HSE concludes that workers feel stress 'when they can't cope with pressures and other issues'. It recommends that 'Employers should match demands to workers' skills and knowledge' as workers can suffer stress if they feel they don't have the skills or time to meet tight deadlines (HSE, 2023).

Poor work–life balance is one of the main stressors that can result in negative impacts to people as well as organisations; it has been associated with substance abuse, alcohol misuse and marital dissatisfaction (Grant-Vallone and Donaldson, 2001). Volatility, uncertainty, complexity and ambiguity (VUCA) are commonly present in construction projects. The negative impact of VUCA, in combination with long – and possibly arduous – working hours, dangerous workplaces, job hazards, intensive labour and conflicting interests, threaten the well-being of project employees (Ling *et al.*, 2017).

According to Rees-Evans (2020), while a moderate amount of stress can assist people in their work, if stress increases too much it can be detrimental to concentration, productivity and mental health. Prolonged excessive stress can cause fatigue and irritability, affecting reasoning, judgement and the ability to make decisions. Ultimately, it can lead to burnout and other serious health problems. The level of pressure is influenced by numerous external factors, such as management style, unrealistic deadlines, excessive workload, bullying in the workplace, lack of control, lack of clarity over job role or being either over-skilled or under-skilled for a specific job (Rees-Evans, 2020).

Worsening mental health can lead to poor risk perception. Workers need both psychological and mental capacities to remain sufficiently concentrated and alert to manage the variety of on-site hazards throughout the day, but capacities are likely to be impaired when the worker suffers from mental ill-health. This can be particularly relevant for site supervisors and managers – the subjects of the research – who are responsible, of course, for controlling sites and therefore have jobs with particular high mental demands (Bentley *et al.*, 2006). Mental and physical health are interlinked. Besides suffering the types of mental ill-health diagnoses described above, the sufferer can become more susceptible to physical risks because work functioning may be critically reduced, which may lead to a higher risk of accidents and physical injuries on the job (Beseler and Stallones, 2010).

Besides physical injuries from accidents due to mental impairment, there are other forms of physical harm which can arise which are considerably more sinister. These are referred to as 'negative coping strategies'. Coping strategies can be categorised as positive (adaptive) or negative (maladaptive) – for instance, alcohol, drug abuse and substance abuse (ADSA) are all used as a diversion for dealing with the effects of strenuous work (Chan *et al.*, 2020). Construction workers turn to ADSA, with substance abuse found to be associated with anxiety (Langdon and Sawang, 2018) and this can contribute to the likelihood of suicide (Rees-Evans, 2020) .

When considering traditional measures of project success, deliverables may be in the form of human, technical or financial resources and deliverables (PMBOK, 2021), and may be different for different projects. Traditionally, there are three fundamental project success criteria: time, cost and quality (TCQ) (Lester, 2021). Beyond TCQ it appears that current evaluation methods for OSC are mostly cost based and not value based. Blismas *et al.* (2006) found that when evaluating OSC against traditional construction methods, ' 'softer' issues such as health, safety, quality, human and environmental factors are almost totally excluded'. Potentially, until evaluation is more holistic and value based rather than cost based, the uptake of OSC will be slow (Blismas *et al.*, 2006).

Current evaluation methods for comparing TC and OSC building solutions do not adequately account for all the main factors that affect cost, while also ignoring the wider benefits and dis-benefits inherent within the different approaches. Current traditional models are largely ignorant of value and therefore cannot 'record' some of the benefits OSC can promote (Blismas *et al.*, 2006).

The main cost benefits in OSC are from indirect cost savings and non-cost value-adding items. However, monetary measures are inadequate for items that cannot be directly attributable to an element, such as health and safety, sustainability or wider human factors (Blismas *et al.*, 2006). OSC offers a promising means to improve the efficiency of construction projects, but the lack of experience and knowledge regarding its use can result in errors in design owing to conflicts and omissions of considerations for OSC projects (Hyun *et al.*, 2022).

It should be noted that early engagement is important for success in OSC. Thompson *et al.* (2021) noted that how 'clients and their teams set their projects up from Strategic Definition stage onwards has a massive impact on the success or otherwise of adopting manufacturing principles and effectively deploying the various modern methods of construction'. Furthermore, 'DfMA [design for manufacture and assembly] is not just a technical solution. It is a philosophy [that] needs to be embedded early in the project' (Thompson *et al.*, 2021).

The site managers who were interviewed for this research were highly experienced in managing both TC and OSC projects, although this is not to suggest that their OSC projects were entirely problem-free. As seen above, though, these are precisely the types of errors that can cause stress, anxiety and other mental ill health issues.

In summary, the literature review indicates that, generally, work is not only beneficial for individuals' well-being, but it can also – dependent on the nature and quality of the work – deliver societal and economic benefits. Indeed, some researchers believe that work is fundamental to humans' well-being. An alternative belief, though, is that the nature of work has changed and that work intensification and the resultant pressure are having a dangerously damaging effect on parts of the workforce (Asquin *et al.*, 2010).

Mental health problems are ubiquitous and affect a significant proportion of the population. Construction workers, though, suffer more ill-health than the general population and the effects can be catastrophic for both the individual and the organisation. Many of the causes of mental ill-health in construction are associated with difficulties that workers face in their roles, and the effects can be compounded by the apparent reluctance of workers to seek help. The stressors that lead to mental ill-health include job demands, lack of communications and lack of clarity in task arrangements, all of which are especially significant as construction projects are generally large and highly complex.

The results of workers' mental ill-health manifest themselves in numerous ways, some of which impact on project performance and others – arguably more importantly – impact on other aspects of workers' health, including their physical health and their propensity towards negative behaviours.

The literature review did not reveal any specific research that compared workers on TC projects compared to OSC project.

11.3. Methodological approach

11.3.1 Research details

The hypothesis of the original research was that there may be benefits in terms of project delivery through OSC and that these benefits might translate into an improved sense of well-being amongst the project participants, especially those managing sites. Alternatively, would taking a significant proportion of activities away from the site and into the factory make management of projects more difficult for the site manager, with a potential negative impact on their well-being?

The literature review identified multiple stressors for site managers and team leaders. It identified that the foremost risk factors to mental illness of construction workers relate to job demand and job control, along with poor work-life balance (Cheung *et al.*, 2022; Nwaogu *et al.*, 2022).

Conducting semi-structured interviews was the preferred method of research for obtaining qualitative data; nevertheless, in the course of their interviews each manager also volunteered quantitative data on, for example, the time saved on projects through implementing OSC. Likewise, the interview of the Head of Health and Wellbeing resulted in both qualitative and quantitative data.

11.3.2 Interviewees

The interviewees (Table 11.1) were selected because they had experience in managing both TC and OSC projects (Table 11.2) and so were able to assess the levels of project success which had been achieved on multiple projects. They were further able to comment on whether OSC had resulted in improved project success and hence could articulate from their own experiences the risks and benefits of both TC and OSC and describe the comparative degree of well-being they derived from both, as well as being able to report on the well-being in the teams they managed.

Table 11.1 Research interviewees' details

Interviewees	SM01	SM02	SM03
Job title	Senior Operations Manager	Project Director	Site Manager
Years in this role	30	Graduated in 2015	20
Employer	Tier-1 construction company, turnover >£ 1billion	Specialist M&E provider/ installer, subsidiary of major Tier-1 construction company.	Medium-sized house-builder
Years with this employer	4	3	4

Source: Author's creation

Table 11.2 OSC project details

Case Studies	CS01	CS02	CS03
Description of project	Construction of 315 apartments plus commercial space, city centre location	Joint project with parent company to build laboratories; total contract value c£130 million, including > £50 million M&E, science park	Development of 75 new homes, semi-rural 'greenfield' site.
MMC categories used at current project	2, 3 and 5	5	2 and 3

Source: Author's creation

11.3.3 Case studies

In the small sample interviewed, the following MMC categories were in use (Table 11.2).

- MMC Category 2 – Premanufactured 2D primary structural systems (e.g. wall panels, floor decks).
- MMC Category 3 – Premanufactured components (non-systemised primary structure) (e.g. balconies).
- MMC Category 5 – Premanufactured non-structural assemblies and sub-assemblies (e.g. boiler cupboards, service risers).

11.3.4 Interviews

The interviews with experienced site managers were 'semi-structured' with the following aims.

- To draw on their knowledge and understanding of the range and scope of OSC and MMC currently in use on residential projects in the UK.
- To further draw on their experiences to understand the benefits of off-site and MMC in delivering project success.
- To draw on their experiences and observations about their own well-being as well as that of other stakeholders on their sites where possible.
- To use the case studies to seek correlation between project success and workers' well-being.

Questions during the actual interviews were restricted to the MMC categories for which the interviewee claimed actual experience, either on their current project or previous projects they had managed. All the interviews were recorded and transcribed, with the written consent of the interviewees. As the interviews were semi-structured, not every question was answered (or, indeed, asked) – for example, there were instances where the interviewees might be speaking about the benefits of an MMC system at some length and, during that conversation, they might articulate multiple advantages. In such cases, there was no point returning to the list of questions to ask about a particular benefit – for example, higher quality – if the respondent had already covered that aspect in general conversation.

A tabulated summary of the interviews is in Table 11.3.

11.4. Results and discussion

11.4.1 Common themes from interviews

All the interviewees were keen advocates of OSC and articulated the following benefits.

- *Programme certainty (time)*. All the interviewees perceived adherence to the project programme to be one of their primary responsibilities and all of them stated that programme certainty was easier to achieve using OSC compared to TC.
- *Cost*. Cost savings were reflected not only through the virtual elimination of waste, but also by shorter build times with consequent reduction in site overheads and 'prelims'.
- *Quality*. Prefabricated components and systems achieve close tolerances resulting in easy, 'right-first-time' assembly, with few defects. Repetitive assembly processes and clear design and assembly documentation mean that the work can be completed by fewer people than in TC, reducing risks.
- *Health and safety*. The health and safety advantages were numerous and benefitted people away from the site as well as on site. All the interviewees spoke about the significantly fewer deliveries required to the site, with positive benefits to nearby residents or road users. Fewer people on site meant fewer incidents; the interviewees spoke of scenarios in TC where multiple trades would be either working in or passing through the same spaces simultaneously, increasing risk of slips, trips and falls.
- *Wellbeing*. The site managers all described how the combination of benefits listed above improved their individual sense of well-being and led to 'happier' teams, too.

Overall, the interviewees were confident in the certainty of programme, cost and quality of their finished products – the typical time, cost and quality measures. More significant was the evidence of healthy, long-term relations with their subcontractors, albeit new entrants' lack of familiarity with the various MMC systems could expose some tensions.

There was an interesting contrast between SM01 and SM02 in terms of how problems or delays might be resolved. The former had the longest experience of OSC out of all the interviewees, having had the opportunity to influence the design of the project and building a team of people with whom he had had extensive previous experience (and enjoyed previous successes). Nevertheless, the subjective impression was that if any individual's or teams' performances fell behind schedule or endured other problems, a display of empathy might not be as apparent as might be hoped – 'they know what's required, it's up to them to sort it out' and '[in those circumstances] I might not be their best friend.' In contrast, SM02 appeared at a personal and corporate level to be more pro-active in supporting the workforce's well-being.

Both SM01 and SM03 had experienced similar difficulties with the in-situ parts of their projects, specifically groundworks. Both had engaged sub-contractors with whom they'd not previously worked and both ended up finding replacements. It is reasonable to assume that the breakdown of those relationships may have involved some acrimony.

11.4.2 Discussion of results

The results of this research indicate that a significant proportion of the general population suffers from mental ill-health, but the construction workforce appears to be especially susceptible, as evidenced most dramatically by the disproportionate rates of suicidality in the industry. While people's ability to work is generally considered to be beneficial for their well-being, the type of work combined with the complexity of projects which are common in construction can often create the conditions which cause workers to feel stressed. If OSC can alleviate some of the stressors then it has the potential to improve the mental health of the workforce and should, therefore, become a material consideration in procuring construction projects.

Objective 1 was to establish the current thinking and research concerning mental health and well-being amongst UK construction workers.

The research demonstrated that poor mental health is prevalent in the construction industry (Sumner, 2016), leading to detrimental effects not only for workers but also for projects (Bevan, 2010; Rees-Evans, 2020). The causes of mental ill-health are wide and varied and affect individuals in different ways, but nevertheless some of the stressors that lead to mental ill-health can be alleviated – if not eliminated altogether – by the project certainty (HSE, 2023), which tends to be inherent in OSC. This enhances workers' well-being and they become better able to deliver their work efficiently; absences – and the resultant costs – will be reduced and physical accidents will be reduced (Beseler and Stallones, 2010).

Objective 2 was to conduct interviews to understand the range and scope of OSC currently in use on residential projects in the UK in the small sample interviewed. Objective 2 was also to understand the risks and benefits of OSC in delivering project success. The interviewees were unanimous in their enthusiasm for the benefits of OSC, especially the project certainty that they felt resulted from OSC. Each felt that using premanufactured components of high quality and delivered on time, in combination with experienced assembly teams who were familiar with the products, resulted in fewer delays, less (or zero) re-work, fewer operatives on site at any one time, and fewer vehicle movements and health and safety incidents. None mentioned any risks; at least not once teams were familiar with the various OSC systems. In summary, the interviewees were able to articulate multiple benefits of OSC in delivering project success with minimal risks.

Objective 2 also sought to understand the effect of OSC on the well-being of the site staff. The research demonstrated that there are many potential stressors arising from delivering large and complex construction projects; uncertainty about job roles and poor work-life balance from excessive workloads can have serious implications for mental health and well-being (Chan *et al.*, 2020; Li *et al.*, 2022). All the interviewees described how they experienced fewer stressors because of the greater project certainty delivered by OSC. Indeed, where stressors were noted, they arose from parts of the project being delivered by way of TC (e.g. substructures) by unfamiliar teams.

Therefore, a picture is formed of a population that suffers from significant mental health and well-being issues in general, and an industry in which the workforce suffers disproportionately due to the common stressors associated with complex construction projects. However, there is evidence that OSC may be able to reduce some of those stressors and hence benefit the workforce.

11.5. Implications of the study

The implication of this study is that the decisions made about how construction projects are procured – either by way of TC or OSC – can potentially affect the well-being and mental health of the workforce engaged to deliver those projects.

The proposal arising from this research is that the well-being of the workforce should be a material consideration when commissioning building projects. This research suggests that there is a positive correlation between the use of OSC solutions and workforce well-being and therefore weighting should be applied to OSC proposals. This study highlights that there is potential to bring about positive outcomes for the construction management workforce by specifying OSC solutions rather than TD, albeit if OSC is to be considered as a potential delivery process, clients and their representatives need to be prepared to adopt OSC as a philosophy that is embedded early in the decision-making process.

In an industry where the incidences of mental ill-health exceed those of the population at large, every opportunity should be taken to specify those solutions which are proven to reduce stressors on the workforce.

11.6. Conclusions

This research set out to assess the extent of mental ill-health in the construction sector by researching the evidence and current thinking on the topic. It also set out to explore, through case study interviews, whether OSC can reduce or eliminate some of the key stressors to the workforce and hence improve their mental health.

The existing research around mental health in construction is extensive and clear; workers in construction – especially those involved in managing construction sites – suffer disproportionately from mental ill-health, due in significant part to the stressors they experience because of the job demands associated with traditional construction.

Against this background, the original purpose of this research was two-fold: firstly, to see whether OSC delivered greater project certainty; and secondly, if so, to explore whether this results in any beneficial effects on the well-being of site management, especially compared to traditional construction. Despite the limited number of interviewees, the responses to the research questions were unequivocal. The experiences of all the interviewees in delivering both TD and OSC had led them to conclude that OSC did indeed deliver greater project certainty against numerous key performance indicators. As a result, all the interviewees had experienced fewer stressors – despite the complexity and size of their current projects – with consequent benefits to their sense of well-being. Furthermore, they also observed more harmonious relationships amongst the site workforces.

REFERENCES

Asquin A, Garel G and Picq T (2010) When project-based management causes distress at work. *International Journal of Project Management* **28(2)**: 166–172.

Bentley TA, Hide S, Tappin D *et al.* (2006) Investigating risk factors for slips, trips and falls in New Zealand residential construction using incident-centred and incident-independent methods. *Ergonomics* **49(1)**: 62–77.

Beseler CL and Stallones L (2010) Safety knowledge, safety behaviors, depression and injuries in Colorado farm residents. *American Journal of Industrial Medicine.*

Bevan S (2010) The Business Case for Employee Health and Wellbeing [online]. http://investorsinpeople.ph/wp-content/uploads/2013/08/The-Business-Case-for-Employee-Health-and-Wellbeing-Feb-2010.pdf.

Blismas N, Pasquire C and Gibb A (2006) Benefit evaluation for off-site production in construction. *Construction Management and Economics* **24(2)**: 121–130.

Boschman JS, Molen HF van der, Sluiter JK and Frings-Dresen MHW (2013) Psychosocial work environment and mental health among construction workers. *Applied Ergonomics* **44(5)**: 748–755.

Boschman JS, van der Molen HF, Sluiter JK and Frings-Dresen M (2013) Improving occupational health care for construction workers: a process evaluation. *BMC Public Health* **13**: 218.

Bryson K and Duncan A (2018) *Mental Health in the Construction Industry Scoping Study*, BRANZ, Porirua, New Zealand.

Chan A, Nwaogu J and Naslund J (2020) *Mental Ill-Health Risk Factors in the Construction Industry: Systematic Review*. National Library of Medicine [online] https://www.ncbi.nlm.nih.gov/pmc/articles/PMC7785059/.

Chapman J, Roche AM, Duraisingam V *et al.* (2020) Exploring the relationship between psychological distress and likelihood of help seeking in construction workers: the role of talking to workmates and knowing how to get help. *Work* **67(1)**: 47–54..

Cheung CM, Bowen P, Cattell K and Davis J (2022) How the well-being of construction professionals mediates the effect of work–life balance on their commitment to the organization. *Journal of Management in Engineering* **38(4)**.

Egan J (1998) *Rethinking Construction: The Egan Report (1998)*. Blackwell Science Ltd, Oxford, UK.

Farmer M (2016) *The Farmer Review of the UK Construction Labour Model.*

Frefer A, Mahmoud M, Haleema H and Almamlook R (2018) Overview success criteria and critical success factors in project management. *Industrial Engineering & Management* **7(1)**: 1–6.

Grant-Vallone EJ and Donaldson SI (2001) Consequences of work-family conflict on employee well-being over time. *Work & Stress* **15(3)**: 214–226.

HSE (2023) Stress and Mental Health at Work, Health and Safety Executive [online]. https://www.hse.gov.uk/stress/ (accessed 27 April 2026).

Hutton EA, Skues JL, Sullivan JA and Wise LZ (2022) Mental health research in the global construction industry: a scoping review using a dual-continuum model of mental health, *Mental Health & Prevention* **28**: 200249.

Hyun H, Kim H-G and Kim J-S (2022) Integrated off-site construction design process including DfMA considerations. *Sustainability* **14(7)**, p. 4084. https://doi.org/10.3390/su14074084

Jiang T, Wumaier G, Li X, Yang X and Liu J (2021) Research on the effects of occupational stress and the DRD2 gene on the psychological health of workers in the Xinjiang Desert Oil Field. *Frontiers in Psychiatry* [online] https://www.scopus.com/inward/record.uri?eid=2-s2.0-85116062342&doi=10.3389%2ffpsyt.2021.737228&partnerID=40&md5=a28ff06e148d50fb405aa02072dd7ec9.

Kamardeen I and Hasan A (2023) Analysis of work-related psychological injury severity among construction trades workers. *Journal of Management in Engineering* **39(2)**.

Kamardeen I and Sunindijo R (2017) Personal characteristics moderate work stress in construction professionals. *Journal of Construction Engineering and Management* **143**.

Keyes CLM (2002) The mental health continuum: from languishing to flourishing in life. *Journal of Health and Social Behavior* **43(2)**: 207–222. https://doi.org/10.2307/3090197.

Keyes CLM (2005) Mental illness and/or mental health? Investigating axioms of the complete state model of health. *Journal of Consulting and Clinical Psychology* **73(3)**: 539–548.

Langdon R and Sawang S (2018) Construction workers' well-being: what leads to depression, anxiety and stress? *Journal of Construction Engineering and Management* **144**.

Lester A (2021) *Project Management, Planning and Control*, 8th ed. Elsevier Science & Technology, Oxford, UK.

Li K, Wang D, Sheng Z and Griffin MA (2022) A deep dive into worker psychological well-being in the construction industry: a systematic review and conceptual framework. *Journal of Management in Engineering* **38(5)**, [online], https://www.scopus.com/inward/record.uri?eid=2-s2.0-85134031810&doi=10.1061%2f%28ASCE%29ME.1943-5479.0001074&partnerID=40&md5=c7c317f8880e42197a15ef9febe09337.

Lim S, Chi S, Lee J, Lee H-J and Choi H (2017) Analyzing psychological conditions of field-workers in the construction industry. *International Journal of Occupational and Environmental Health* **23(4)**: 261–281. https://doi.org/10.1080/10773525.2018.1474419

Ling FYY, Ning Y, Chang YH and Zhang Z (2017) Human Resource Management Practices to Improve Project Managers' Job Satisfaction. *Engineering, Construction and Architectural Management* **25(5)**: 654–669. https://doi.org/10.1108/ECAM-02-2017-0030

Lingard H and Francis V (2004) The work-life experiences of office and site-based employees in the Australian construction industry. *Construction Management and Economics* **22(9)**: 991–1002.

McCord J, McCord M, Davis PT, Haran M and Rodgers WJ (2015) Understanding delays in housing construction: evidence from Northern Ireland. *Journal of Financial Management of Property and Construction* **20(3)**: 286–319.

Nwaogu JM, Chan APC and Naslund JA (2022) Measures to improve the mental health of construction personnel based on experts' opinion. *Journal of Management in Engineering* **38(3)**, 04022018. https://doi.org/10.1061/(ASCE)ME.1943-5479.0001045

Pinto JK, Dawood S and Pinto MB (2014) Project management and burnout: implications of the Demand–Control–Support model on project-based work. *International Journal of Project Management* **32(4)**: 578–589.

PMBOK (2021) *A Guide to the Project Management Body of Knowledge (PMBOK® Guide) – Seventh Edition and The Standard for Project Management (ENGLISH)*, 7th edn, Project Management Institute [online], https://learning.oreilly.com/library/view/a-guide-to/9781628256642/cover.xhtml.

Razak MIA, Khoiry MA, Badaruzzaman WHW and Hussain AH (2022) DfMA for a better industrialised building system. *Buildings* **12(6)** [online] https://www.mdpi.com/2075-5309/12/6/794.

Rees-Evans D (2020) *Understanding Mental Health in the Built Environment*. Chartered Institute of Building, Bracknell, UK.

Seligman ME (2011) *Flourish: A Visionary New Understanding of Happiness and Well-Being*. Simon and Schuster.

Stocks SJ, McNamee R, Carder M and Agius RM (2010) The incidence of medically reported work-related ill health in the UK construction industry. *Occupational and Environmental Medicine* **67(8)**: 574–576. https://doi.org/10.1136/oem.2009.053595.

Sunindijo RY and Zou PXW (2012) Political skill for developing construction safety climate. *Journal of Construction Engineering & Management* **138(5)**: 605–612.

Thompson M, Sinclair D, Tait A and Ostime N (2021) *DfMA Overlay to the RIBA Plan of Work*. Royal Institute of British Architects.

Waddell G and Burton AK (2006) *Is Work Good for Your Health and Well-being?* Department for Work and Pensions, London, UK.

Westerhof GJ and Keyes CLM (2010) Mental illness and mental health: the two continua model across the lifespan. *Journal of Adult Development* **17(2)**: 110–119.

World Health Organisation (2023) *Mental Health Fact Sheet*. World Health Organisation [online], https://www.who.int/news-room/fact-sheets/detail/mental-health-strengthening-our-response (accessed 27 April 2026).

Table 11.3 Tabulated summary of semi-structured interviews (continued on next page)

Topic	SM01	SM02	SM03
OSC Benefits			
Speed	Claimed that despite an [earlier] project starting later than a nearby project (which was being delivered by way of TC), his project was completed more quickly. Recognised shortened programme duration, but also programme certainty. Cat 2 and 5 manufacturing took place while foundations being constructed.	Early 'at risk' design and procurement gave programme certainty despite potential material shortages/cost increases because of political instability in Europe.	The company's (MMC Category 2) system is panellised, deliveries are made to site and the panels are immediately craned into place without delay (and no need for stockpiling). '*The biggest benefit is the quickness that we can get people into the houses.*'
Quality	Claimed that an [earlier] project was handed over 'defect-free'. Accuracy in manufacture of Cat 2 wall panels meant that they were easily assembled by small teams; the teams were well informed and since the work was to a large extent repetitive the team members were knowledgeable and confident in their work.	Quality improved by '*all consultants and designers are working on the same model, live on the cloud.*'	As a measure of the quality of their wall panels, the company was seeking to persuade its kitchen supplier that site visits to measure each house before beginning fabrication of kitchen units were unnecessary as the houses were being built to very fine tolerances.
Cost	Eliminated scaffolding; shorter contract period, less labour on site, reduced prelims.	Cost certainty achieved through early procurement based on 'at risk' detailed designs to produce accurate material schedules.	Significant wastage from cutting of plasterboard on site, but company exploring more pre-cut plasterboard. Otherwise, system delivers cost certainty.

Logistics	Fewer vehicle movements, especially materials deliveries. Also, fewer people on site at any one time compared to TC - estimated roughly 30 people on site compared to more than 100 if it were a TC project.	4D software employed to assess logistics.	Fewer vehicle movements, especially materials deliveries. Also, fewer people on site at any one time compared to TC. Researcher observed very little site traffic despite a delivery of wall panels taking place during one visit.
Health and Safety	OSC meant fewer people on site, resulting in less waste and rubbish on site, which helped reduce accidents, especially slips, trips and falls.	Health and Safety benefits arise from certainty about all aspects of the project which OSC provides.	Estimated 50% reduction in H&S incidents, due to a) fewer people on site b) less rubble and material stocks c) company installs infrastructure such as roads and paths very early in the project.
Other	None	By designing a prefabricated plant room, the project was effectively 'unlocked' as the original design was, SM02 felt, fraught with risk around this issue.	

Table 11.3 Continued

Topic	SM01	SM02	SM03
OSC Drawbacks			
Speed	None	None – the company owns the manufacturing facility.	Currently the company's system requires scaffolding to be installed and this can't take place until the external walls are erected. The company is investigating pre-erecting free-standing scaffold before the walls are delivered.
Quality	None	None – the company owns the manufacturing facility.	None – the company owns the manufacturing facility.
Cost	None	None – the company owns the manufacturing facility.	None – the company owns the manufacturing facility.
Logistics	None	None – the company owns the manufacturing facility.	Specialist teams deliver and erect the wall panels; the vehicles involved do not require special consents as they are not over-sized loads. The main constraint on delivery and installation is any occurrence of high winds on site.
Health and Safety	Described a serious accident on a previous site, implying no room for complacency.	None.	None
Other	None	None.	

Project planning considerations			
Main TC topics	Foundations built using TC had caused delays to programme.	Location of this site means that subcontractors are drawn to London – and away from this project - so the company felt it may face difficulty in recruiting local companies as subcontractors.	The foundations and all substructure works are still delivered by TC and require high levels of accuracy to achieve level walls and floor slabs from which the panels can be erected. There have been instances where this has not been achieved. The company also prefers to install all the infrastructure on 'day one' – e.g. roads, paths, paved areas and so on – which results in significant early expenditure (albeit resulting in a 'tidy' and safe site). Kitchen suppliers still insist on pre-measuring all properties before committing to manufacture, but the company is hoping to avoid this in most cases once a level of trust and experience has been reached with the supplier.
Main OSC topics	SM01's previous OSC experiences led him to be involved with detailed designs and allowed him opportunities to rationalise and standardise wherever possible.	SM02 didn't believe that OSC *necessarily* makes life easier for construction industry workers. Could be helped by additional training (e.g. by way of Supply Chain Sustainability School).	The erection of the wall panels can be adversely affected by high winds, so weather data are critical. Coordinating the erection and removal of scaffolding is essential.
Main differences between TC and OSC	Opportunity for earlier engagement led to positive outcomes on OSC projects (note: this may have been true even for TC projects).	SM02 was also an advocate for early engagement with designers and suppliers and he personally had been afforded those opportunities.	Much of the success of the OSC elements depends on the quality achieved by the TC teams to install all the infrastructure and substructure works.

Table 11.3 Continued

Topic	SM01	SM02	SM03
Project Success KPIs			
Speed	Predicted to be achieved, despite delays in (TC) substructure works.	Predicted to be achieved.	A significant proportion of the houses this company provides are to Social Housing clients, which means speed of delivery is important (and is being achieved). This benefits the client by providing much-needed accommodation for its tenants, but also the company through improved cashflow. The ability to make the properties water-tight very quickly means the following trades can commence work quickly.
Quality	'[OSC] *is just more efficient. I find it's a controlled environment, so you get the quality, absolutely second to none.*'	Being achieved.	Substantial steel sections and innovative connections result in very accurate erections with few quality issues. The high thermal performance – and hence very low running costs – are appreciated by both public sector and private sector customers.
Cost	Achieved c£5m cost savings through rationalisation and standardisation.	Achieved through minimising waste and early procurement.	Significant investment has been made into designing and manufacturing the system.

Logistics	Simplified	Digital design enabled 4D simulations	Delivery and erection are in the hands of the company's own specialist logistics teams. There are generally fewer tradespeople – and hence fewer general vehicle movements – than with TC.
Health and Safety	Improved (but see elsewhere caution against complacency).	Even when working in areas of complexity – typical of the type of work they were undertaking – he felt that the teams on site were surrounded by sufficient expertise to support their work and meanwhile they were benefitting from a low-risk environment.	Generally, the subcontractors on site have long-standing relationships with the company and the knowledge and experience they have gained helps to reduce H&S incidents.
Other	Client satisfaction and the opportunity for repeat business.		Clients very satisfied.

Table 11.3 Continued

Topic	SM01	SM02	SM03
Impact of project performance on well-being – behaviours			
Subcontractors / suppliers	Opportunities for continuing relationships with existing subcontractors. However, 'new' subcontractors sometimes did not become familiar enough quicky enough with the MMC systems and will not be engaged again in the future. *'Everyone's empowered.'* *'Don't blame anybody. As a team you work it out if there's a problem.'*	As long as subcontractors and suppliers *'come to you with the right skills and knowledge and abilities'*… [they can be comfortable with] … *time pressures and a limited time to set up the project' […] we knew we were getting people with the right capabilities from the beginning.'*	*'We work as a team. It's a good atmosphere. Everybody's happy because they're earning money.'*
Management	SM01's OSC projects perceived by management as exemplar projects.	SM02's employer was aware of the issue of mental health in the industry and was proactive in devising strategies.	SM03's OSC projects perceived by management as exemplar projects.
Client	Clients satisfied that cost savings can 'unlock' projects; clients happy with certainty of project delivery.	Client happy to fund early procurement once 'at-risk' early design tabled and global political crises articulated.	Good feedback from customers about high quality finishes and low running costs.
Other stakeholders	Client's funders pleased with project certainty.	Direct employees were redeployed if they were unable to develop the appropriate skills to work with OSC.	The OSC system has been welcomed by local authority planning departments.

Impact of project performance on well-being – personal well-being			
Confusion? E.g. lack of information, incorrect or conflicting information.	None	None, although the company recognised that not every team member was suitable for the project: *"When you are asking someone to do something, you've got to really understand strengths, weaknesses, blind spots.'*	The teams are made aware of what was required of them and were assured of project certainty. If queries *did* arise they were mostly from new team members, unfamiliar with the system.
Anxiety? E.g. worries about missing deadlines.	*'We know how in reality to build it. We know we are just clear what we need when we need it.'*	*'The pressures* [that arise] *when* [people are] *given too much to do and not enough time to do; at a stroke, almost,* [OSC takes] *all those stresses and worries away'*.	*'We deliver when we say we're going to deliver. We're not knocking people back every week because the weather's delayed us or anything like that.'*
Anger	If disputes *were* to arise, they would be most likely to occur should there be any delays on site; in this event, SM01 might not – in his own words - *'be their best friend at that point … I get frustrated .by poor performance, poor quality.'*	None	None
Speed? (Project falling behind programme)	*'I say to the guys, even when we've had a bad day: keep looking ahead, the sun's shining over there.'*	None – the company owns the manufacturing facility.	No concerns.

Table 11.3 Continued

Topic	SM01	SM02	SM03
Quality? (Defects appearing)	Some quality issues with (Cat 3) balconies. Resolved through redesign and minor rework.	None – the company owns the manufacturing facility.	No concerns.
Price? Budget over-runs)	None	None – the company owns the manufacturing facility.	No concerns.
Logistics? Difficulties in transportation, craneage and assembly)	None	None – the company owns the manufacturing facility.	No concerns, except in high winds.
Health and Safety? (Incidents on site)	None (on current site)	None	Very few reported.
Other?	*'The teams certainly on modern methods of construction* [sites] *are a lot happier.'*	*We run a very high performing, high expectation, high-demand* [in terms of results] *team, but we actually then run a really flexible work–life balance strategy.* The company has a strategy for regular personal development reviews (PDR) and team members were supported in many ways such as home working from time to time, compassionate leave, cost-of-living support and so on. Furthermore, SM02 was keen at all times to celebrate success with the team.	

Overall, describe the benefits or harm to your sense of well-being that working in OSC causes.	Felt comfortable during work hours as well as during out-of-work hours, with little cause for concern about teams' performance and few conflict-inducing disputes. This was explained by the fact that OSC delivered cost certainty and programme certainty; the quality was high and they had developed long-term relationships with the teams involved in assembly.	*'I go home worried, but worried in a nice way, because I care about what we do and the legacy that we leave and our reputation for doing a good job.'*	*'I have to say, from a well-being point of view, this is probably the best company I've worked for. They really do have a very good ethical view towards their staff and look after you really well'.*

Source: Author's creation

Part III

Global case studies of MMC in action

Emmanuel I Daniel and Fidelis A Emuze
ISBN 978-1-83708-437-1
https://doi.org/10.1108/978-1-83708-436-420261013

Chapter 12
Sustaining off-site construction: integrated strategies for long-term viability

Tamiris Capellaro Ferreira[1] and Nwakaego Onyenokporo[1]

[1]*University of Huddersfield, Huddersfield, United Kingdom* 0000-0002-4563-2429; 0000-0003-3462-8878

Abstract

Off-site volumetric construction (OSVC) has been promoted in the UK as one of the modern methods of construction (MMC) categories to address housing shortages, productivity gaps and net-zero 2050 targets. Yet, despite the government's support, the sector has witnessed a series of high-profile business failures, raising questions about its viability. This study aimed to identify and analyse the challenges that have led UK OSVC firms into administration and to propose strategies for their long-term viability. To achieve that, a multiple case study strategy with six UK OSVC firms was employed. Four of these firms entered administration, allowing comparative insights between resilient and failed business models. Comparative analysis revealed that firms entered administration due to a combination of volatile demand, capital intensity and policy dependency, compounded by overconfidence and escalation of commitment at the managerial level. Successful firms have sustained slow but stable growth through gradual investment and the implementation of integrated managerial and design strategies, mitigating structural vulnerabilities. The study advances the understanding of business sustainability in the OSVC sector by linking institutional and behavioural theories to practical challenges. It analyses firm collapse and outlines strategic measures for policy-makers and industry stakeholders seeking to stabilise and scale the OSVC sector.

Keywords: modern methods of construction (MMC); volumetric modular; off-site construction; bankruptcy; sustainability; long-term viability

12.1. Introduction

12.1.1 Background

Historically, construction has been one of the most inefficient industries, with projects frequently experiencing cost overruns and scheduling challenges (Razkenari *et al.*, 2020).

Off-site construction (OSC) has emerged as an alternative approach to enhance efficiency compared with traditional construction (Zheng *et al.*, 2024). OSC is recognised for its potential to improve outcomes related to time, cost, quality, efficiency, safety and sustainability (Mai *et al.*, 2025). It can be categorised into two subtypes: volumetric (including pods, modules and complete buildings) and non-volumetric (encompassing components, panels and foldable structures) (Ginigaddara *et al.*, 2022).

Government initiatives in countries such as Japan, China, Singapore, Hong Kong and the United Kingdom (UK) have highlighted the importance of OSC, particularly for the residential sector (Razkenari *et al.*, 2020). The UK has been selected for this research due to its longstanding tradition in OSVC, which emerged post-World War I as a solution to housing shortages (Hammad *et al.*, 2019). Recently, in response to the national target of constructing 300 000 homes annually, the UK government has indicated the MMC as a viable pathway to address housing supply, labour shortages and productivity challenges (UK Parliament, 2024). MMC encompasses a diverse array of non-traditional building systems, broadly classified into seven categories (MCHLG, 2019; UK Parliament, 2024).

Due to the nature of their activities, volumetric MMC (Category 1) can deliver superior quality and sustainability but carries the highest insolvency risk due to capital intensity; panelised and hybrid systems (Categories 2 and 3) offer greater flexibility with moderate sustainability gains; while process-led innovations (Categories 6 and 7) remain least risky but also least transformative. Long-term viability thus depends on aligning external enablers (policy stability, finance mechanisms, demand pipelines) with internal capabilities (managerial learning, risk management and operational flexibility). Due to its potential to drive transformative advancements within the industry (Razkenari *et al.*, 2020) and high insolvency risk, OSVC (or MMC Category 1) was the focus of this study.

12.1.2 Scope of the research

Despite government incentives, many off-site manufacturers struggle to remain competitive (Ayinla *et al.*, 2025). There are several academic studies regarding the challenges in adopting OSC, but very few on how to keep OSC firms financially viable in the long term. Razkenari *et al.* (2020) observe that one of the key challenges for OSC is the design inflexibility together with high initial costs and substantial capital investments in factories. The need for advanced machinery and strategically located facilities demands significant upfront investment, often leading to cash-flow difficulties and business instability. Taylor (2022) reports that in the UK, between 2010 and 2018, a total of 193 OSC companies became either dormant, dissolved or in liquidation, but does not discuss the reason in depth nor strategies for the companies to succeed in the market.

While OSC is viewed as central to achieving the UK's housing and net-zero 2050 targets, financial fragility among manufacturers has cast doubt on its long-term viability (Lowe, 2022). OSC firms appear particularly vulnerable due to high capital investments and dependence on consistent demand. Evidence suggests that off-site manufacturers have continuously faced challenges in performing on a scale that enables their survival, going into bankruptcy within a few years of setting up (Ayinla *et al.*, 2025). The collapse of major UK OSC firms such as Ilke Homes, Legal & General Modular Homes and House by Urban Splash between 2022 and 2023 raised concerns about the sector's resilience. Homes England invested in MMC companies such as Ilke Homes and House by Urban Splash, but the companies eventually went bankrupt and government bodies were not paid, resulting in losses (UK Parliament, 2024). This prompted the House of Lords Built Environment Committee, in 2024, to urge the UK Government to develop a coherent, evidence-based strategy to support MMC adoption and sustainability (UK Parliament, 2024).

This study, therefore, addresses a critical gap: understanding why off-site volumetric construction companies in the UK face failure despite the sector's policy importance and sustainability potential and what they could have done differently. This chapter aims to identify and analyse the challenges that have led OSVC firms into administration and to propose strategies for their long-term viability. Using multiple case studies of failed and successful companies, this study seeks to extend existing knowledge on OSC business sustainability. The findings are expected to inform industry stakeholders and policymakers on measures that could enhance firm resilience and ensure that OSC and MMC can contribute effectively to the UK's housing, productivity and carbon-reduction objectives.

12.2. Perspectives of OSC in the UK

MMC, particularly OSC, has been promoted in the UK as a key enabler of faster, greener and more efficient project delivery, capable of addressing the housing and labour shortages while reducing the sector's carbon footprint (Ashayeri *et al.*, 2022; Goulding *et al.*, 2017; Gunawardena and Mendis, 2022). Despite this promise, the medium-to-long-term viability of many MMC firms remains uncertain. The sector's successes and failures can be understood by distinguishing between external drivers (policy, finance and market context) and internal drivers (management and operational capability), which interact through complex institutional and resource dependencies.

12.2.1 External drivers of success and failure

Government initiatives such as the Affordable Homes Programme (AHP) have required housing associations to use MMC for 25% of developments to secure funding (UK Parliament, 2024). These incentives have stimulated uptake but also exposed firms to market volatility and policy dependency. Nevertheless, limited standardisation, high capital costs and insurance difficulties have discouraged many investors and developers from patronising the OSC manufacturers (UK Parliament, 2024).

The policy-market interface has produced uneven outcomes: while commercial applications (e.g. student housing) have been relatively successful, residential developers remain hesitant due to higher perceived costs and fragmented supply chains (Brennan *et al.*, 2024; UK Parliament, 2024). According to institutional theory, firms' dependence on public incentives and regulatory legitimacy constrains strategic flexibility (DiMaggio and Powell, 1983), while resource dependency theory highlights how inadequate demand pipelines weaken firms' ability to maintain factory throughput and manage cash flow risks (Pfeffer and Salancik, 1978). Interruptions between projects can create severe cashflow pressures for OSVC manufacturers, who must maintain staff and equipment even when production pauses.

This financial strain increases the risk of insolvency, a problem more common in the MMC sector than in traditional construction (Brennan *et al.*, 2024). This risk is further aggravated by conventional payment mechanisms, where clients release funds only after onsite inspection of completed work. Since volumetric manufacturers must procure materials and produce entire modules upfront, such payment structures expose them to significant cashflow vulnerabilities (Jalil *et al.*, 2015).

12.2.2 Internal drivers and managerial behaviour

Business history is replete with examples of innovative firms failing to sustain themselves long term. While some literature attributes failure to external factors, studies such as Park *et al.* (2020) and Brown (2024) emphasise internal factors, particularly managerial dysfunction. Brown (2024) identifies behaviours such as myopia, groupthink and escalation of commitment, where managers persist with failing strategies due to optimism bias, reputational concerns or agency pressures. Subjective expected utility theory (SEUT) explains how inexperience and overconfidence lead decision-makers to overlook risks, a pattern relevant to the UK OSC sector. Self-presentation theory (SPT) and self-evaluation under threat (SEUT) further suggest that managers under scrutiny may 'save face' by maintaining poor courses of action, especially when reliant on investor approval.

Lowe (2022) conducted case studies in modular construction companies operating across the residential and commercial sectors, identifying key factors undermining their sustainability in the UK. Internally, these challenges include poor financial management, non-compliance of designs with building regulations, overproduction relative to market demand and insufficient project pipelines to maintain factory efficiency. Lowe (2022) argues that the limited success of modular construction, despite the considerable attention it has received in recent years, reflects deeper structural issues. The sector continues to grapple with high upfront costs, regulatory compliance difficulties and a reliance on steady demand in an inherently volatile market.

The UK Parliament (2024) attributes the collapse of modular companies such as Ilke Homes, House by Urban Splash and Legal & General to individual business decisions, insufficient order books and wider economic challenges following Covid, together with the increase in interest rates since the autumn of 2022. According to Lowe (2022), Swan, a pioneer in modular development within the housing sector in the UK, was found to be non-compliant with regulatory standards by the Regulator of Social Housing. A similar issue was encountered in the House by Urban Splash, whose collapse was attributed to factories' under-utilisation and design defects, which eroded customer confidence (Lowe, 2022). The Government has also noted that customers are hesitant to place large orders with unprofitable manufacturers, fearing they may cease operations (UK Parliament, 2024).

12.3 Methodological approach and case study description

This research adopted a qualitative empirical methodology aimed at identifying patterns emerging from the data. The approach was grounded in case study research, an established empirical strategy that enables in-depth investigation of contemporary real-world practices within their contextual boundaries (Yin, 2003). This approach was considered suitable due to the exploratory and practice-based nature of the topic, focusing on OSVC companies in the UK. Case studies are particularly relevant when investigating complex organisational processes (Gibbert *et al.*, 2008). A known limitation of this method is its restricted generalisability (Brown, 2024; Gibbert *et al.*, 2008), but this was mitigated through the inclusion of multiple case studies and the application of theoretical replication logic (Yin, 2003).

Between 2022 and 2023, six case studies were conducted to provide comparative insights on companies' production and design strategies. The companies were intentionally selected to represent diverse scales, production capacities and market sectors within the UK OSVC industry, strengthening the analytical generalisation of findings (Gibbert *et al.*, 2008). The sample included firms operating in residential, commercial, healthcare and educational sectors, enabling cross-case comparison of different design and production strategies.

By 2025, four of the six companies studied had filed for bankruptcy, providing an additional opportunity to analyse the factors contributing to business failure in the sector. As direct contact with former employees was not possible due to company dissolution, supplementary analyses were conducted using publicly available sources, including trade publications and business news reports. Table 12.1 summarises the key characteristics of the companies analysed in the case studies.

Table 12.1 Companies analysed in the case studies

Identification	Sectors of Operation	Foundation year	No. of Employees	Production (modules/ year)	Status
Company A	Educational, residential and commercial	2007	51–200	380 to 630 modules/year	Bankrupt in 2023
Company B	Residential	2016	51–200	500 modules/ year	Bankrupt in 2024
Company C	Residential	2017	501–1000	2020 modules/ year	Bankrupt in 2023
Company D	Healthcare, construction, industrial and commercial	1984	11–50	250 modules/ year	In operation
Company E	Residential	2016	501–1000	500 to 750 modules/ year	Bankrupt in 2024
Company F	Commercial	2013	11–50	500 to 750 modules/ year	In operation

Source: Authors' creation

12.3.1 Data collection and analysis

Data were collected through a combination of site observations, semi-structured interviews and document analysis, allowing triangulation across multiple evidence sources. Site observations were conducted during factory visits to examine production workflows, automation level and layout configurations. A total of 37 interviews with managers were carried out, representing 1746 minutes of discussion to explore organisational strategies, project delivery models and production challenges. Documents such as company reports were reviewed as one more source of information.

The interviews followed a semi-structured protocol and were only recorded when authorised, but notes were taken during all the interviews. The data was subsequently transcribed and coded. Observational data were recorded through field notes and photographs. A qualitative content analysis was employed to interpret and categorise data into key themes related to production and design strategies, and to the challenges in OSVC. These themes informed the cross-case comparison and the identification of patterns among companies with different operational models.

Ethical approval for the research was obtained from the University of Huddersfield Research Ethics Committee. Confidentiality and anonymity were strictly maintained, with company names and individual identities anonymised.

12.4. Results and discussion

This section reports the case study findings for the six UK OSVC firms (Companies A - F). Each case is presented considering *(a)* business model and market position, *(b)* design and production process and *(c)* financial outcome and resilience trajectory (continued operation or insolvency). This section then compares the cases and identifies recurring risk factors and resilience capabilities in the sector. Section 12.4.3 discusses the long-term sector viability, linking strategic recommendations to specific empirical observations from the cases and to recent policy and market developments.

12.4.1 Case study findings

12.4.1.1 Company A

BUSINESS MODEL AND MARKET POSITION

Company A (founded 2007) operated as an integrated delivery firm, combining design, module manufacture and on-site assembly. It acted as project delivery manager when required and therefore could deliver the entire build programme. It operated across residential, education and commercial markets, with residential work representing around 60% of sales. Its clients were mainly institutional buyers, such as government departments, local authorities and developers, rather than individual end users. This positioned Company A within the higher-volume procurement environment.

DESIGN AND PRODUCTION PROCESS

Company A generally received fully developed designs intended for traditional construction and adapted them for volumetric manufacture. The main factory layout was positional: modules were kept largely in one place while materials and labour moved to them. Steel chassis were initially assembled in a cellular sub-area (Figure 12.1), then moved to their final position in the factory. Automation levels were low and many processes resembled traditional site-based construction.

The company could produce 2.5 modules per day. However, highly customised projects reduced productivity to 1.5 modules per day. This implied an annual requirement to output approximately 380 to 630 modules to sustain throughput targets.

FINANCIAL OUTCOME AND RESILIENCE TRAJECTORY

After a period of rapid expansion, including acquisition of a new 9300m^2 manufacturing base in Hull and winning places on major public sector procurement frameworks, Company A entered administration in October 2023. At its peak, it employed 85 staff and reported turnover approaching £30 million in April 2022, but this growth was debt-leveraged and exposed the business to project delays and input cost inflation (Built Offsite, 2023; Laister, 2023; Prior, 2023). Their debts rose to approximately £7.1 million. Company filings warned that commodity price volatility, political and regulatory shifts and programme slippage posed material risks to delivery (Laister, 2023). In resilience terms, the firm pursued aggressive scale-up but did not achieve sustained cost absorption across the pipeline to service its debt.

12.4.1.2 Company B

BUSINESS MODEL AND MARKET POSITION

Company B was focused on residential modular housing. Originally founded in 2016, it underwent judicial reorganisation in 2021 and relaunched at a new site under a different name. As a result, during the case study the firm was still rebuilding internal capability and processes. Interviewees described fast but disorganised growth, high staff turnover and instability in production routines. In September 2022, the company had 84 employees and was actively recruiting. The company's customer base consisted largely of private developers and housing associations.

Figure 12.1 Chassis assembly

Source: Author's personal archive.

DESIGN AND PRODUCTION PROCESS

Company B usually received completed designs from clients and adapted them for modular manufacture. During initial visits to the factory, it used a positional layout. The firm subsequently began transitioning to a line-based layout with defined workstations and repeatable steps. In October 2022, that transition was underway; however, four projects were being produced simultaneously, causing delays in implementation. The stated aim of moving to a flow line was to increase repeatability and reduce bespoke variation from project to project, improving the factory's efficiency and productivity. Automation levels were low to moderate: mainly timber cutting machinery and some handling equipment, such as hydraulic jacks and overhead cranes.

FINANCIAL OUTCOME AND RESILIENCE TRAJECTORY

Company B entered administration in April 2024. The company had struggled with cash flow, had attempted but failed to sell the business and could not meet short-term liabilities. The firm's history reflects sectoral fragility: it had acquired the assets of a previous modular producer that failed during Covid despite government-backed finance (Banks, 2024; Built Offsite, 2024). Although unaudited accounts for the year ending November 2022 indicated net assets of £8.2 million, the firm continued to face liquidity pressure and lacked a stable, forward-funded pipeline of work. Upon inquiry, the firm claimed capacity to deliver 650 homes in 2024 with ambitions of over 1000 homes in 2025, but these projections did not convert into secured, cash-backed orders. This illustrates a recurring pattern across the failing firms: stated capacity ramp-up ahead of actual secured demand.

12.4.1.3 Company C

BUSINESS MODEL AND MARKET POSITION

Company C, founded in 2017, provided off-site modular housing for distribution throughout England and Wales. The company received the first direct investment from the UK Government in a modular housing factory for expansion and was repeatedly described as an exemplar MMC firm (Battersby, 2023a; Roofing Today, 2023). Its manufacturing facility opened in November 2018 and the factory automation project was completed in 2020, aiming to achieve the production of 3000 homes per year by 2023. The firm's model relied on supplying standardised housing types at industrial scale and its main clients were local authorities and developers.

DESIGN AND PRODUCTION PROCESS

Unlike Companies A and B, Company C did not primarily deliver bespoke designs. Instead, it offered a catalogue of nine pre-designed house types, with limited finish options. Each house type consisted of stacked volumetric modules (one module per floor). Customers selected their design options before the production process started. The production system was a flow line with three main lines and two auxiliary lines (for floors and ceilings). Workstations were time-balanced across the line and real-time progress data were displayed on factory screens. Automation was medium; CNC cutting and automated lifting/handling were integrated with semi-skilled manual tasks. Overhead cranes minimised manual lifting and digital work instructions were accessible at each station. Figure 12.2 illustrates some factory floor equipment.

Figure 12.2 Equipment present on the factory floor of Company C: (a) a suction-cup machine positioning wooden sheets onto panel frames, (b) an operator using a CNC lathe and (c) an overhead crane lifting a lightweight steel panel structure.

Source: Author's personal archive

FINANCIAL OUTCOME AND RESILIENCE TRAJECTORY

Despite rapid growth, high-profile funding injections (including £60 million in 2021 and £100 million in 2022 from private investors) and political backing, Company C entered liquidation in 2023 with over £319 million owed to creditors (Battersby, 2023a; 2023b; Offsite Hub, 2023). Homes England was the largest creditor at an estimated £68.7 million debt (Battersby, 2023b). The investor group withdrew funding over concerns around cash burn and the gap between production capacity and actual sales pipeline. In 2020–2021 the firm generated turnover of £57 million–£64 million but still posted operating losses of over £30 million per year, due partly to very high fixed costs associated with its automated plant (Roofing Today, 2023). Company C highlights a paradox in policy-led scaling: state endorsement and capital injections enabled rapid expansion of capacity, but demand, planning approvals and land supply did not scale at the same pace (Butler, 2023).

12.4.1.4 Company D

BUSINESS MODEL AND MARKET POSITION

Company D is a family-owned volumetric manufacturer active since 1984. Unlike Companies A, B and C, which were largely focused on residential delivery at scale, Company D specialises in highly customised, sector-specific permanent buildings for healthcare, education and commercial clients, as well as temporary accommodation for sites. Its core client base includes government departments, NHS bodies and private developers in the health sector.

DESIGN AND PRODUCTION PROCESS

The company offers complete customisation services, meeting the needs of each client. Designs are typically provided by the client in traditional form and then converted by Company D's in-house design team into modular buildable units. The production system is predominantly positional, with modules assembled in fixed bays in the main building, supported by cellular sub-facilities for timber cutting and panel assembly. Automation is low; most work is manual, although timber cutting is mechanised and standardised cassette elements are pre-manufactured to accelerate assembly.

FINANCIAL OUTCOME AND RESILIENCE TRAJECTORY

Company D remains solvent and has grown. For the financial year ending September 2024, it reported turnover of £23.31 million, a 74% increase from the previous year. It has secured a position on an NHS framework, which enables it to bid for public sector projects.

12.4.1.5 Company E

BUSINESS MODEL AND MARKET POSITION

Company E, founded in 2016, manufactured modular homes and apartments for the residential sector. Its model relied on batch orders of at least 50 homes per client, which was justified by crane installation costs and site logistics. Its customers were primarily developers and large-scale residential investors. The firm sought to operate as a volume housebuilder using off-site manufacture, supported by significant external finance from institutional investors.

DESIGN AND PRODUCTION PROCESS

Company E offered 29 pre-designed house and apartment types. While the designs were pre-established, the finish choices were flexible. The factory layout followed a space-constrained U-shaped line, with central sub-areas organised in a cellular configuration for parallel kit-of-parts preparation. Automation was medium; early-stage production leveraged CNC machinery for timber cutting, nailing and panel assembly, with human operators supervising and initiating each cycle. Later-stage finishing work, including fit-out, was largely manual and resembled traditional on-site construction.

FINANCIAL OUTCOME AND RESILIENCE TRAJECTORY

Company E went into administration in November 2024. Between 2020 and 2023, it accumulated heavy losses, including a reported £46.3 million loss before tax for the year ending October 2023 (PBC Today, 2025). Although the firm secured substantial investment (including backing from Homes England) and announced plans for a new large-scale factory, cash burn and stalled pipeline undermined viability (Gerrard, 2024). The failure was attributed to the sluggish consumer demand, to the rise in raw material costs and to the escalating interest rates (Built Offsite, 2025; PBC Today, 2025). By August 2024, one of its major investors wrote down a £25 million stake and expansion plans were cancelled at significant cost. Redundancies followed and forward production effectively ceased (Gerrard, 2024). The firm reported no income for certain divisions and rising administrative overheads, indicating structural underutilisation of capacity.

12.4.1.6 Company F

BUSINESS MODEL AND MARKET POSITION

Company F delivers volumetric modular schemes primarily in high-demand, institutionalised and repeatable asset classes: build-to-rent housing, student accommodation, co-living and hotels. Rather than competing directly with speculative for-sale housing, Company F positions itself in sectors where demand is driven by long-term operators and investors with predictable occupancy models.

DESIGN AND PRODUCTION PROCESS

Company F produces precision-engineered modules. Individual fully fitted modules are manufactured on a production line, then transported to site and craned into place. On-site work therefore focuses on stacking, connection and final finish. This reduces vehicle movements, waste, theft risk and overall site duration compared to traditional construction. The firm emphasises that reduced on-site labour, faster programme delivery and embodied carbon savings form part of its competitive offer to clients.

FINANCIAL OUTCOME AND RESILIENCE TRAJECTORY

Company F is profitable. It reported turnover of £167 million and pre-tax profit of £11 million in the financial year ending August 2021. Turnover rose to approximately £187 million in 2024 and the forward pipeline was described as 'sizable' (Endole, 2024; Lowe, 2022). Management explicitly links this performance to two narratives: *(a)* contributing to national housing need and *(b)* supporting decarbonisation goals at scale (Lowe, 2022). Unlike Companies A, B, C and E, Company F demonstrates that profitability in modular delivery is achievable under certain conditions: a disciplined product strategy, reliable repeat-client demand and credible evidence of regulatory and safety performance.

12.4.1.7 Case studies summary

Table 12.2 synthesises key dimensions across all six companies. 'Resilience trajectory' is interpreted as either collapse (administration/liquidation) or continued operation and profitable growth.

Table 12.2 Case studies key dimensions (continued on next page)

Company	Design Approach	Factory Layout	Automation Level	Client Type	Resilience Trajectory
A	Adapts client designs for modular manufacture	Positional	Low	Public bodies and developers	Bankrupt in 2023 – after debt-funded expansion
B	Adapts client designs for modular manufacture	Transition from positional to line-based	Low/ Medium	Developers and housing associations	Bankrupt in 2024 – cash flow failure and insufficient demand
C	Catalogue of nine house types	Line-based	Medium	Local authorities and developers; backed by government investment	Bankrupt in 2023 – cash flow failure and insufficient demand
D	Adapts client designs for modular manufacture. Standardised cassettes	Positional/ cellular sub-areas	Low	NHS/Public sector	In operation

Table 12.2 Continued

Company	Design Approach	Factory Layout	Automation Level	Client Type	Resilience Trajectory
E	Batch orders	U-shaped/ Line-cell hybrid	Medium	Large developers and investors; backed by government investment	Bankrupt in 2024 – after cumulative losses and cancelled expansion
F	Repeatable modules	Line-based	Medium	Institutional investors/ operators	In operation

Source: Authors' creation

12.4.2 Cross-case discussion

Across the six firms, several patterns emerge, as discussed below.

12.4.2.1 Demand and pipeline risk

The main structural weakness in the failed firms (A, B, C, E) is not related to lack of technical capability but lack of stable, cash-backed demand at the scale required to absorb fixed factory costs. Companies C and E invested heavily in physical capacity, automation and headcount on the assumption that the pipeline would materialise. In practice, inflows of work lagged behind expansion, leading to underutilised factories and unsustainable cash burn. Companies A and B showed a similar pattern on a smaller scale by forecasting hundreds of units per year without securing a forward order book sufficient to stabilise cash flow. By contrast, Companies D and F anchor themselves in sectors where demand is both repeatable and procurement-driven (healthcare, build-to-rent/student accommodation), which creates predictable throughput and smoother revenue profiles.

12.4.2.2 Product strategy and standardisation

Companies A and B were largely reactive, adapting client designs originally intended for traditional construction. This approach preserves customer flexibility but erodes manufacturability as every project becomes a prototype, reducing throughput and preventing learning-curve effects. Company C closed this gap by forcing customers to choose from a defined catalogue with limited options. Company E also moved towards catalogued designs. Company F and, to an extent, Company D use standardised elements, allowing for customisation. They standardise module parts (e.g. wall panels) and repeatedly deploy them. This supports testing (e.g. fire testing) and accreditation and strengthens the business case to clients and insurers.

12.4.2.3 Regulatory assurance and credibility

Insurers, lenders and building control bodies have expressed caution around MMC because traditional building regulations are not specifically geared to manufactured systems

(UK Parliament, 2024). In that environment, firms must prove compliance at system level. Company F invested in extensive large-scale fire testing and accreditation work, in the absence of fully standardised MMC guidance. Company D, likewise, collaborates with contractors, conducts fire tests and uses building information management (BIM)-led coordination to demonstrate safety and regulatory compliance. This is not just technical but commercially critical. It reassures risk-averse public clients (e.g. the NHS) and enables access to frameworks that guarantee workload. In contrast, some firms, including high-profile residential modular ones, have faced design and compliance issues that damaged client confidence and slowed adoption (Lowe, 2022; UK Parliament, 2024).

12.4.2.4 Managerial maturity and financial discipline

Company D has operated since 1984 and Company F since 2013. Both have had time to develop managerial systems, governance, cost control and client relationships. Younger firms often attempt to have rapid capacity expansion funded by equity injections, headline growth targets and political endorsements without first building robust delivery management, risk controls and cash discipline. Company C in particular shows the risk of financing-led expansion without matching operational maturity. Although it attracted over £100 m in investment and political backing as an MMC exemplar, it still collapsed within six years, leaving substantial debts (Battersby, 2023b; Roofing Today, 2023). The UK Parliament (2024) notes that most large modular housing manufacturers only emerged after 2015, implying a sector-wide lack of embedded experience.

12.4.2.5 Strategic positioning relative to policy and sustainability

Both resilient firms align themselves with wider public priorities. Company D positions off-site modules as a way to rapidly deliver NHS capacity with minimal disruption, directly addressing public service resilience. Company F frames volumetric delivery as a pathway to net-zero carbon, reduced site traffic and shorter build times for high-demand rental and student accommodation markets. This alignment is reinforced by emerging policy moves such as NHS England's expectation that large health projects include high proportions of MMC and central government ambitions to increase MMC use to achieve carbon and productivity goals (UK Parliament, 2024).

12.4.3 Integrated strategies for the long-term viability of off-site construction

This section consolidates the empirical patterns identified above into four strategic pillars for the long-term viability of OSVC. Each pillar links observed practice in the case studies to broader policy developments.

12.4.3.1 Achieving economies of scale through pipeline consolidation

Since 2021, the government strategy has emphasised the 'Five-S' approach (scale, standardisation, safety/assurance, soft levers and stimulation of demand) (UK Parliament, 2024). From the case studies, it is clear that without demand, unit costs cannot fall and firms are forced into debt-funded speculative scaling, which has repeatedly proven fatal.

A central failure mode in Companies A, B, C and E was underutilised capacity that led to severe cash burn. The government has been focusing on coordinated pipeline creation by way of procurement mandates and public frameworks – for example, NHS England (2025) has introduced targets for high MMC content (e.g. 70% MMC for new builds and 50% MMC for refurbishments projects over £25 million). Nonetheless, as per the institutional theory, firms should not rely solely on public incentives and regulatory legitimacy (DiMaggio and Powell, 1983).

According to the UK Parliament (2024), it is unclear why the private sector is not driving demand for MMC if it offers significant benefits. The Government should consider whether there are policy barriers or if businesses are overly cautious and develop further mandates for a minimum use of Categories 1 and other specific categories.

12.4.3.2 Standardisation, regulation and assurance as enablers of market trust

One of the barriers for residential-focused firms is insurance and warranty acceptance. Building regulations are not MMC-specific, which creates uncertainty for insurers and building control when assessing modular systems (UK Parliament, 2024). While the insurance industry and other stakeholders have called for MMC-specific guidance in the Approved Documents, there is a lack of clear direction, leading to a fragmented approach among manufacturers.

Warranty and insurance providers indicate they can support MMC products but face challenges due to complex processes and insufficient data (UK Parliament, 2024). The extended time needed to secure warranties and the reluctance of insurers to accept current building regulations as adequate are hindering the growth of MMC homes. The government could update guidance and encourage insurance providers to work collaboratively to gather necessary data.

The Government has committed to establishing a universal specification for key components in MMC to enhance interoperability, mitigate supply chain risks and achieve economies of scale (UK Parliament, 2024). Despite this pledge made in 2019, a Publicly Available Specification PAS 8700:2025 has just been issued by the British Standards Institute (BSI) in April 2025. A specific standard for MMC will help with the issue of not meeting building regulations or fear from insurance companies.

This direction aligns with what was observed in Companies D and F. Both firms heavily invest in demonstrable compliance: fire tests, digital coordination and repeatable cassette or module types. This ensures that the final product meets existing building regulations and can be eligible for insurance and warranties. This in turn gives institutional clients (such as NHS trusts, in the case of Company D) confidence to procure at scale. In that way, accreditation is treated as an asset and a sales tool. Lowe (2022) highlighted that two failed modular companies (Swan and House by Urban Splash) were found non-compliant with regulatory standards.

12.4.3.3 Governance, managerial discipline and phased growth

The insolvencies show a pattern of 'premature scaling' (Joseph *et al.*, 2023), significant capital expenditure on factories, technology and headcount ahead of confirmed work. Company C is the clearest example. It was heavily capitalised, publicly endorsed as an MMC exemplar and forecast production in the thousands of homes per year. Yet it entered liquidation, because investor funding was withdrawn once it became clear that sales and planning approvals were lagging behind fixed cost commitments (Battersby, 2023b; Offsite Hub, 2023). Company E followed a similar path, announcing a major new factory and attracting investors, but ultimately reporting cumulative losses and cancelling expansion plans (Gerrard, 2024; PBC Today, 2025).

By contrast, Company D grew within a defined market niche and focused on repeat work, while Company F concentrated on high-demand markets. Both firms exhibit traits associated with 'phased scaling' (Joseph *et al.*, 2023): tightening product scope, aligning delivery with stable demand and reinvesting profits into capability rather than pursuing speculative expansion. That shows that resilience depends on managerial discipline, scaling capacity only when pipeline, accreditation and cash flow justify it. This requires internal governance maturity that many younger MMC firms have not yet developed. The fact that only a handful of the largest modular manufacturers in residential delivery were active before 2015 (UK Parliament, 2024) reinforces the immaturity of sector managerial systems.

12.4.3.4 Portfolio diversification

To address the underutilisation of factories and optimise return on initial capital investment, OSVC manufacturers can consider diversifying their work portfolio to create a continuous pipeline of projects. This strategy aligns with portfolio management theory, which emphasises spreading risk across different business units or markets to stabilise cash flow and maximise long-term returns (Brzęczek, 2020).

Both Companies D and F demonstrate effective portfolio management in practice. Company F focuses on student accommodations and buy-to-rent properties, contrasting with the traditional buy-to-sell model pursued by many developers and MMC manufacturers. Similarly, Company D targets the healthcare sector and site offices and accommodation, securing a consistent workflow for its manufacturing facilities. This approach reduces market volatility risk. Both companies further enhance their operational stability by maintaining strong client relationships and a reliable pipeline of upcoming projects, reflecting principles of resource-based theory, where leveraging unique capabilities and strategic assets contributes to sustainable competitive advantage (Kozlenkova *et al.*, 2014).

12.5. Implications for practice, policy and research

The findings from the six UK OSVC firms reveal systemic tensions between ambition and operational maturity, and between policy intent and market readiness. While the cases highlight recurring risk factors such as premature scaling, demand volatility and compliance uncertainty, they also point to resilience capabilities rooted in managerial discipline, standardisation and sectoral fit. The implications, therefore, span practical lessons for industry, priorities for public policy and directions for future research.

12.5.1 Implications for practice

From a practitioner perspective, the evidence underscores that resilience in modular manufacturing depends less on technical innovation and more on organisational governance, market focus and cash discipline. Firms such as Companies D and F illustrate that sustainable performance emerges from incremental, demand-led scaling, coupled with tight product–process integration and strong client relationships. By contrast, the insolvencies of Companies A, B, C and E demonstrate the dangers of speculative capacity expansion. Practitioners should therefore prioritise phased scaling strategies, where capacity growth follows verifiable pipeline. It is also recommended to diversify client portfolios and invest in accreditation, as demonstrable compliance enhances insurer and client confidence.

12.5.2 Implications for policy

At the policy level, the cases expose the limits of top-down stimulus without synchronised demand creation and regulatory assurance. State-backed investment, as seen in Companies C and E, enabled rapid factory expansion but did not guarantee throughput or market trust. Sustainable sectoral growth requires policies that balance scale with stability. Policy measures should therefore include clear regulatory pathways, as an MMC-specific guidance within the Building Regulations, and insurance and warranty collaboration frameworks, enabling insurers to reduce perceived risk. Beyond technical and regulatory levers, policy should also recognise that MMC adoption is a systemic issue: planning regimes, land supply constraints and fragmented finance models all interact with manufacturing viability.

12.5.3 Implications for research

Future work should be expanded to include other categories of MMC. Additionally, although improvement strategies have been proffered, future work should develop an effective business model or decision-making framework to guide the management of MMC companies, encouraging their long-term viability and sustaining the broader off-site construction sector. In addition, comparative studies across sectors could reveal how procurement models and risk perceptions shape MMC viability.

12.6. Conclusions

This chapter aimed to identify and analyse the challenges that have led OSVC UK firms into administration and to propose strategies for their long-term viability. This was achieved through six comparative case studies. The multi-case study design provided a comprehensive understanding of the diversity and challenges of the UK modular construction sector. The analysis demonstrated that long-term viability is determined by the dynamic alignment of external and internal capabilities.

Externally, policy instability, clients' hesitancy and volatile market created structural fragility. Internally, companies face challenges related to designs not meeting industry standards or building regulations, insufficient work pipeline, limited managerial experience and inadequate risk governance. The most resilient firms were those able to reconfigure their business models to manage cashflow volatility, diversify their client portfolios and maintain regulatory and operational discipline.

The cross-case analysis also shows that the viability of Category 1 volumetric construction in the UK is not primarily a question of technical capability. Instead, resilience depends on five interlocking conditions, namely

(a) stable demand that can keep factories utilised
(b) standardised product platforms that are insurable, certifiable and repeatable
(c) credible evidence of regulatory compliance, particularly around fire safety and building performance
(d) governance structures that scale capacity only when pipeline and cashflow justify it
(e) strategic alignment with long-term public mandates.

Firms lacking these conditions (A, B, C, E) expanded faster than their assured pipeline and entered administration or liquidation. Firms that embedded these conditions (D, F) remain solvent, profitable and able to attract further work. This suggests that future MMC policy should focus less on subsidising capacity in isolation and more on underwriting accredited, standardised solutions within guaranteed procurement frameworks that translate policy goals into bankable, repeatable demand.

Theoretically, the study contributes by shedding light on a topic underexplored in the literature, identifying the reasons for failure of UK OSVC companies. The study advances understanding of business sustainability in the OSVC sector by linking institutional and behavioural theories to practical challenges faced by companies. Analysing why they failed and proposing strategies for their long-term viability, this research contributes not only to the development of the literature in the area, but also practically, by providing company managers with insights that could help other companies to succeed in such a volatile market.

This study also provides insights for policymakers seeking to develop stable, evidence-based strategies to help the implementation of the MMC and consequently the development of a more efficient and sustainable construction sector.

It is acknowledged that there were limitations to the research. The dissolution of four case companies constrained opportunities for follow-up interviews, leading to secondary data sources and analysis. Despite inherent limitations related to access and generalisability, the methodological design ensured sufficient rigour and transparency to support the study's analytical and theoretical contributions. However, future research should incorporate direct stakeholder interviews to deepen understanding of resilience mechanisms in the OSVC sector.

Acknowledgement

This work was supported by the Coordination for the Improvement of Higher Education Personnel – Brazil (CAPES) – Financing Code 001.

REFERENCES

Ashayeri I, Goulding J, Heesom D *et al.* (2022) Implementation of a novel data-driven approach to optimise UK offsite housing delivery. *IOP Conference Series: Earth and Environmental Science* **1101(4)**: 1–11.

Ayinla K, Cheung F and Towe B (2025) Time-driven activity-based costing (TDABC) model for performance optimization of offsite manufacturing methods housebuilding. *Construction Management and Economics* **43(8)**: 637–656.

Banks C (2024) Timber-frame modular builder on brink of administration. *Construction News* [online], 18 March. https://www.constructionnews.co.uk/financial/administrations/

timber-frame-modular-builder-on-brink-of-administration-18-03-2024/ (accessed 10/07/2025).

Battersby M (2023a) Ilke Homes enters liquidation leaving £319m debts unpaid. *Housing Today* [online], 20 October. https://www.housingtoday.co.uk/news/ilke-homes-enters-liquidation-leaving-319m-debts-unpaid/5125875.article (accessed 9/7/2025).

Battersby M (2023b) Modular builder Ilke Homes enters liquidation with £319m in unpaid debts. *Building Design* [online], 22 October. https://www.bdonline.co.uk/news/modular-builder-ilke-homes-enters-liquidation-with-319m-in-unpaid-debts/5125887.article (accessed 09/07/2025).

Brennan J, Vokes C, Massey T *et al.* (2024) *Volumetric Modular Construction Research.* Ministry of Housing, Communities & Local Government [online]. https://www.gov.uk/government/publications/volumetric-modular-construction-research (accessed 11/07/2025).

Brown RS (2024) Escalation of commitment and bankruptcy: the case of Katerra Industries. *Journal of Management Policy & Practice* **25(4)**.

Brzęczek T (2020) Optimisation of product portfolio sales and their risk subject to product width and diversity. *Review of Managerial Science* **14(5)**: 1009–1027.

Built Offsite (2023) Modular Builder M-AR Rescued from Administration. [online] 1 December. https://builtoffsite.com.au/news/modular-builder-m-ar-rescued-from-administration/ (accessed 11/07/2025).

Built Offsite (2024) Timber-Frame Modular Builder Lighthouse on the Brink of Administration. [online] 24 March. https://builtoffsite.com.au/news/timber-frame-modular-builder-lighthouse-on-the-brink-of-administration/ (accessed 11/07/2025).

Built Offsite (2025) Declining demand hits UK modular builder TopHat as losses climb to £53.8 million (AUD $101.4 million). [online] 25 January. https://builtoffsite.com.au/news/declining-demand-hits-uk-modular-builder-tophat-as-losses-climb-to-53-8-million-aud-101-4-million/ (accessed 09/07/2025).

Butler R (2023) Ilke Homes falls into administration. *Construction Wave* [online], 30 June. https://constructionwave.co.uk/2023/06/30/ilke-homes-falls-into-administration/ (accessed 09/07/2025).

DiMaggio PJ and Powell WW (1983) The iron cage revisited: institutional isomorphism and collective rationality in organizational fields. *American Sociological Review* **48(2)**: 147–160.

Endole (2024) *Tide Construction Limited – Company Profile.* https://open.endole.co.uk/insight/company/08656212-tide-construction-limited (accessed 31/07/2025).

Gerrard N (2024) UK firm TopHat to wind down modular housing operation. *Construction Briefing* [online], 22 October. https://www.constructionbriefing.com/news/uk-firm-tophat-to-wind-down-modular-housing-operation/8039781.article (accessed 11/07/2025).

Gibbert M, Ruigrok W and Wicki B (2008) What passes as a rigorous case study?. *Strategic Management Journal* **29**: 1465–1474.

Ginigaddara B, Perera S, Feng Y and Rahnamayiezekavat P (2022) Development of an offsite construction typology: a Delphi study. *Buildings* **12(20)**: 1–21. https://doi.org/10.3390/buildings12010020.

Goulding JS, Rahimian FP, Arif M, Akintoye A and Faraj I (2017) Offsite manufacturing in architecture, engineering and construction: an unfolding story. *Welcome to Delegates, IRC 2017*: 862.

Gunawardena T and Mendis P (2022) Prefabricated building systems – design and construction. *Encyclopedia* **2(1)**: 70–95.

Hammad AW, Sutrisna M and Zaman A (2019) Towards optimising OSM implementation in construction projects. In *Offsite Production and Manufacturing for Innovative Construction: People, Process and Technology*. Routledge, Abingdon, pp. 189–222.

Jalil AABD, Nuruddin AR and Mydin AO (2015) A new procurement method for housing projects implementing IBS modular system. In *International Conference on Advances in Civil and Environmental Engineering 2015*, pp. 1–8.

Joseph G, Aboobaker N and Zakkariya KA (2023) Entrepreneurial cognition and premature scaling of startups: a qualitative analysis of determinants of start-up failures. *Journal of Entrepreneurship in Emerging Economies* **15(1)**: 96–112.

Kozlenkova IV, Samaha SA and Palmatier RW (2014) Resource-based theory in marketing. *Journal of the Academy of Marketing Science* **42(1)**: 1–21.

Laister D (2023) Administrators called to Hull modular construction firm. *Business Live* [online], 30 October. https://www.business-live.co.uk/manufacturing/administrators-called-hull-modular-construction-28077682 (accessed 09/07/2025).

Lowe T (2022) Why are we struggling to make modular work?. *Building* [online]. https://www.building.co.uk/focus/why-are-we-struggling-to-make-modular-work/5118795.article (accessed 18/07/2025).

Mai TP, Doan DT and Ghaffarianhoseini A (2025) Utilizing multiskilled resources in addressing labor shortage issues in off-site construction: benefits, challenges and best practices. *Journal of Construction Engineering and Management* **151(4)**: 04025014.

MCHLG (2019) Modern Methods of Construction: Introducing the MMC Definition Framework [online]. https://www.buildoffsite.com/content/uploads/2019/04/MMC-I-Pad-base_GOVUK-FINAL_SECURE-1.pdf (accessed 02/02/2025).

NHS England (2025) NHS Modern Methods of Construction Assessment Tool User Guide [online]. https://www.england.nhs.uk/long-read/nhs-modern-methods-of-construction-assessment-tool-user-guide/ (accessed 31/07/2025).

Offsite Hub (2023) Ilke Homes sinks into administration with most of firm's 1,100 staff set to lose their jobs. [online] 30 June. https://www.offsitehub.co.uk/ilke-homes-sinks-into-administration-with-most-of-firms-1100-staff-set-to-lose-their-jobs/ (accessed 09/07/2025).

Park J, Lew Y and Park B (2020) An organismic approach to multinational enterprise failure. *Management Decision* **59(1)**: 18–31.

PBC Today (2025) TopHat run at a loss for last four years, reveals new report. [online] 21 January. https://www.pbctoday.co.uk/news/mmc-news/tophat-run-at-loss-for-last-four-years-reveals-new-report/147544/ (accessed 11/07/2025).

Pfeffer J and Salancik RG (1978) The external control of organizations: a resource dependence perspective. *Academy of Management Review* **4**: 521–532.

Prior G (2023) Modular contractor M-AR files administration notice. *Construction Enquirer* [online], 19 October. https://www.constructionenquirer.com/2023/10/19/modular-contractor-m-ar-files-administration-notice/ (accessed 11/07/2025).

Razkenari M, Fenner A, Shojaei A, Hakim H and Kibert C (2020) Perceptions of offsite construction in the United States: an investigation of current practices. *Journal of Building Engineering* **29**: 101138.

Roofing Today (2023) Modular Constructor Ilke Homes Administration Forces Supplier Debt Statement. [online] 22 June. https://roofingtoday.co.uk/modular-constructor-ilke-homes-administration-forces-supplier-debt-statement/ (accessed 09/07/2025).

Taylor MD (2022) A definition and valuation of the UK offsite construction sector: ten years on. *International Journal of Construction Management* **22(15)**: 2877–2885.

UK Parliament (2024) *Modern Methods of Construction – What's Gone Wrong?* [online]. Built Environment Committee. https://committees.parliament.uk/work/7950/modern-methods-of-construction-whats-gone-wrong/ (accessed 11/07/2025).

Yin RK (2003) *Case Study Research: Design and Methods*, 3rd ed. Sage Publications, Thousand Oaks, CA, USA.

Zheng R, Li Z, Li L *et al.* (2024) Proposing a lean-optimized scheduling model of mixed-flow prefabricated component production in off-site construction. *Journal of Construction Engineering and Management* **150(8)**: 04024088.

Modern Methods of Construction: Foundations, Processes and Applications

Emmanuel I Daniel and Fidelis A Emuze
ISBN 978-1-83708-437-1
https://doi.org/10.1108/978-1-83708-436-420261015

Chapter 13
Enhancing onsite performance through lean construction processes and approaches: evidence from Europe

William Power

Lean Construction Practitioner, Ireland 0000-0001-5791-846X

Abstract

This chapter aims to evaluate the integration of lean construction methods and combine various approaches to improve project delivery. The objectives of this study are to examine how Categories 6 and 7 of national modern methods of construction (MMC) definition frameworks can be realised on-site by the application of the Last Planner System, Takt and Scrum concepts and principles. This qualitative research uses a case study design on eight projects in Ireland, the UK and Europe. The studies include high-density and low-density residential buildings, a commercial distribution centre, a marine port extension and a pharmaceutical project. Most are new-build constructions; one involves decommissioning a redundant pharmaceutical manufacturing facility and another examines improving design collaboration on a new pharmaceutical plant. Results indicate improved schedule alignment, reduced activity duration and more proactive quality and safety planning. Implementing and combining the Last Planner® System, Takt and Scrum present opportunities to enhance on-site production performance through their theoretical and practical concepts.

Keywords: lean construction; Last Planner System; Takt; Scrum; agile

13.1. Introduction

Despite considerable advancements in off-site product innovation and development in recent decades, there is still widespread dissatisfaction with construction's on-site productivity and efficiency (Barbosa *et al.*, 2017). The sector experiences substantial inefficiency and rework stemming from a proliferation of non-value adding activities and institutionalised waste embedded within traditional delivery processes and models (Oakland and Marosszeky, 2017;

Sarhan *et al.*, 2017). Construction, globally, suffers from challenging relationships, poor productivity and substantial wastefulness; thus costing too much and taking too long (Farmer, 2016; Oakland and Marosszeky, 2017).

Increased adoption of MMC presents numerous potential advantages for professionals within the industry, as well as for society. MMC facilitates the creation of a more sustainable built environment, enhances working conditions and improves productivity within the construction sector (HM Government, 2022; Sweeney, 2024). Furthermore, it enables the expedited delivery of high-quality homes compared to traditional on-site construction methods which can contribute to addressing one of society's biggest concerns: the rapid supply of affordable housing (Sweeney, 2024).

Construction contributes to excessive environmental pollution, production of physical waste, energy consumption, as well as wasteful productive time in its processes (Liu *et al.*, 2020). An Irish Government report (Oireachtas, 2022) estimated that 37% of Ireland's carbon emissions emanate from the built environment – 14% originate from the carbon in building materials and their transport, maintenance, repair and disposal and the other 23% are from heating, lighting and cooling buildings. While O' Hegarty *et al.* (2025) claim there is no broad link between MMC and reduced embodied carbon, Sweeney (2024, p. 11) suggests '… the main benefit of MMC is its potential environmental impact'. Improving construction productivity while increasing efficiency and reducing waste has the potential to drive sustainable development objectives and will positively contribute to UN Department of Economic and Social Affairs Sustainable Development Goal 12.

Construction and project management has predominantly relied on the critical path methodology (CPM) over the past 60 years, but its traditional use has been critiqued for failing to address the needs of production management (Ballard and Tommelein, 2021). Academics and practitioners alike have debated the shortfalls of CPM and have suggested alternative methodologies such as lean, Last Planner System (LPS®), agile, Scrum and Takt for the management of project controls, scheduling, planning and execution (Ballard and Tommelein, 2021; Koskela *et al.*, 2019).

Lean construction is proposed as an antidote to many of the problems inherent in construction project management (Ballard *et al.*, 2020). While lean construction originated as a set of countermeasures to specific problems in construction, it has evolved by developing its own methodologies and by adopting and adapting methods and principles of lean production (Koskela *et al.*, 2019). Integration of different scheduling and planning techniques seeks to address shortfalls in both traditional and innovative methods with the specific objective of increased planning reliability and construction productivity. Enhancing on-site management practices creates a superior environment for incorporation of off-site components on to the project.

National MMC definition frameworks (DHLGH, 2023; HM Gov, 2022; MHCLG, 2019) identify seven categories of MMC – categories one to five relate to off-site manufacturing while six and seven refer to on-site productivity and process improvements. This research presents evidence from on-site implementation of lean construction methods that aim to increase on-site efficiency, work-flow visibility and planning reliability, thus applying to MMC categories six and seven. The objectives of this research are to examine how categories six and seven can be realised on-site by the application of LPS, Takt and Scrum concepts and principles. This chapter refers to case studies in pharmaceutical, residential, commercial and infrastructural projects that adopted collaborative planning methods to improve on-site process improvement.

13.2. Perspectives on lean construction approaches to on-site construction planning

The Construction Playbook was published by the UK Government in 2022 and

> 'focused on getting projects and programmes right from the start. Whether the delivery or refurbishment of a school, hospital or major infrastructure project, the principles and policies in this Playbook will transform how we assess, procure and manage public works projects and programmes' (p.2).

The Construction Playbook sets out what contracting authorities and supply chain should do to facilitate its adoption. While MMC and collaborative working are noted, there is no specific mention of lean practices or philosophies that could address categories six and seven. A recent report on the Irish construction industry (DPER, 2024) posits supporting and developing a robust construction sector is a key objective for regional and national economic development – adoption of lean methodologies like LPS and building information management (BIM) are specifically highlighted in the presented studies.

A production management process specific to construction exists in the form of LPS that achieves reliable flow of work through increasing planning reliability, improving production performance and creating predictable workflow in both design and construction operations (Daniel and Pasquire, 2017). The core functions of LPS are master/milestone schedule, phase/pull planning, look ahead and make-ready process, commitment/weekly work planning, daily huddles/coordination, and learning and action (Ballard and Tommelein, 2021; Daniel *et al.*, 2015).

Recent years have seen an academic-led advancement of Takt planning studies in lean construction. Researchers have developed application methodologies for utilising Takt concepts to enhance construction planning and production management. Takt time applied in construction is a work structuring system to align the production rates of trades by pacing work sequentially through planned zones creating continuous workflow, reliable hand-offs and an opportunity to continuously improve the production system (Frandson *et al.*, 2013). Takt time regards 'space' as a resource to be considered when planning construction projects and designing production operations. Frandson *et al.* (2013) suggest the difference between Takt time planning and other location-based planning methods is the balance between 'work waiting on workers' and 'workers waiting on work'. Takt time has been used successfully in home building, modular home manufacturing and highway construction (Fiallo and Howell, 2012).

Leading lean construction researchers have suggested looking outside of construction to other sectors – for example, the software and IT sectors for inspiration towards improving construction's value offering.

Agile focuses on early and repeated delivery of value and continuous improvement of the processes used to deliver the product while consistently adapting to customer needs (Layton *et al.*, 2020). Agile emerged from the software industry in the early 1990s as a response to the need to modernise project management and has dominated the software and IT sector since publication of the agile Manifesto (Layton *et al.*, 2020; Sutherland and Sutherland, 2014). Scrum of Scrums is a scaled agile technique that offers a way to connect multiple teams who need to work together to deliver complex solutions. It helps teams develop and deliver through transparency, inspection and adaptation, at scale (Layton *et al.*, 2020).

Scrum is the most widely used agile framework and is a flexible, adaptable, empirical, productive and iterative method that uses the ideas of industrial process control theory for the development of software systems (Sanchez and Nagi, 2001). The Scrum framework uses incremental steps daily, as well as on a weekly/monthly basis to complete critical work scope (Streule *et al.*, 2016). Case studies exist of agile and Scrum enhancing design performance (Demir and Theis, 2016) but there is a paucity of examples showing Scrum assisting construction execution. Despite advancements from recent studies there is still a research gap between the alignment of short-term, medium-term and long-term planning.

13.3. Methodological approach

This qualitative research utilises mixed-methods and multiple case studies to address the aim and examine outcomes from the application of LPS, Takt and Scrum on construction projects. A qualitative strategy is suitable for this study based on its attributes identified by Naoum (2012) and Bryman (2012). It enables research in a real-life social setting to understand the current phenomenon, aligning with the study's aims.

Methods include interviews, case studies, action research, observations and participative inquiry (focus groups). Following the initial literature review, eight distinct case studies were conducted. The accumulation of case studies is important for construction's body of research as Flyvbjerg (2006, p.42) asserts that 'a discipline without a large number of thoroughly executed case studies is a discipline without systematic production of exemplars, and that a discipline without exemplars is an ineffective one'. The eight case projects are described in Table 13.1.

The researcher participated as a 'lean facilitator' on all the projects, however the degree of involvement varied from assisting twice weekly on some projects to as little as once every six weeks on other projects. This classified the researcher as being an 'insider' – a privileged position according to Byrne *et al.* (2015), as the researcher is a member of the community being studied.

Reflexivity and research rigour was addressed in accordance with Olmos-Vega *et al.* (2022) to neutralise the influence of personal subjectivity. Bias was mitigated by compliance with research ethics codes, independent co-researchers and distinct sources of evidence to increase

Table 13.1 Projects examined in this study

Case	Value (€M)	Sector	Description	Reference
A	€460 million	Pharmaceutical	Agile in design development	Power *et al.* (2021a)
B	€240 million	Pharmaceutical	Facilitated LPS implementation in pharmaceutical construction	Power *et al.* (2021b)
C	€310 million	Pharmaceutical	Commissioning and qualification (C&Q) – LPS and agile concepts in pharmaceutical commissioning	Power *et al.* (2021c)
D	€175 million	Demolition and decommissioning	Benefits from effective lookahead planning	Power *et al.* (2023)
E	€26 million	Commercial	Scrum complementing LPS	Power *et al.* (2022)
F	€32 million	Residential	Takt enhancing LPS in housing and apartment construction	Power *et al.* (2024a)
G	€44 million	Residential	Integrating LPS, Takt and Scrum to create a comprehensive management system	Power *et al.* (2024b)
H	€310 million	Marine infrastructure	LPS improving schedule adherence on new jetty berth project	Awaiting publication.

Source: Author's creation

construct reliability. The units of analysis in the case study are industry specific recognised metrics. Planned per cent complete (PPC) is utilised in LPS and Takt as a measure of the success of the planning process. Reasons for non-completion (RNC) is another metric in LPS and Takt that serves to categorise and analyse the root causes of missed task execution (Ballard and Tommelein, 2021).

Scrum utilises metrics in burndown (a graphic representation of two factors: the amount of work a team has already completed in each sprint and the amount of time that the team has left to complete that work) and user story points (units of measurement used to estimate the effort required to complete a defined work element) (Layton *et al.*, 2020). Results were triangulated after analysis of interviews, documentation review, observations and focus groups findings.

13.3.1 Literature selection

A systematic integrative literature review encompassing construction, lean construction, LPS, Takt, Scrum and agile was conducted in accordance with Snyder (2019). A time range from

January 2010 to April 2023 (13+ years) was selected as this was considered appropriate to capture the latest and most recent lean research findings and trends. Selected literature was read, critically analysed and emerging themes were evaluated by content analysis in accordance with Creswell and Poth (2016).

The literature review also examined lean construction conceptual models and frameworks in peer-reviewed journals, construction conference papers (International Group for lean Construction (IGLC) and Association of Researchers in Construction Management (ARCOM)) and books.

13.3.2 Sampling

Non-probability sampling principles, in accordance with Bryman (2012), are adopted in the study. To fully support the aim of the research, in accordance with Saunders *et al.* (2016), a purposeful self-selection technique focused on individuals who agreed to participate in the study once they had met specified criteria. The criteria used in selecting the respondents include

(*a*) being directly involved on the project planning for at least 3 months
(*b*) working at a minimum of supervisor level on the project
(*c*) having undertaken training on the specific practice (LPS, Takt, Scrum).

13.3.3 Data collection strategy

The research adhered to the four principles of case study data collection as outlined by Yin (2018), utilising multiple sources of evidence, creating a case study database, maintaining a chain of evidence and exercising caution when using data from electronic sources. The methodology included interviews with key personnel, four focus group sessions, analysis of project documentation, performance data and systematic observational studies from each project, as follows.

- *Interviews*. In qualitative research with an exploratory nature, in-depth or semi-structured interviews should be used (Saunders *et al*., 2016). Semi-structured interviews were selected as they have the advantage that important themes and key questions are covered, but at the same time, questions may be reordered, added or omitted due to the organisational context where the interview takes place (Saunders *et al*., 2016). The purpose of the interviews was to elicit in-depth, personalised data on participants opinions, thoughts and experiences that cannot be obtained through other methods. Forty-nine in-depth interviews were conducted across the eight case projects over a three-year period. Interviewees comprised seven client directors, fourteen project managers, twenty-six supervisors and two lean coaches.
- *Structured observation*. The lean facilitator maintained an 'observation research diary' across the project's duration. This involved recording behaviours, attitudes, moods and observations from morning huddles, planning workshops, meetings, walks, reviews and problem-solving sessions relating to the process implementation. Mason (2002) posits that researchers should go to the natural settings where activities occur and observe what people 'really' do in those settings. According to Leicht *et al*. (2010)

observational studies are particularly useful in construction research as it constitutes so many social interactions; researching how teams work and interact offers understanding into how people lead and make decisions. This method is justified for its consistency and objectivity as it gathers data that is difficult to obtain through surveys or interviews alone.

- *Focus groups*. Following the interviews and analysis of the observation research diary, focus group interviews were conducted on four of the case studies. Interviewing a focus group is often used as an explorative tool in qualitative research, where a group of interviewees is asked about their opinions and perceptions on a specific theme or topic (Saunders *et al*., 2016). Focus groups are justified as they are valuable for understanding group dynamics, observing non-verbal cues and providing flexibility to explore new avenues as they arise.
- *Documentation study*. Comprehensive documentation data was available, comprising schedule and budget reports, revisions of schedules, product goal and road map visuals, pull plans, contract documents, weekly LPS PPC reports, marked up weekly work plans, RNC, Scrum burndown charts, Trello boards data, constraints logs and the lessons learned log. By relying on several independent sources of evidence, the researchers were able to increase the construct reliability of the research (Yin, 2018). Utilising documentation study allowed the researchers to provide a transparent and reproducible record of the research process, give proper credit to original sources and allow others to understand, verify and build upon the work.

13.3.4 Data analysis

The interviews were transcribed and analysed using a thematic analysis approach, as suggested by Braun and Clarke (2006). Emerging data was organised into different themes, and inferences drawn from the emerging themes were checked by triangulation against the literature review findings and against other sources to check their reliability and integrity. Following Creswell and Poth (2016), data analysis involved: organising raw data (transcripts, field notes, images); reading all data; coding by hand or computer; generating themes/descriptions from codes; interrelating themes; and interpreting their meaning. The researcher hand-coded and analysed the data, along with collecting and writing up the findings.

13.4. Results and discussion

This section describes, based on the case studies, the key findings from the data collection instruments employed in the research and how implementation of LPS, Takt and Scrum can advance categories six and seven of MCC. Both theoretical and practice findings are presented.

Analysis of the literature confirms a clear gap in the ability of both CPM and traditional construction management theory to underpin the complete execution of construction projects. Table 13.2 summarises the key elements of the theoretical findings.

Literature asserts CPM and traditional construction management is not equipped to adequately incorporate trade contractors planning inputs to develop near-term lookahead plans. This results in value and efficiency loss on construction projects and underpins the urgent need for process and productivity improvement as espoused in MMC categories six and seven.

Table 13.2 Summarised theoretical elements required for construction delivery

Method	Key theoretical elements: shortfall and/or advantage
CPM and traditional construction management	Forms a network of linked activities that portrays the overall project scope. Intends the initiation and direction of action before it takes place. Uses mathematical modelling and quantitative models. Does not address operations production or incorporate the view below subcontract level.
Lean construction theory – transformation, flow, and value	Seeks smooth flow with minimum variation and addresses waste reduction. Transforms inputs to outputs, strives for continuous flow to improve output rate of the system, and adds value in the form of quality and internal and external customers' needs.
Last Planner System	Looks for reduction of variability through constraint-free tasks, process performance measurement (PPC), examining reasons for task non-completion, creation of an available backlog, using make-ready and lookahead planning.
Takt Planning	Satisfying customer demand rate. Maintain resource use while adjusting zone size and pace of crews. Uses production science to establish continuous flow.
Scrum	Empiricism and lean thinking founded on transparency, inspection and adaptation. Observing and experiencing actual results and status. Accepting change and making change readily. Accommodates planning, execution and control theory.

Source: Author's creation

13.4.1 Case A

An effective design process is a critical input to the smooth flow of construction tasks. Case A introduced Scrum and agile concepts on a pharmaceutical new-build project in the Netherlands. It required design and support services inputs from local offices and regional offices in Ireland, thus necessitating enhanced coordination across different countries and cultures.

Design teams were organised as 'self-performing' Scrum teams using the Scrum of Scrums framework to progress their work packages. Each design discipline had individual Scrum workspaces and specifically worked on predefined and sized batches of work ensuring frequent release of deliverables. This ensured early inspection and alignment with the next-customer's requirements for satisfaction. Figure 13.1 presents the layout of the Scrum of Scrums implementation on the project.

Cumulative stacking of weekly delays resulted in everything becoming urgent, rather than identifying and focusing on what was the critical priority. As part of the Scrum intervention, blockers to the highest priority tasks were identified and escalated for resolution. Figure 13.1 presents how this collaboration and escalation functioned.

Figure 13.1 Scrum of Scrums managing design process and escalations

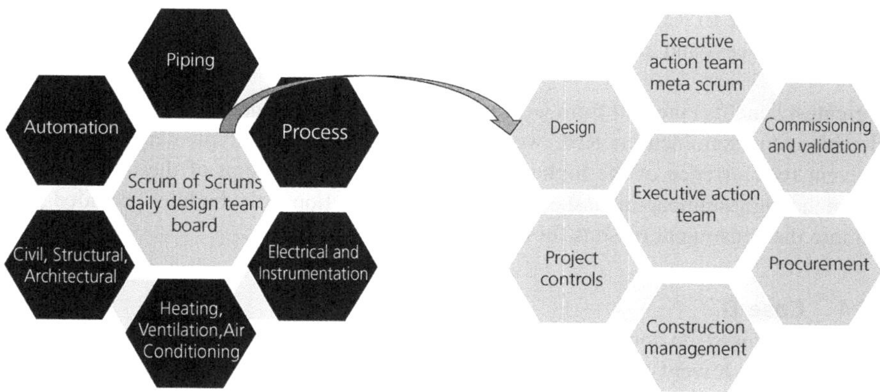

Source: Author's creation

Teams became familiar with the process and, with increased clarity and visibility, could focus wholly on their own deliverables and the number of missed dates began to fall to a weekly average of 28, an 85% improvement. Weekly 'blockers raised' fell to an average of 19, a 66% improvement. As teams and Scrum Masters recognised patterns of repetitive blockers arising, the input providers made enhancements to ensure recurring blockers were permanently resolved. As teams and Scrum Masters recognised patterns of repetitive blockers, input providers made enhancements to ensure these were permanently resolved. Team velocity increased as members developed a better understanding of agile concepts and the Scrum framework. The number of tasks undertaken by the teams expanded from 220 in week one to 355 in week 12, an increase of 62%.

13.4.2 Case B

The introduction of modern methods of construction can be challenging, as changes to embedded traditional processes may encounter resistance. Facilitated change can overcome such resistance, and Case B examines a facilitated implementation of LPS on the construction phase of a pharmaceutical expansion in Ireland. As critical schedule milestones were being missed on the project, and as key handover dates loomed closer, the client agreed to position a dedicated LPS facilitator full-time on the project. LPS was implemented at the project level and the facilitator became the 'external enabler' as referred to in Daniel and Pasquire (2017). Results showed a 12% increase in PPC and schedule reduction from 31 to 27 weeks accruing from the facilitator's coaching and interventions.

13.4.3 Case C

On many large capital projects, high levels of complexity create productivity bottlenecks that can slow handovers between phases, thus leading to further downstream schedule impact (O'Connor and Mock, 2019). Study C examined LPS implementation from the construction phase into the C&Q phase and utilised LPS complemented with Scrum concepts.

From an average of 55% PPC over the first seven weeks of the implementation, productivity steadily rose between weeks 8 to 17 averaging 69%. From week 18 to week 40, PPC stabilised at an average of 77%. This improved stability provided reliability and predictability in the planning process and as interim milestones were achieved, client confidence improved.

The greatest benefits emerged from weekly collection and examination of the RNC data. The leadership team examined the RNC weekly and mitigations were implemented immediately to prevent reoccurrence of the highest impact RNCs. Visualisation of the data channelled focus on the highest impacting issues. Accurate categorisation at the huddles provided greater detail into the constituent reasons for the highest impact RNC.

13.4.4 Case D

This study focused on lookahead planning, a critical but often overlooked function of LPS. The research went 'back to basics' and explored how improved lookahead planning could enhance delivery on a demolition and rehabilitation project. Outcomes showed more emphasis should be placed on work structuring, first run studies and constraints management. A first run study showed a 35% increase in concrete crusher output and an applied constraints management process showed over 100% increase in constraints raised and closed on time. Visualised workflow through defined zones enhanced communication with the entire team.

13.4.5 Case E

LC literature extolls the virtues of LPS however, numerous 'blind-spots' have also been highlighted (Aslam *et al.*, 2020; Ballard and Tommelein, 2021). Study E reported on a 20-week implementation where Scrum complemented LPS on a food distribution centre expansion that sought to achieve a critical building weathertight milestone. The team had been using LPS but were still experiencing problems at crew level relating to coordination and workface organisation. Scrum complemented the LPS at crew level daily interfaces and average PPC from week 27 to week 46 was 87%; this was an 11% PPC increase from the first 26 weeks average PPC. Comparison of RNC from post-week 27 data (with Scrum) with pre-week 27 data (without Scrum) pointed to a reduction in scheduling and coordination issues at interface handoffs between crews.

13.4.6 Case F

Previous studies have suggested combining LPS and Takt to address shortfalls however, there is a paucity of examples showing how this would work. Study F combined Takt planning with LPS on two residential projects in Dublin, Ireland. One was a low-density development with 40 housing units in 15 individual blocks and a high-density element of 8 duplex units with an apartment block of 24 units over 4 floor levels. The other was a high-density residential development on a greenfield urban site. The overall project had 1100 apartments and duplex units in 10 individual blocks varying from 3 to 6 floors in height. This study focuses on 2 duplex blocks, each with 8 apartments on the ground floor with 8 x 2 storey houses overhead and an apartment block with 46 apartments over 5 floors. The application of Takt to complement LPS resulted in a 13% increase in PPC on the first project and 12% increase on the second. Improved cycle time showed an average of 11% schedule reduction since the implementation commenced.

13.4.7 Case G

The success from applying Scrum and Takt to enhance the LPS implementation in case F, prompted a follow-on study to synthesise all three methods into a single construction planning system. Study G was conducted on a new-build residential complex within a development that had been under construction for 18 months. The overall project consisted of 1800 residential units comprising apartments and two storey duplex units. This research focused on three blocks, six storeys high, with 12 apartments per floor – a total of 216 units to be handed over as a single phase.

The first nine weeks of the implementation focused on LPS alone and the average PPC was 72% per week. Takt planning was added and up to week 18 PPC had improved to an average of 83% per week. Scrum was introduced to assist the constraints process at week 12 and results became evident by week 18. From week 18 to week 40 the average PPC raised to consistently above 90% and averaged at 93% per week. This is a critical finding and points to the importance of the diligence required to ensure an effective constraints identification and resolution process. This in turn offered value in the form of stabilised inputs, increased Tasks Make Ready (TMR) and resulting improved commitment planning with higher PPC. The impact of Scrum assisting the constraints and TMR process brought an 11% PPC increase – when PPC is already in the mid-80s it is usually challenging to achieve additional PPC gains. Therefore, the 11% gain is significant. Utilising LPS, Takt and Scrum in an integrated Hybrid agile-lean Production Planning and Control System (Power and Sinnott, 2025) resulted in weekly task completion rising from the industry average of 54% to a consistent and reliable 93%, a 39% increase in planning performance effectiveness over traditional methods.

Case H - Studies exist of implementing LPS on infrastructural projects: Daniel *et al.* (2016) examined highways in UK and Daliri *et al.* (2021) reported on the Minnevika railway bridge project in Norway. However, there is a dearth of examples from marine infrastructure projects. Study H examined the implementation of LPS on a marine project for the creation of a new berth quay wall along the main Dublin port channel and a 290m jetty with nine marine platforms. By implementing a 12-week lookahead and a diligent constraints identification and resolution process, task execution performance increased from less than 60% pre-implementation to consistently over 86% weekly. Advance identification of tasks with incomplete inputs allowed postponement and subsequent reallocation of expensive marine plant to work on alternative 'ready' tasks.

13.4.8 How lean methods improved on-site performance

The results from the presented case studies illustrate how implementation of LPS, Takt and Scrum concepts can improve construction execution on projects. Literature highlights there is a paucity of specific instruction relating to how construction production should be planned and managed. This leads to ad hoc practices and negates the improvement offered by most aspects of MMC as site-based processes are not sufficiently modernised and 'ready' to maximise the advantage offered by MMC innovations. PMI PMBOK® Guides do not offer prescriptive detail on how construction operations should be executed; it rests with teams and management at practice level to apply processes to deliver construction projects. Historically, tacit knowledge would have been developed and retained within organisations but since the introduction of management contracting and the 'hollowed-out' firm, such competency has been lost in the construction domain (Green, 2023). Therefore, a distinct process gap emerges which tends to be filled by the aforementioned ad hoc practices that have no clear theoretical or practical foundation. Lean construction offers a solution to this issue, with detail and definition being advanced with LPS, Takt and Scrum. In case A, concepts of agile and Scrum ensured the team were always working on the highest priority items, ensuring focused attention on the value-generating deliverables. The role of the LPS facilitator in case B ensured all functions of LPS were effectively implemented, leading to enhanced project performance by following a distinct series of planning routines.

Flow and Value theories are brought to the crew level of operations by LPS. By approaching construction planning at the milestone level and then breaking it down into distinct more granular planning phases, LPS brings structure by performing specific functions to accomplish the purpose of planning and control (Daniel and Pasquire, 2017; Hamzeh *et al.*, 2016). The resulting actions of using make-ready and lookahead planning to create constraint-free tasks, measuring process performance through PPC, examining reasons for task non-completion, ensuring improvements are implemented and the creation of an available backlog, apply the theory to practice. Cases C and D ensured an effective lookahead planning process assisted identification and removal of constraints to create a smooth workflow for commissioning and demolition teams. This develops an environment where prepared tasks are available to the crews at the right time order and place to allow production progress. Striving to eliminate variation and variability is a distinct affront to the traditional way construction operations have been executed. Case H increased delivery performance and plant utilisation by ensuring a ready backlog of constraint-free tasks was available to allow consistent execution.

The introduction of Takt theory to construction planning creates a different mindset to that experienced in traditional construction project management. Planning now addresses customer demand rate – on the construction project the internal customer may be the next trade in line. As small work-batch and zone sizes are introduced, trades can become nervous about continuity of work should variation occur. This requires trust and confidence in the planning and commitment process while also having the fallback of an LPS workable backlog to buffer any breakdown in the Takt plan. Maintaining resource use while adjusting the zone size and pace of crews is a key component of Takt planning (Frandson *et al.*, 2013). This balances crew size for durations of the Takt plan and when a level of maturity arises in the implementation, stability ensues in production, achievement and PPC. Cases F and G combined LPS and Takt and the scientific production concepts introduced increased task execution by focusing on small batch sizes and preparing tasks to be worked on. Takt introduces mathematical and production science concepts to establish continuous flow. Strategic buffer placement and under-utilisation of crews is a major difference from traditional thinking and this can be one of the principal challenges encountered during a Takt implementation. Some clients examined compensation for under-utilisation of crews and resulting output restriction to guarantee resource availability. Resultingly, the pursuit of stability and reliability will encounter obstacles, but workarounds are possible by negotiation and compromise.

Scrum theory works with any project, provided the work to be done can be listed and prioritised. Scrum's empiricism and lean thinking introduces transparency which lets clients decide where money is best allocated and allows changes, inspection and adaptation as the project proceeds. While traditional project and construction management thinking seeks to resist change, Scrum differs. Scrum welcomes change, makes change and is equipped to work through the effects of change to reach the most optimal outcome for all parties (Layton *et al.*, 2020). Scrum allows the team to recognise and develop course-correcting issues and solutions early. Scrum was effectively used in case A to manage the design coordination process on a large design project. Scrum complemented LPS on case E and both LPS and Takt on case G to manage the constraints identification and removal process. Like LPS, Scrum observes daily and weekly progress and by experiencing actual results and status, can offer rapid solutions to unexpected variation and variability. This concept is critical to ensuring Takt and LPS plans have the best opportunity of staying on track. Scrum accommodates planning, execution and control theory and therefore becomes a secure 'container' where LPS and Takt can function within. Figure 13.2 presents how Scrum, LPS and Takt can work together in a synthesised production management system.

Figure 13.2 The hybrid agile lean production system

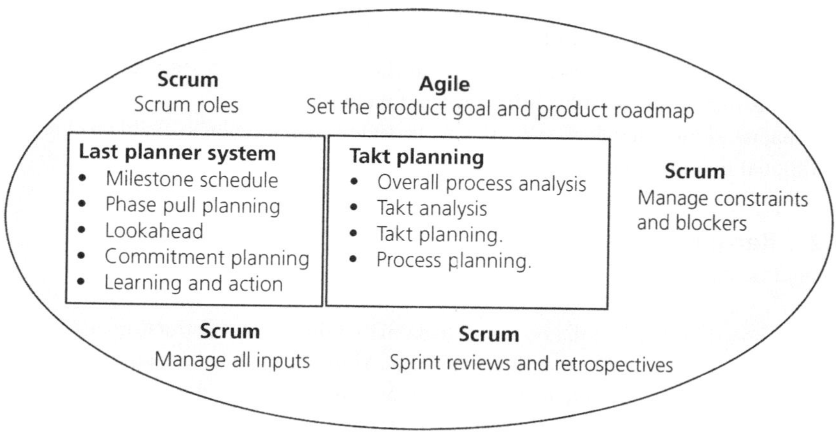

Source: Power and Sinnott (2025)

Several distinct factors emerged that supported the successful application of the lean methods. These have also been highlighted in previous research (Daniel and Pasquire, 2017; Hamzeh *et al.*, 2016). Visible leadership support is essential to support the necessary modification to work practices and overcome resistance to change. This should also be reflected from the procurement stage in the contract conditions, where a collaborative working culture is promoted rather than traditional adversarial practices. A clear sense of urgency or 'need' must be communicated to the team and stakeholders regarding the importance of the lean implementation. Understanding the desired behavioural and social context to support collaborative engagement and planning is critical and it may be necessary to engage a competent Facilitator to create and initiate this environment. Effective training and coaching on the lean methods is also a key enabler of success as it can be easy for teams to abandon implementations early prior to witnessing improvements.

13.5. Implications for practice and research

Based on the research outcomes, the following recommendations are proposed for the construction sector to complement faster adoption of MMC practices. As the MMC community progresses Categories one to five, it is important to engage and align with the lean construction community to ensure rapid and effective adoption of practices to ensure Categories six and seven are progressed in parallel. Lean construction methods are proven on construction projects and their integration with MMC creates a stronger delivery model and solution for many of construction's embedded issues. Future research should focus on aligning the MMC and lean construction approaches to maximise their potential and benefits.

13.5.1 Recommendations for the construction sector

The recommendations of the study for the construction sector are as follows.

- Construction's representative bodies should promote and fund-support the required change in behavioural and working practices required for wider lean construction adoption and integration of MMC.
- Construction projects and site management should be educated, trained and certified on lean construction philosophy, principles and methods.
- Companies should develop strategic lean implementation plans to enable a change from traditional to a more integrated lean approach to project execution.

13.5.2 Recommendations for construction clients

The recommendations of the study for construction clients are as follows.

- Clients should recognise the potential from mandating lean construction and its methods in the construction delivery process. Aligning MMC and lean construction practices will bring advantage. Contracts should incentivise and accommodate more collaborative and less adversarial working practices.
- Repeat clients of construction should invest in comprehensive learning and knowledge-building from lessons and errors that occur in project delivery. Implementing lean construction methods can ensure known lessons will be addressed.

- Clients, projects, companies and practitioners often attempt their own techniques at implementing lean. Clients should develop their own 'single best way' of creating a sustainable lean implementation that offers value at all levels.

13.5.3 Recommendations for academic research

The recommendations of the study for future academic research are as follows.

- Future studies should examine further refinement of the integration of lean planning methods and experiment with technological options to digitise construction operations planning. Identifying variation and variability should be improved with artificial intelligence and an integrated planning system should prioritise creating 'ready' tasks for work teams. Enhancing existing production planning and control processes will contribute to more efficient and effective construction execution.
- PMI has adopted agile concepts in PMBOK® 7th edition and researchers should examine how the concepts of agile and Scrum can enhance construction production operations.
- Evaluating issues MMC experience on projects should initiate lean problem-solving exercises to develop a future-state combined MMC and lean construction production system.

13.6. Conclusions

The objectives of this research are to examine how MMC categories six and seven can be realised on-site by the application of Last Planner System, Takt and Scrum concepts and principles. A considerable contribution has been made to prevailing construction practices by conducting eight studies across the residential, pharmaceutical, marine and commercial construction sectors.

The study found that introducing Takt to LPS demonstrated how both can work together and lead to enhanced outcomes. Previously, the industry was reliant on literature assessments of how both complement one another. Introducing agile and Scrum onto live construction projects is a major step forward in trying to address the variation in inputs that blights traditional construction planning. This research presented how Scrum can manage all inputs and offers a simple solution to constraints identification and management, identified as a shortfall in traditional practice.

Full benefits of MMC cannot be achieved without considerable change in the way construction projects are planned and managed. Deep adoption of lean construction practices will ensure MMC can penetrate the construction sector at greater speed and effectiveness. Wider adoption of lean construction will create a 'pull' process with increased preparedness to receive MMC solutions. This will reduce some of the barriers that exist in the current 'push' system of introducing MMC onto ill-prepared people, processes and projects.

Acknowledgements

The author would like to acknowledge the contributions of their co-researchers on the case studies – Dr Derek Sinnott, Dr Patrick Lynch, Chris Solorz and Aidan Mullin. The author also recognises the contributions of their colleagues in Arcadis, Cork Operations, Ireland – Aiden O Dwyer, Eddie Doyle and David McCarthy.

REFERENCES

Aslam M, Gao Z and Smith G (2020) Development of innovative integrated last planner system (ILPS). *International Journal of Civil Engineering* **18(6)**: 701–715.

Baker SE, Edwards R and Doidge M (2012) *How many qualitative interviews is enough? Expert voices and early career reflections on sampling and cases in qualitative research*. Working paper. https://eprints.ncrm.ac.uk/id/eprint/2273/ (accessed 27 April 2026)

Ballard G and Tommelein I (2021) 2020 Current Process Benchmark for the Last Planner System of Project Planning and Control. *Lean Construction Journal*: 53–155.

Barbosa F, Woetzel J and Mischke J (2017) *Reinventing Construction: A Route of Higher Productivity*. McKinsey Global Institute.

Braun V and Clarke V (2006) Using thematic analysis in psychology. *Qualitative Research in Psychology* **3(2)**: 77–101.

Byrne E, Brugha R, Clarke E, Lavelle A and McGarvey A (2015) Peer interviewing in medical education research: experiences and perceptions of student interviewers and interviewees. *BMC Research Notes* **8(1)**: 1–11.

Creswell J and Poth C (2016) *Qualitative Inquiry and Research Design: Choosing Among Five Approaches*. Sage Publications, Thousand Oaks, CA, USA.

Daliri S, Young BK and Lædre O (2021) Last Planner® System on The Minnevika Bridge Project. *Proc. 29 th Annual Conference of the International Group for lean Construction (IGLC29)* (Alarcon LF and González VA (eds)), Lima, Peru, pp. 757–766, 10.24928/2021/0167, online at iglc.net.

Daniel E and Pasquire C (2017) *Last Planner System Path Clearing Approach (LPSPCA): An Approach to Guide; Clients, Main Contractors and Subcontractors in the Implementation of the LPS*. Research report for external body. Nottingham: Nottingham Trent University.

Daniel E, Pasquire C and Dickens G (2015) Exploring the implementation of the Last Planner® System through IGLC community: Twenty-one years of experience. In *Proceedings of the 23rd Annual Conference of the International Group for lean Construction*. Perth, Australia, 153–162.

Daniel E, Pasquire C and Dickens G (2016) Exploring the factors that influence the implementation of the last planner system on joint venture infrastructure projects: a case study approach. In: *Proc. 24th Ann. Conf. of the Int'l. Group for lean Construction*, Boston, MA, USA. pp. 23–32.

Demir S and Theis P (2016) agile design management–the application of Scrum in the design phase of construction projects. In *Proceedings of the 24th Annual Conference of the International Group for lean Construction*, (IGLC). Boston, MA, USA.

DHLGH (2023) *Modern Methods of Construction, Introductory Guide. Modern Methods of Construction – A Keyway of Helping to Address Ireland's Housing Needs*. Department of Housing, Local Government and Heritage. https://www.gov.ie/en/publication/e5e78-modern-methods-of-construction-introductory-guide/.

DPER (2024) *Project Ireland 2040, Build 2022: Construction Sector Performance and Capacity*. Department of Public Expenditure and Reform, https://www.gov.ie/en/publication/ceb9d-build-2024-construction-sector-performance-and-capacity/.

Farmer M (2016) *Modernise or Die: Time to Decide the Industry's Future. The Farmer Review of the UK Construction Labour Model.* Construction Leadership Council, London, UK.

Fiallo M and Howell G (2012) Using production system design and Takt time to improve project performance. In *Proceedings of the 20th Annual Conference of the International Group for lean Construction (IGLC 20)*, San Diego, CA, USA.

Flyvbjerg B (2006) Five misunderstandings about case-study research. *Qualitative Inquiry* **12(2)**: 219–245.

Frandson A, Berghede K and Tommelein I (2013) Takt time planning for construction of exterior cladding. In *Proceedings of the 21st Annual Conference of the International Group for lean construction*, Fortaleza, Brazil. pp. 21-2.

Green SD (2022). Modern methods of construction: reflections on the current research agenda. *Buildings and Cities* **3(1)**: 653–662, 10.5334/bc.265.

Green S (2023) *Making Sense of construction improvement.* Routledge, Abingdon, UK.

Hamzeh F, Kallassy J, Lahoud M and Azar R (2016) The first extensive implementation of lean and LPS in Lebanon: results and reflections. In *Proc. 24th Annual Conference of the International Group for lean Construction*, Boston, USA, pp. 33–42.

HM Government (2022) The Construction Playbook. *Cabinet Off* **1**: 1. [Google Scholar]

Koskela L, Ferrantelli A, Niiranen J, Pikas E and Dave B (2019) Epistemological explanation of lean construction. *Journal of Construction Engineering and Management* **145(2)**: 04018131.

Layton M, Ostermiller S and Kynaston D (2020) *agile Project Management for Dummies.* John Wiley and Sons, Hoboken, NJ, USA.

Leicht R, Hunter S, Saluja C and Messner J (2010) Implementing observational research methods to study team performance in construction management. *Journal of Construction Engineering and Management* **136(1)**:76–86.

Liu J, Yi Y and Wang X (2020) Exploring factors influencing construction waste reduction: a structural equation modeling approach. *Journal of Cleaner Production* **276**: 123185, 10.1016/j.jclepro.2020.123185.

Liu M, Ballard G and Ibbs W (2010) Workflow variation and labor productivity: case study. *Journal of Management in Engineering* **27(4)**: 236–242.

Mason J (2002) *Qualitative Researching.* 2nd edn. Sage Publications, London, UK.

MHCLG Modern methods of construction, Introducing the MMC Definition Framework. Minist. Hous. Communities Local Gov. (2019). https://www.cast-consultancy.com/wp-content/uploads/2019/03/MMC-I-Pad-base_GOVUK-FINAL_SECURE.pdf (accessed 02/3/ 2025).

Naoum S (2012) *Dissertation Research and Writing for Construction Students.* Routledge, London, UK.

O'Connor J and Mock B (2019) Construction, commissioning and startup execution: problematic activities on capital projects. *Journal of Construction Engineering and Management* **145(4)**: 04019009.

O'Hegarty R, McCarthy A, O'Hagan J, *et al.* (2025). Understanding the embodied carbon credentials of modern methods of construction. *Buildings and Cities* **6(1)**: 70–89, 10.5334/bc.515.

Oakland J and Marosszeky M (2017) *Total Construction Management: lean Quality in Construction Project Delivery*. Taylor & Francis, London, UK, 10.4324/9781315694351.

Oireachtas (2022) *Embodied Carbon in the Built Environment, Report, Joint Committee on Housing, Local Government and Heritage*. https://data.oireachtas.ie/ie/oireachtas/committee/dail/33/joint_committee_on_housing_local_government_and_heritage/reports/2022/2022-10-14_report-on-embodied-carbon-in-the-built-environment_en.pdf (accessed 29/3/2025).

Olmos-Vega F, Stalmeijer R, Varpio L and Kahlke R (2022) A practical guide to reflexivity in qualitative research: AMEE Guide No. *149. Medical Teacher* **45(3)**: 241–251, 10.1080/0142159X.2022.2057287.

Power W and Sinnott D (2025). Integrating agile, Scrum and LPS principles: the development of a Hybrid agile lean (HAL) production planning & control system. In *Proceedings of the 33rd Annual Conference of the International Group for lean Construction* (IGLC33) (Seppänen O, Koskela L and Murata K (eds.)), pp. 1–10, 10.24928/2025/0254.

Power W, Sinnott D and Lynch P (2021a) Accelerating construction project delivery using agile's integrated system of principles and methods. In Thirty-Eighth Annual Conference ARCOM, Glasgow, Scotland, p. 481.

Power W, Sinnott D and Lynch P (2021b) Evaluating the efficacy of a Dedicated Last Planner® System Facilitator to enhance construction productivity. *Construction Economics and Building*, **21(3)**: 142–158.

Power W, Sinnott D and Mullin A (2021c) Improving commissioning and qualification delivery using Last Planner System. *Lean Construction Journal* 36–52.

Power W, Sinnott D and Lynch P (2022) Scrum complementing last planner system – a case study. In *Proceedings of the 30th Annual Conference of the International Group for lean Construction*, Edmonton, Canada, pp. 175–186, 10.24928/2022/0120.

Power W, Sinnott D and Lynch P (2024a) Takt complementing Last Planner® System on residential Construction projects. *Construction Economics and Building* **24(4/5)**, 10.5130/AJCEB.v24i4/5.8846.

Power W, Sinnott D and Lynch P (2024b) Synthesizing Last Planner® System, Takt and Scrum theory and practice to enhance construction project delivery. *Lean Construction Journal*, 16–28, 10.60164/3xymkr1ws.

Power W, Sinnott D, Lynch P and Solorz C (2023) An evaluation of the lookahead planning function in Last Planner® System. In *Proceedings of the 31st Annual Conference of the International Group for lean Construction*, Lille, France, pp. 1337–1349, 10.24928/2023/0217.

Sanchez L and Nagi R (2001) A review of agile manufacturing systems. *International of Production Research* **39(16)**: 561–600.

Sarhan S, Pasquire C and King A (2017) The concept of Institutional Waste within the construction industry: a potential theoretical framework. *Lean Construction Journal* **14(1)**: 12–24.

Saunders M, Lewis P and Thornhill A (2023) Research Methods for Business Students. 9th edn. Pearson, Harlow, UK.

Snyder H (2019) Literature review as a research methodology: an overview and guidelines. *Journal of Business Research* **104**: 333–339.

Streule T, Miserini N, Bartlomé O, Klippel M and García De Soto B (2016) Implementation of Scrum in the construction industry. *Procedia Engineering* **164**: 269–276, 10.1016/j.proeng.2016.11.619.

Sutherland J and Sutherland J (2014) *Scrum: The Art of Doing Twice the Work in Half the Time*. Crown Currency.

Sweeney R (2024) *Modern Methods of Construction: barriers and benefits for Irish housing*. Chartered Institute of Building, Dublin, Ireland.

Yin RK (2018) *Case Study Research and Applications: Design and Methods*. 6th edn. Sage Publications, Thousand Oaks, CA, USA.

Emmanuel I Daniel and Fidelis A Emuze
ISBN 978-1-83708-437-1
https://doi.org/10.1108/978-1-83708-436-420261017

Chapter 14
Revolutionising construction in China: lean, off-site and digital innovations

Yanqing Fang[1], Emmanuel I. Daniel[2] and Shang Gao[3]

[1]*Tianjin University of Finance and Economics, Tianjin, China* 0000-0003-2989-9737
[2]*The University of Wolverhampton, Wolverhampton, United Kingdom* 0000-0002-5675-1845
[3]*The University of Melbourne, Melbourne, Australia* 0000-0002-4161-5592

Abstract

The construction industry in China has experienced significant changes, driven by the adoption of modern methods of construction (MMC), including building information modeling (BIM), lean construction (LC) and offsite construction techniques. These approaches, both individually and combined, have played a key role in tackling issues such as inefficiencies, high costs, environmental impacts and fragmented processes commonly seen in traditional construction practices. This chapter examines the connections between LC, digitalisation through BIM and offsite construction methods, emphasising their use and influence on improving construction processes in China. Based on insights from two questionnaire surveys, main findings show that BIM adoption is more advanced than LC in China's prefabricated housing projects. BIM has helped improve design, visualisation and construction coordination. Meanwhile, lean practices like total quality management and standardisation have shown potential in boosting production efficiency and reducing waste. However, LC remains less developed, with tools like the Last Planner System and just-in-time gaining limited recognition. Many digital construction technologies are strongly linked to LC capability indicators. The chapter provides strategic advice for construction professionals to adopt digital technologies and LC practices. Construction companies are encouraged to strategically invest in LC methods and digital tools to better adapt to the changing industry landscape.

Keywords: BIM; lean construction; off-site construction; digitalisation; modern methods of construction

14.1. Introduction

As a developing country with a large population and a substantial housing stock, China has experienced sustained and rapid growth in residential construction over the past two decades. In line with the '14th Five-Year Urban Renewal Plan', China is in the late acceleration phase of rapid urbanisation. During this period, various new housing construction related issues and challenges have emerged (Wuni *et al.*, 2021). The Chinese Government has introduced the 'high-quality housing' concept, which sets higher standards for housing quality, efficiency, sustainability and management processes. This has drawn the construction industry and academia's attention to modern construction methods (MMC) that can enhance construction standards and housing quality.

China's construction sector has undergone significant changes, mainly due to the adoption of MMC, including building information modelling (BIM), lean construction (LC) and off-site construction technologies. Their use has been proven in studies to effectively tackle challenges like inefficiency, high costs, environmental impact and process fragmentation in traditional construction (Daniel *et al.*, 2024a). As a modern industrial production model, off-site construction technology is considered a promising solution for achieving higher-quality, innovative products and improved management processes (Wu *et al.*, 2021a).

The concept of off-site construction is implemented by breaking down the entire structure into multiple prefabricated components that are manufactured in a controlled factory setting and subsequently transported to the construction site for final assembly (Zaalouk *et al.*, 2023). Off-site construction ensures better product quality due to standard quality control and supervision in factory settings, improving production efficiency and shortening construction time (Goh and Goh, 2019; Shahpari *et al.*, 2020). Consequently, off-site construction technology is increasingly developed to meet sustainability and China's housing demands (Ji *et al.*, 2019). In China, it is widely used in prefabricated construction, especially in housing.

The Chinese Government has set an ambitious target for the prefabricated construction industry: approximately 30% of all new buildings are expected to be prefabricated within the next 10 years (CPC Central Committee and the State Council of China, 2016). Some developed areas in China, like Shanghai, have achieved higher prefabrication standards. The current state of China's prefabricated housing construction (PHC) industry is decentralised and diversified, making coordination difficult (Zhai *et al.*, 2014).

BIM is a paragon of construction digital innovation and a robust tool for boosting off-site construction tech (Li *et al.*, 2018a). Integrating BIM with off-site construction technology can save resources and reduce unnecessary rework (Ji *et al.*, 2019). LC represents the application of lean principles within the construction industry. It offers a method to clearly define value, minimise unnecessary interference to enhance efficiency and achieve more with less. Lean philosophy focuses on aligning products more closely with user needs while eliminating waste and creating value (Fang and Gao, 2024).

LC has proven effective in smoothing processes, improving coordination and tackling the fragmentation in off-site construction tech (Heravi and Firoozi, 2017). Recent studies have focused on actively integrating BIM and LC in prefabricated construction (Gbadamosi *et al.*, 2019). BIM enhances communication among stakeholders (Wang *et al.*, 2021), while LC addresses inherent construction process issues (McHugh *et al.*, 2019). 'KanBIM', created by Sacks *et al.* (2010), combines with Last Planner System (LPS) – the best-known lean construction technique – on the BIM platform to shorten feedback cycles. Nascimento *et al.*

(2017) integrated BIM and LC to optimise planning and control, thereby reducing construction time.

BIM, LC and off-site construction have all attracted significant attention in both academic and industrial circles. In China, some people have already begun to explore the integration of any two of these three practices – for example, some scholars have conducted research on the application of BIM in China's prefabricated construction field, covering multiple aspects such as comprehensive benefits (Ji *et al.*, 2019), implementation barriers (Wu *et al.*, 2021b), energy consumption (Xu *et al.*, 2019) and carbon emissions (Li *et al.*, 2021).

However, relatively speaking, although Gao and Low (2014) have mentioned that LC practices may be more suitable for prefabricated environments, research focusing on the application of LC in China's prefabricated construction is still relatively scarce. Some scholars have proposed management systems and models for the integration of LC and BIM in China's prefabricated construction (Chen and Ban, 2020; Li *et al.*, 2020), as well as individual case analyses of the integrated application of LC and BIM in prefabricated construction (Yao, 2017).

At present, the status of LC and BIM integration in China's prefabricated construction is still unclear and most existing research remains at the conceptual stage. Therefore, it is necessary to have a thorough understanding of the implementation status of LC and BIM integration in China's PHC, as well as the factors that affect their integration in PHC. Therefore, the research question of this chapter is: *What is the level of adoption of key MMC (BIM, off-site construction, LC) in China?*

This study conducts two surveys. The first focuses on the application status of LC and BIM in China's PHC and their effects on project delivery. The second examines the adoption levels of major construction digital technologies (such as BIM, cloud computing and Internet of Things (IoT)) in China and their correlation with LC capability indicators. The study aims to clarify the adoption and development trends of BIM, off-site construction technology and LC as key components of MMC, identify imbalances in technology adoption and provide data support for the promotion of MMC.

Meanwhile, the correlation analysis reveals the synergistic effects between digital technologies and LC practices, which help accurately assess the adoption level of LC in China's construction industry and promote the actual progress of China's construction industry in the field of MMC.

14.2. Perspectives and concepts

In China, MMCs are increasingly becoming a key driving force for the transformation and upgrading of the construction industry. MMC encompasses a variety of cutting-edge technologies, such as LC, off-site construction and digital technologies centred on BIM (Sánchez-Garrido *et al.*, 2023; Forbes and Ahmed, 2010; Pan *et al.*, 2007). These methods and technologies have significantly improved the quality and sustainability of construction while enhancing efficiency. This section focuses on the research background of LC, off-site construction and BIM in China.

14.2.1 LC in China

LC is a project management philosophy that focuses on reducing waste and enhancing value. It emphasises the use of methods such as continuous process improvement, enhanced team collaboration and just-in-time delivery techniques throughout the entire construction project life cycle to improve efficiency and maximise stakeholder value (Daniel *et al.*, 2024b; Lean Construction Institute, 2012).

In China, LC has been highly valued and actively supported by national policies as an essential means of promoting the transformation and upgrading of the construction industry. In 2016, the 'Several Opinions on Further Strengthening Urban Planning, Construction and Management' issued by the Central Committee of the Communist Party of China and the State Council explicitly proposed that within about 10 years, the proportion of prefabricated buildings in new constructions should reach 30% and that LC should be used to accelerate the industrialisation of construction (Tong and Niu, 2016). This provided a clear direction for the development of LC.

In 2020, the Ministry of Housing and Urban-Rural Development and other departments further clarified the requirements for promoting LC in the 'Several Opinions on Accelerating the Development of New-Type Construction Industrialisation', including optimising construction techniques and methods, and innovating construction organisation. In addition, Jiangsu Province listed LC as the first advocated development method in the 'Jiangsu Construction Action Outline 2025' and promoted its implementation through policy support and demonstration projects. These policy documents not only provide a basis for the promotion of LC in China but also lay a solid foundation for its wide application in the construction industry.

However, it has been shown that LC in China still faces many problems, especially the insufficient promotion of LC tools (such as the limited application of LPS and just-in-time (JIT)), rigid organisational culture and the lack of standardisation (Fang and Gao, 2024). Some scholars have explored the successful application of LC in specific projects through case studies and put forward corresponding suggestions for improvement (Xing *et al.*, 2021). Overall, although China has achieved specific results in the field of LC, there are still many challenges to be overcome to achieve its comprehensive promotion and in-depth application.

14.2.2 Off-site construction in China

Off-site construction refers to the practice of manufacturing components in a factory environment and then transporting them to the site for assembly. Specific methods include modular construction, prefabrication, design for manufacturing and assembly (DfMA) and volumetric construction (Daniel and Oshodi, 2023; Daniel *et al.*, 2024b). The application of prefabrication in China has gained high recognition from the government due to its significant advantages in reducing waste, lowering environmental pollution and promoting the modernisation of the construction industry (Fang *et al.*, 2023).

As early as the late 1980s, China proposed the concept of 'residential industrialisation'. In 1998, the National Center for the Promotion of Residential Industrialisation was established and, in the following year, the 'Several Opinions on Promoting the Modernisation of the Residential Industry and Improving the Quality of Residences' was issued.

In recent years, the Chinese Government has been vigorously promoting the development of prefabricated modular construction based on off-site construction and has introduced a series of supportive policies – for example, the 'Several Opinions on Promoting the Development and Reform of the Construction Industry' in 2014 clearly pointed out that promoting construction industrialisation is one of the key measures to transform the industry's development model (Jiang *et al.*, 2018). In addition, the Ministry of Housing and Urban-Rural Development (MOHURD) identified demonstration cities, industrial bases and projects for prefabricated construction in two batches in 2017 and 2020 (Fang *et al.*, 2023).

Although China's prefabricated modular construction has entered a period of rapid development, its off-site construction system is still in a key stage of transformation from traditional production models to modern industrial models. It faces many urgent problems to be solved, including the lack of a standardisation system, uneven production technology and product quality, weak information management, insufficient integration of the entire project process, relatively high costs and insufficient understanding of off-site construction-based prefabricated construction by management (Zhai *et al.*, 2014).

14.2.3 BIM in China

In China, BIM technology has made significant progress in terms of policy, theory and practice.

In terms of policy, the Ministry of Housing and Urban-Rural Development of China issued the 'Guiding Opinions on Promoting the Application of Building Information Modelling' in 2015, which clarified the development goals and implementation paths of BIM technology in the construction industry. It provided a policy basis for the promotion of BIM technology.

In terms of theoretical research, scholars have carried out extensive and in-depth research on BIM technology – for example, Wang *et al.* (2023) proposed a conceptual framework that explains the mechanism of using BIM to promote dispute management throughout the life cycle of construction projects. Tu *et al.* (2023) developed a system dynamics model to investigate the implementation level of BIM in the construction phase.

In terms of practical application, many construction companies have actively explored the implementation of BIM technology in real-world projects. Overall, policy guidance, theoretical support and practical exploration promote each other and jointly drive the rapid development of BIM technology in China's construction industry.

However, the promotion of BIM technology in China still faces some challenges, such as capital-related factors and insufficient support from owners (Wu *et al.*, 2021b). Nevertheless, with the continuous maturation of the technology and the gradual expansion of the market, the application prospects of BIM technology in China's construction industry are still broad and it is expected to inject new vitality into the high-quality development of the industry.

14.2.4 Research gaps

Currently, there is a relative lack of empirical discussion in the research on LC and BIM. Although the BIM and LC integration framework proposed by Sacks *et al.* (2010) is pioneering, it is essentially a conceptual model that still needs to be validated through empirical data. Similarly, Maraqa *et al.* (2021) quantitatively assessed the benefits of BIM and LC.

However, these studies are mainly based on some particular LPS tools and have not been carried out in the context of PHC. Although previous studies have explored the application of LC in PHC (Goh and Goh, 2019; Höök and Stehn, 2008; Innella *et al.*, 2019) and the application of BIM in PHC (Gbadamosi *et al.*, 2019; Ji *et al.*, 2019) separately, research that simultaneously involves LC, BIM and PHC is still in a void. In China, the application level of MMC and construction digital technologies urgently needs to be thoroughly examined through empirical research. This study aims to understand the level of adoption of MMC in China, particularly focusing on BIM, off-site construction and LC.

14.3. Methodological approach

14.3.1 Research design

To satisfy the research question, questionnaire surveys were used. Surveys are the most used data-collection method for conducting this kind of research – for example, Zhai *et al.* (2014) investigated the barriers and challenges to off-site production adoption in China by surveying 110 construction professionals in the country. Wu *et al.* (2021a) also surveyed to obtain data on the transaction costs of prefabricated housing in China. Similarly, this chapter explores the level of adoption of key MMC (BIM, off-site construction, LC) in China through surveys.

This study involves two rounds of surveys. The first round aims to clarify the current application status of LC and BIM in China's PHC, as well as the application effects of BIM and LC on PHC project delivery. The second round of surveys is designed to determine the adoption levels of major construction digital technologies in China and their correlation with LC capability indicators. The analysis of the adoption levels of major construction digital technologies in China can provide insight into the current adoption status of BIM and off-site construction within China's construction industry, situated within a broader digital technology context.

Through comparison, the position and importance of BIM and off-site construction in the digital transformation of the construction industry can be clarified, which in turn allows for a more accurate assessment of their adoption levels and development trends as key components of MMC. Moreover, this comparison can also help identify potential imbalances in technology adoption during the digital transformation of China's construction industry, providing a more comprehensive perspective and data support for in-depth research on the promotion and application of MMC, and thus better addressing the research question.

Correlation analysis reveals the interactions and synergies between digital technologies (such as BIM, cloud computing and Internet of Things) and LC practices, helping to more accurately assess the adoption level of LC in China's construction industry. By understanding how these technologies support or promote LC, a more comprehensive perspective on the research question can be provided, clarifying the role of digital technologies in driving the overall development of MMC and thus better grasping the actual progress of China's construction industry in the field of MMC.

The first survey was made possible through the assistance of a collaborating institution. The sampling frame comprised a list of approximately 400 businesses representative of active participants in China's prefabrication industry, thus forming a representative survey population. Key demographic data showed that most respondents were between 21 and 40 years old (96.06%), held a bachelor's degree or higher (92.91%) and had over five years of experience in the construction sector.

The respondents, who represented a range of roles including developers, designers, contractors and suppliers, were drawn from organisations across 23 Chinese provinces and municipalities to ensure geographic diversity and minimise regional bias. Over half of these respondents were from organisations employing more than 500 people.

For the second survey, the sampling frame was constructed using a purposive sampling strategy initiated through the first author's professional network in Tianjin, followed by snowball sampling to expand the respondent pool. The target population of the second survey encompassed all Chinese construction companies that had adopted digital technologies in their construction projects. The inclusion criteria were specifically designed to identify such companies, with the additional requirement that respondents must hold management positions. This ensured that participants possessed sufficient knowledge of organisational practices and technology adoption to provide informed and reliable responses.

The respondents were predominantly front-line managers such as project managers and supervisors (63.44%), with 30.11% middle managers and 6.45% top executives; 43.01% possessed over 10 years of experience in the construction industry, 56.99% worked in state-owned enterprises, while 43.01% were from non-state-owned entities and organisations were almost evenly split between large firms (over 500 employees, 50.54%) and SMEs (49.46%), with most firms engaged in residential (26.88%) and public projects (44.09%), followed by industrial (10.75%), infrastructure (11.83%) and other project types (6.45%).

Participation in the surveys was voluntary. The first survey sent out 273 invitations and collected 127 valid responses, while the second survey sent out 1050 invitations and collected 93 valid responses. The response rates were 46.5% and 8.9%, respectively. The difference in response rates was due to the first survey being conducted with the help of a partner organisation that had a detailed list of respondents. Although the total number of responses for both surveys was not large, the sample size was still acceptable, similar to studies by Munianday *et al.* (2022) and Zhu and Cheung (2017), which also relied on about 100 participants.

14.3.2 Measurement and data analysis

To investigate the implementation of BIM and LC at different stages of PHC projects, PHC projects were divided into several stages: operation and maintenance, logistics and transportation, production and manufacture, and design and construction. To understand the application effects of LC tools and BIM functions in PHC projects, the main tools of LC were selected, including visual management, total quality management, standardisation, 5S and safety, Kaizen, JIT, mistake proofing, value stream mapping, LPS and early involvement of stakeholders (Tzortzopoulos *et al.*, 2020).

The main functions of BIM include visualisation, online/electronic object-based communication, rapid generation and evaluation of construction plan alternatives, collaboration in design and construction, rapid generation of multiple design alternatives, automated generation of drawing and documents, reuse of model data for predictive analyses, and maintenance of information and design model integrity.

To understand the correlation of BIM and off-site construction with other major construction digital technologies and LC, this study adopted 11 measurement indicators of LC capability by Fang *et al.* (2025), including

- timely production of customer-satisfied products (LC1)
- efficient operation of standard modules (LC2)
- keep flow continuity (LC3)
- standard operating procedures to ensure stable product quality (LC4)
- exploitation of existing knowledge and technology (LC5)
- resilience (LC6)
- buffering (LC7)
- flexible response to customer customisation needs (LC8)
- variation toleration (LC9)
- continuous improvement (LC10)
- establish long-term cooperation (LC11).

Construction digital technologies mainly include

- BIM (CD1)
- common data environment (CD2)
- unmanned aerial systems (CD3)
- cloud-based project management (CD4)
- augmented reaility and virtual reality (CD5)
- artificial intelligence (CD6)
- cybersecurity (CD7)
- big data and analytics (CD8)
- blockchain (CD9)
- laser scanners (CD10)
- robotics and automation (CD11)
- sensors (CD12)
- IoT (CD13)
- workers with wearable sensors (CD14)
- actuators (CD15)
- additive manufacturing (CD16)
- off-site construction (CD17)
- equipment with sensors (CD18).

Data collected through questionnaires were analysed and evaluated using percentages, means and correlation analysis methods by way of SPSS.

14.4. Results and discussion

14.4.1 Implementation of LC and BIM in prefabricated housing construction

According to the survey results, the implementation of BIM and LC in PHC reveals a clear adoption trend. Many respondents (55.12%) reported using both methodologies in their projects. A significant proportion (40.94%) implemented BIM alone, while only a small fraction (3.15%) adopted LC alone. In aggregate, BIM was employed in over 95% of the projects surveyed, starkly contrasting with LC's adoption rate of less than 60%. This indicates that BIM is a far more prevalent tool in PHC than in LC. Only a negligible minority (0.79%) of projects utilised neither approach.

The questionnaire investigated when the implementation of BIM and LC occurs across different stages of PHC projects. As shown in Figure 14.1, the adoption of LC appears to be more mature in production and manufacturing, as well as logistics and transportation stages. In contrast, BIM has seen more comprehensive implementation in design, construction, operation and maintenance. LC is particularly effective during the production and manufacturing stages, where it facilitates process optimisation and cost control through lean management practices.

The prevalence of BIM application is notably high during the design phase, with the construction phase recognised as pivotal for maximising BIM's value. Moreover, a multitude of studies have highlighted the diverse benefits of BIM implementation in both design and construction phases (Tu *et al.*, 2023).

Figure 14.1 Implementation phase of BIM and LC in PHC projects

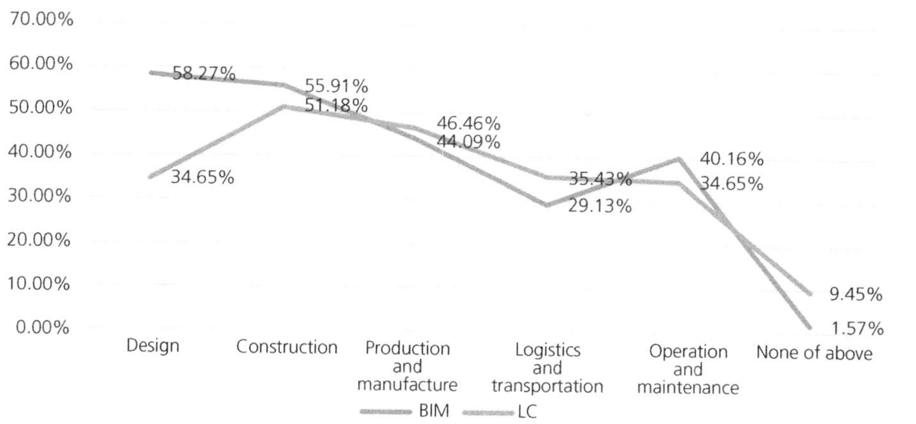

Source: Authors' creation

14.4.2 The effect of applying BIM and lean in supporting the delivery of PHC projects

Figure 14.2 presents participants' evaluation of the effect of LC and BIM in supporting the delivery of PHC projects. Visual management scored 4.35, the highest-rated LC tool. Total quality management (4.34) and standardisation (4.21) follow as tools identified by participants as more likely to achieve a better result in supporting the delivery of PHC projects. 5S and safety ranked fourth, indicating that respondents value it. This is followed by continuous improvement.

However, LC's representative tools JIT, mistake proofing, value stream mapping and the LPS are identified as not effective enough in supporting the delivery of PHC projects and early involvement of stakeholders is regarded as the least effective and, as described above, its proportion of application is also the lowest. The results also indicate that knowledge of certain systematic lean tool concepts is not widely disseminated in China's construction industry.

The mean of all BIM functions in PHC projects is above four points, indicating that the effect of applying BIM functions is highly recognised. The first three most effective functions for PHC projects are visualisation, online/electronic object-based communication, and rapid

Figure 14.2 The effect of applying LC tools and BIM functions in PHC projects

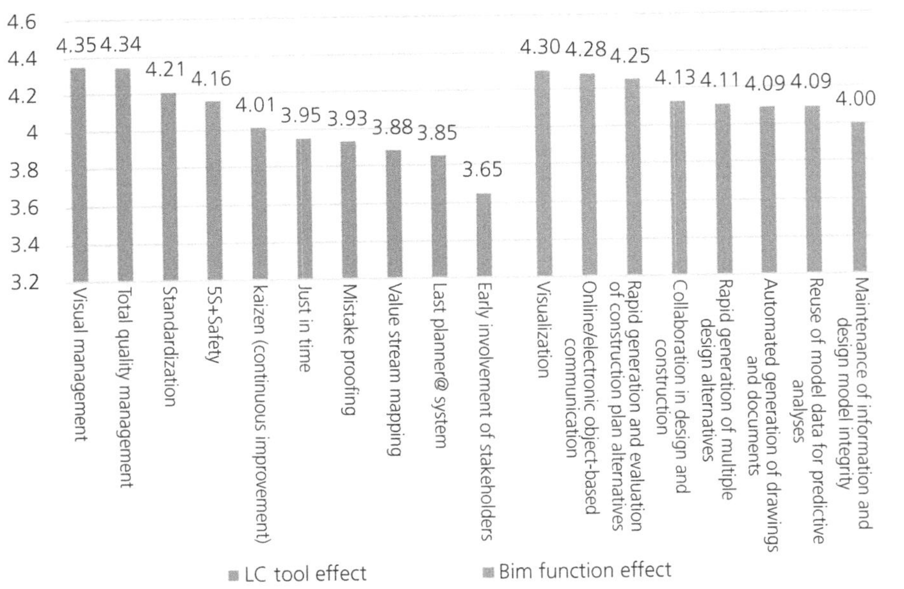

Source: Authors' creation

generation and evaluation of construction plan alternatives. Thus, BIM is most recognised for its effectiveness in visualisation, communication and planning. Collaboration in design and construction takes fourth place, showing that the effectiveness of BIM in integrated design and construction is also well recognised.

14.4.3 Adoption levels of BIM and off-site construction with other major construction digital technologies in China

The results show the adoption status of construction digital technologies in China's construction industry. As Table 14.1 indicates, BIM (3.75) has the highest mean score of the digital tools and off-site construction (3.68) ranks second. The mean scores of most technologies are between three and four, indicating that respondents' attitudes towards most technology adoption levels are neutral to agreeable.

Table 14.1 Level of adoption of main construction digital technologies

Items	Mean	Rank	Sig. (Wilcoxon signed rank test)
CD1: BIM	3.75	1	0.000*
CD2: Common data environment	3.25	3	0.083
CD3: Unmanned aerial systems	3.15	4	0.301
CD4: Cloud-based project management	3.15	4	0.283
CD5: AR/VR	3.14	6	0.300
CD6: Artificial intelligence	3.09	8	0.572
CD7: Cybersecurity	3.08	9	0.503
CD8: Big data and analytics	3.08	9	0.582
CD9: Blockchain	2.15	18	0.000*
CD10: Laser scanners	3.05	12	0.409
CD 11: Robotics and automation	2.98	15	0.939
CD 12: Sensors	3.14	6	0.267
CD 13: IoT	3.06	11	0.552
CD 14: Workers with wearable sensors	3.04	13	0.874
CD 15: Actuators	2.89	16	0.568
CD 16: Additive manufacturing	2.39	17	0.000*
CD 17: Off-site construction	3.68	2	0.000*
CD 18: Equipment with sensors	3.03	14	0.792

Note: *Significant at p-value < 0.05.
Source: Authors' creation

Only four technologies scored below three. The Wilcoxon signed rank test showed that BIM, off-site construction, blockchain and additive manufacturing show statistical significance in terms of adoption. These findings directly address our research question: 'What is the level of adoption of key MMC in China?'. Likely driven by government policies, the significant adoption levels of BIM and off-site construction suggest that these technologies are being widely recognised in construction projects in China.

The lower adoption levels of blockchain and additive manufacturing indicate that these technologies are still in the early stages of implementation and may require further development and industry education to achieve broader adoption. Fang *et al.* (2023) pointed out that it has been a long journey for BIM to penetrate the industry, although there are still roadblocks in the continuous promotion.

Similarly, the prefabricated construction technology is heavily promoted by the Chinese Government, aided by many incentive measures. Although the application of additive manufacturing (3D printing) has begun to emerge, it only occurs at present in experimental testbed projects.

Blockchain is a concept that has started to receive traction over the past two years and researchers are now diligently exploring its industry-scale applications. For practitioners, this indicates that investing in BIM and off-site construction may yield significant operational improvements. Conversely, the lower adoption of blockchain and additive manufacturing suggests that companies may need to focus on building internal capabilities and understanding of these emerging technologies to capitalise on their potential benefits fully.

14.4.4 Correlation of major construction digital technologies and LC capability indicators

As Table 14.2 shows, many correlation coefficients between construction digital technologies and LC indicators are greater than 0.3, showing greater than moderate correlations, which indicates that the application of construction digital technologies is closely related to LC capability. The study, therefore, suggests that emphasis should be placed on mutual facilitation between construction digital technologies and LC capability in practice.

There is a significant and positive correlation between BIM and the timely production of customer-satisfying products (LC1) and the efficient operation of standard modules (LC2), respectively. This is because BIM integrates all project information into a digital platform where the project team can identify clashes and variations early, leading to better standardised module operations. It can also accurately predict schedules that improve just-in-time production (Jeong *et al.*, 2016).

Off-site construction (CD17) presents a strong correlation in the effective implementation of modular construction (LC2), which is synergistic with off-site construction's own characteristic of producing standardised prefabricated components in factories (Jin *et al.*, 2018). Although the application of blockchain technology (CD9) and 3D laser scanning technology (CD10) is not yet commonly adopted, these are still positively correlated with most LC capability indicators.

Table 14.2 Correlation between main construction digital technologies and LC capability measure items

Items	CD1	CD2	CD3	CD4	CD5	CD6	CD7	CD8	CD9	CD10	CD11	CD12	CD13	CD14	CD15	CD16	CD17	CD18
LC1	0.254*	0.173	0.199	0.172	0.118	0.054	0.180	0.228*	0.230*	0.144	0.225*	0.171	0.139	0.246*	0.298**	0.288**	0.129	0.292**
LC2	0.330**	0.306**	0.232*	0.215*	0.271**	0.208*	0.247*	0.289**	0.203	0.224*	0.228*	0.147	0.199	0.242*	0.291**	0.147	0.304**	0.173
LC3	0.170	0.186	0.127	0.303**	0.191	0.131	0.173	0.255*	0.230*	0.354**	0.392**	0.208*	0.338**	0.448**	0.391**	0.316**	0.206*	0.402**
LC4	0.162	0.244*	0.110	0.210*	0.084	0.093	0.224*	0.204	0.226*	0.207*	0.252*	0.045	0.200	0.275**	0.282**	0.218*	0.063	0.281**
LC5	0.094	0.113	0.006	0.214*	0.104	0.122	0.066	0.061	0.222*	0.213*	0.302**	-0.022	0.293**	0.277**	0.351**	0.274**	0.065	0.267**
LC6	0.178	0.175	0.132	0.298**	0.183	0.231*	0.214*	0.194	0.285**	0.306**	0.292**	0.150	0.277**	0.344**	0.348**	0.173	0.218*	0.360**
LC7	0.166	0.226*	0.138	0.229*	0.246*	0.228*	0.192	0.181	0.335**	0.363**	0.337**	0.262*	0.272**	0.417**	0.336**	0.223*	0.229*	0.381**
LC8	0.118	0.138	0.082	0.161	0.113	0.099	0.109	0.079	0.179	0.336**	0.282**	0.135	0.260*	0.304**	0.239*	0.159	0.193	0.290**
LC9	0.174	0.224*	0.155	0.207*	0.231*	0.133	0.177	0.196	0.198	0.221*	0.239*	0.150	0.288**	0.220*	0.248*	0.142	0.245*	0.279**
LC10	0.118	0.243*	0.146	0.226*	0.174	0.169	0.227*	0.233*	0.238*	0.301**	0.203	0.217*	0.156	0.286**	0.267**	0.190	0.135	0.335**
LC11	0.022	-0.018	-0.079	0.083	0.096	0.084	0.048	-0.027	0.374**	0.264*	0.308**	0.091	0.184	0.235*	0.197	0.235*	-0.03	0.199

Note: ** Correlation is significant when the confidence (two tails) is 0.01. * Correlation is significant when the confidence (two tails) is 0.05.
Source: Authors' creation

It is worth noting that there are weak negative correlation trends between familiar data environment (CD2), unmanned aerial systems (CD3), big data and analytics (CD8), off-site construction (CD17) and establishing long-term cooperation relationships with suppliers and clients (LC11). This may be because establishing long-term partnerships often requires significant upfront costs, which potentially increases the company's financial pressures in the short term. Moreover, it might take some time before the positive outcomes of such alliances fully materialise.

The correlation coefficients between utilisation of existing knowledge and advanced technology (LC5) and sensors (CD12) are also negative, but none of these negative correlation coefficients are significant.

14.5. Managerial implications for practice

14.5.1 Comprehensive application of the MMC method

Although BIM and LC each have their own advantages in different stages and aspects of PHC projects, only a few enterprises have implemented both technologies simultaneously. This suggests that managers should pay attention to the comprehensive application of MMC technologies. The adoption of LC is more mature in the production and manufacturing, as well as logistics and transportation stages. At the same time, the application of BIM is more widespread in the design and construction stages. This indicates that different technologies may play different roles in different stages of the PHC project life cycle.

Managers should reasonably arrange and apply the corresponding technologies according to the specific stages and needs of the project to ensure efficient operation and integration in each stage.

14.5.2 Differences in the effectiveness of MMC methods

For lean tools with high scores (such as visual management, total quality management and standardisation), managers should further enhance the application of these tools to ensure they play the maximum role in projects – for example, reducing waste through standardised processes and improving project quality through total quality management.

For high-scoring BIM functions (such as visualisation, online/electronic object-based communication and rapid generation and evaluation of construction plan alternatives), managers should ensure that these functions are fully utilised throughout the entire project life cycle to increase project transparency and decision-making efficiency.

For lean tools that perform poorly (such as JIT, mistake proofing, value stream mapping and LPS), it is necessary to analyse the reasons for their insufficient application in projects and take measures to improve them – for example, to strengthen LC education (Alves *et al.*, 2012), providing training and sharing case studies to enhance the team's understanding and application capabilities of these tools.

14.5.3 Differences in the adoption levels of construction digital technologies

The adoption levels of BIM and off-site construction are relatively high in China's construction industry, indicating that these technologies have gained widespread recognition and application within the sector. Managers should continue to promote the in-depth application of these technologies to ensure they deliver maximum benefits in projects. Some technologies are still in the development stage of application within the industry. Managers should assess the potential value of these technologies – for example, the adoption levels of blockchain and additive manufacturing (3D printing) are currently low, but these technologies hold significant potential. Managers should keep an eye on the development trends of these emerging technologies and explore their application in projects at the right time to maintain competitiveness.

14.5.4 The synergy of LC and the adoption of construction digital technologies

By examining the correlation between LC capabilities and digital technologies, relevant collaborative strategies can be formulated – for example, enhancing the synergy between BIM and LC capabilities can lead to significant benefits. Through the BIM platform, project teams can identify potential issues earlier and adjust plans promptly, thereby improving client satisfaction. By leveraging BIM's schedule forecasting capabilities, JIT production can be achieved, enhancing the efficiency of modular construction.

Some technologies, however, are still under-adopted and require further promotion. Promoting the integration of off-site construction and modular construction can leverage the characteristics of off-site construction, such as the production of standardised prefabricated components in factories, to improve the efficiency and quality of modular construction. By optimising production processes and logistics management, the timely supply and installation of prefabricated components can be ensured.

Pilot applications of blockchain technology in key projects can be used for supply chain management and quality traceability (Sheng *et al.*, 2020). For technologies with low adoption levels but positive correlations with most LC capability indicators – such as blockchain and 3D laser scanning – active exploration of their applications in projects is essential; for instance, using 3D laser scanning technology for on-site quality inspection and progress monitoring can ensure that construction quality meets design requirements (Rebolj *et al.*, 2017).

14.6. Conclusions

This study, through questionnaire surveys, reveals the current implementation status of BIM and LC in China's PHC projects and explores the adoption of construction digital technologies and their correlation with LC capabilities in China's construction industry. The results show that the adoption levels of BIM and off-site construction technologies are the highest. Many construction digital technologies are significantly correlated with LC capability indicators. This indicates that the management implications of construction digital technologies and LC capabilities require a holistic approach.

This study makes several theoretical contributions. It contributes to the knowledge system of BIM and LC in the field of PHC. Firstly, through empirical research, it investigates the current status of BIM and LC in China's PHC projects, filling the gap in empirical studies of BIM and LC in the field of prefabricated construction. Secondly, it empirically demonstrates

China's adoption levels of construction digital technologies. Thirdly, it deepens our understanding of the interrelationships between the various measurement indicators of LC capabilities and construction digital technologies.

The study also provides managerial insights. The findings can offer practical insights for construction practitioners and companies seeking to adopt construction digital technologies and LC. Construction firms can make strategic decisions about investing in LC practices and adopting digital construction technologies to adapt to a changing environment.

Some limitations in this study are worth further exploration in the future. Firstly, the collection of data from the two surveys was based on the experience of experts and derived from relatively small samples. Future research should invite more experts to participate. Secondly, some factors that showed statistically insignificant and negative correlations prompted us to identify potential contributors to these results and conduct further studies. Thirdly, given that the application of some technologies in the construction field is still in its infancy, their interactions with other technologies may not yet be fully mature, thus necessitating continued research.

REFERENCES

Alves TCL, Milberg C and Walsh KD (2012) Exploring lean construction practice, research and education. *Engineering, Construction and Architectural Management* **19(5)**: 512–525, 10.1108/09699981211259595.

Chen J and Ban L (2020) lean management model for promoting prefabricated building lean construction based on building information model. *Science and Technology Management Research* **40(10)**: 196–205.

CPC Central Committee and State Council of China (2016) Several Opinions of the CPC Central Committee and the State Council on Further Strengthening Administration on Urban Planning and Construction Issued. http://www.gov.cn/zhengce/2016-02/21/content_5044367.htm (accessed 28/04/26).

Daniel EI and Oshodi O (2023) lean-offsite-simulation nexus for housing construction: a state-of-the-art review of the existing knowledge. *Construction Innovation* **23(5)**: 994–1017.

Daniel EI, Konara AN and Oshodi O (2024b) Exploring the use of lean, offsite and simulation in housing delivery: evidence from the UK. *Buildings* **14(6)**: 1642, 10.3390/buildings14061642.

Daniel E, Babalola A, Oladinrin O, Obi L and Oshodi O (2024a) Where lean construction and offsite construction meet: a bibliographic scientometric analysis. Construction Innovation: Information, Process, Management, 10.1108/CI-08-2023-0210.

Fang Y and Gao S (2024) lean construction in China: a review. In *Routledge Companion to Creativity and the Built Environment*, 1st edn., Routledge, Abingdon, UK.

Fang Y, Gao S and Tortorella G (2025) Exploring the use of lean construction ambidexterity and Industry 4.0 to achieve sustainable performance. *International Journal of lean Six Sigma*, 10.1108/IJLSS-09-2024-0211.

Fang Y, Gao S, Jiang Y and Li S (2023) BIM and lean construction in prefabricated housing construction in China. *International Journal of lean Six Sigma* **14(7)**: 1329–1353.

Forbes LH and Ahmed SM (2010) *Modern Construction: lean Project Delivery and Integrated Practices*. CRC Press.

Gao S and Low SP (2014) Barriers to lean implementation in the construction industry in China. *Journal of Technology Management in China* **9(2)**: 155–173.

Gbadamosi AQ, Mahamadu AM, Oyedele LO *et al*. (2019) Offsite construction: developing a BIM-Based optimizer for assembly. *Journal of Cleaner Production* **215**: 1180–1190.

Goh M and Goh YM (2019) lean production theory-based simulation of modular construction processes. *Automation in Construction* **101**: 227–244.

Heravi G and Firoozi M (2017) Production process improvement of buildings' prefabricated steel frames using value stream mapping. *The International Journal of Advanced Manufacturing Technology* **89(9)**: 3307–3321.

Höök M and Stehn L (2008) Applicability of lean principles and practices in industrialized housing production. *Construction Management and Economics* **26(10)**: 1091–1100.

Innella F, Arashpour M and Bai Y (2019) Lean methodologies and techniques for modular construction: chronological and critical review. *Journal of Construction Engineering and Management* **145(12)**: 04019076.

Jeong W, Chang S, Son J and Yi J-S (2016) BIM-integrated construction operation simulation for just-in-time production management. *Sustainability* **8**: 1106.

Ji Y, Chang S, Qi Y *et al*. (2019) A BIM-based study on the comprehensive benefit analysis for prefabricated building projects in China. *Advances in Civil Engineering* **2019**: 3720191.

Jiang R, Mao C, Hou L, Wu C and Tan J (2018) A SWOT analysis for promoting off-site construction under the backdrop of China's new urbanisation. *Journal of Cleaner Production*. https://doi.org/:225-234.

Jin R, Gao S, Cheshmehzangi A and Aboagye-Nimo E (2018) A holistic review of off-site construction literature published between 2008 and 2018. *Journal of Cleaner Production* **202**: 1202–1219.

lean Construction Institute (2012) What Is lean Construction? https://leanconstruction.org/lean-topics/lean-construction/ (accessed 10/12/2025).

Li T, Li Z, Li L and Li L (2020) Research on construction process management of prefabricated building based on LC-BIM. *Construction Economy* **41(7)**: 38–42.

Li XJ, Lai JY, Ma CY and Wang C (2021) Using BIM to research carbon footprint during the materialization phase of prefabricated concrete buildings: a China study. *Journal of Cleaner Production* **279**: 123454.

Maraqa MJ, Sacks R and Spatari S (2021) Quantitative assessment of the impacts of BIM and lean on process and operations flow in construction projects. *Engineering, Construction and Architectural Management* **28(8)**: 2176–2198.

McHugh K, Dave B and Craig R (2019) Integrated lean and BIM processes for modularised construction – a case study. *27th Annual Conference of the International Group for lean Construction*, IGLC 2019, pp. 227–238.

Munianday P, Radzi AR, Esa M and Rahman RA (2022) Optimal strategies for improving organizational BIM capabilities: PLS-SEM approach. *Journal of Management in Engineering* **38(3)**: 04022015.

Nascimento DLDM, Sotelino ED, Lara TPS, Caiado RGG and Ivson P (2017) Constructability in industrial plants construction: a BIM-lean approach using the Digital Obeya Room framework. *Journal of Civil Engineering and Management* **23(8)**: 1100–1108.

Pan W, Gibb AGF and Dainty ARJ (2007) Perspectives of UK housebuilders on the use of off-site modern methods of construction. *Construction Management and Economics* **25(2)**: 183–194, 10.1080/01446190600827058

Rebolj D, Pučko Z, Babič NČ, Bizjak M and Mongus D (2017) Point cloud quality requirements for Scan-vs-BIM based automated construction progress monitoring. *Automation in Construction* **84**: 323-334.

Sacks R, Koskela L, Dave B A and Owen R (2010) Interaction of lean and building information modeling in construction. *Journal of Construction Engineering and Management* **136(9)**: 968–980.

Sánchez-Garrido AJ, Navarro IJ, García J and Yepes V (2023) A systematic literature review on modern methods of construction in building: an integrated approach using machine learning. *Journal of Building Engineering* **73**: 106725.

Shahpari M, Saradj FM, Pishvaee MS and Piri S (2020) Assessing the productivity of prefabricated and in-situ construction systems using a hybrid multi-criteria decision making method. *Journal of Building Engineering* **27**: 100979.

Sheng D, Ding L, Zhong B *et al.* (2020) Construction quality information management with blockchains. *Automation in Construction* **120**: 103373.

Tong YM and Niu ZW (2016) The Construction Industry Must Upgrade and Transform to Promote Green Development through lean Construction. Chinese Government Website, April 12, 2016. https://www.gov.cn/zhengce/2016-04/12/content_5063120.htm.

Tu B, Zuo J, Chang RD *et al.* (2023) A system dynamic model for assessing the level of BIM implementation in construction phase: a China case study. *Engineering, Construction and Architectural Management* **30(4)**: 1321–1343.

Tzortzopoulos P, Kagioglou M and Koskela L (eds.) (2020) *Lean Construction: Core Concepts and New Frontiers*. Routledge.

Wang J, Zhang S, Fenn P *et al.* (2023) Adopting BIM to facilitate dispute management in the construction industry: A conceptual framework development. *Journal of Construction Engineering and Management* **149(1)**: 03122010.

Wang W, Gao S, Mi L *et al.* (2021) Exploring the adoption of BIM amidst the COVID-19 crisis in China. *Building Research and Information* **49(8)**: 930–947.

Wu H, Qian QK, Straub A and Visscher H (2021a) Stakeholder perceptions of transaction costs in prefabricated housing projects in China. *Journal of Construction Engineering and Management* **147(1)**: 04020145.

Wu P, Jin R, Xu Y *et al.* (2021b) The analysis of barriers to BIM implementation for industrialized building construction: a China study. *Journal of Civil Engineering and Management* **27(1)**: 1–13.

Wuni IY, Wu Z and Shen GQ (2021) Exploring the challenges of implementing design for excellence in industrialized construction projects in China. *Building Research and Information*: 1–15.

Xing W, Hao JL, Qian L, Tam VW and Sikora KS (2021) Implementing lean construction techniques and management methods in Chinese projects: a case study in Suzhou, China. *Journal of Cleaner Production* **286**: 124944.

Xu Z, Wang S and Wang E (2019) Integration of BIM and energy consumption modelling for manufacturing prefabricated components: a case study in China. *Advances in Civil Engineering* **2019**: 1609523.

Yao K (2017) *On the Application of BIM-Based Lean Management Model in PC Building Project*. HuaQiao University.

Zaalouk A, Moon S and Han S (2023) Operations planning and scheduling in off-site construction supply chain management: scope definition and future directions. *Automation in Construction* **153**: 104952.

Zhai X, Reed R and Mills A (2014) Factors impeding the offsite production of housing construction in China: an investigation of current practice. *Construction Management and Economics* **32(1–2)**: 40–52.

Zhu L and Cheung SO (2017) Harvesting competitiveness through building organizational innovation capacity. *Journal of Management in Engineering* **33(5)**: 04017020.

Emmanuel I Daniel and Fidelis A Emuze
ISBN 978-1-83708-437-1
https://doi.org/10.1108/978-1-83708-436-420261018

Chapter 15
Institutional barriers to the adoption of precast concrete structures in developing countries

Danny Murguia[1] and Edson Zanabria[1]

[1]*Department of Engineering, Pontifical Catholic University of Peru* 0000-0003-1009-4058; 0009-0006-3223-2772

Abstract

Precast concrete structures have been widely acknowledged as a key enabler of more efficient and sustainable industrialised construction. While this construction technology has been extensively adopted in many developed countries, its uptake in developing nations remains limited. This is primarily attributed to the continued dominance of traditional in-situ construction, the availability of low-cost labour and the presence of complex institutional blockers. Using the lens of isomorphism, this research aims to reveal the key institutional barriers to the adoption of precast components and propose a framework to inform policy-making. A qualitative approach was adopted, using data from 40 semi-structured interviews with construction professionals and structural engineers from the architecture, engineering and construction (AEC) industry of Peru. Twelve coercive, mimetic and normative factors were identified. AEC professionals highlighted a lack of authoritative technical standards, insufficient investment in research and development, low client demand, and uncertainty and resistance to change as the primary institutional barriers to adoption. The findings emphasise the role of institutional fragmentation and weak sectoral leadership in perpetuating these barriers. This research proposes an industry-level responsibility matrix to inform policymakers and industry stakeholders seeking to industrialise the construction sector in developing markets.

Keywords: adoption of innovations; developing countries; industrialised construction; institutional theory; precast structures

15.1. Introduction

Industrial construction using precast structural systems offers significant performance improvements in productivity, sustainability and quality in the AEC sector (Akinradewo *et al.*, 2023; Kosbar *et al.*, 2023; Saad *et al.*, 2025). As such, precast concrete structures represent a strategic opportunity to transform the way buildings are planned, designed and built (Chourasia *et al.*, 2023; León *et al.*, 2024). Countries such as Japan, the United States and New Zealand have progressively incorporated these systems into building construction (Bhatta *et al.*, 2022; Zhao and Takahashi, 2025; Zhou *et al.*, 2024). The focus on innovation and the limited availability of labour in such economies have driven its adoption (Maqbool *et al.*, 2024).

Despite significant housing and infrastructure needs in developing countries, the adoption of precast structures remains limited due to cultural, institutional and organisational barriers (Ali *et al.*, 2025; Oti-Sarpong *et al.*, 2022). Across diverse contexts, including Colombia (León *et al.*, 2024), Chile (Ortega *et al.*, 2023), Ethiopia (Akinradewo *et al.*, 2023), Ghana (Guribie and Alkanam, 2021), Egypt (Ali *et al.*, 2023), India (Suliman and Rankin, 2021) and Saudi Arabia (Luo *et al.*, 2024), studies converge in applying technology adoption or diffusion frameworks, yet consistently neglect institutional and regulatory barriers. To address this global gap, the present study examines the barriers to the adoption of precast structural systems from an institutional perspective.

In Peru, precast concrete structures in building construction are limited to a few applications, such as staircases and lattice planks, which are often used as substitutes for traditional formwork and seldom integrated into structural design. The country's high seismic risk further complicates adoption, given limited standards and building codes, as well as concerns about the reliability of connections between precast components. Furthermore, structural engineers are particularly wary, as connection systems may not meet the structural integrity requirements. Existing research shows that steel modular connections are widely studied, whereas concrete connections are less so (Chourasia *et al.*, 2023). This lack of trust and supporting research presents a key barrier to the broader use of precast systems. Peru, therefore, provides a critical case for examining the institutional barriers to precast adoption in developing countries, particularly in contexts characterised by high seismic risk and regulatory uncertainty.

The technology–organisation–environment (TOE) framework (Tornatzky and Fleischer, 1990) is widely used to examine innovation adoption within organisations (Hong *et al.*, 2023; Suliman and Rankin, 2021). While technology benefits and organisational changes are pivotal for sustained adoption, this study argues that barriers in the early stages of adoption are primarily institutional. Small institutional shifts can, in turn, influence organisational and technological factors. Therefore, this paper adopts institutional theory, specifically, institutional isomorphism (DiMaggio and Powell, 1991), to frame the analysis.

Institutional theory focuses on the enduring structures, including rules, norms and routines, that guide social behaviour (Scott, 2004). According to institutional isomorphism, organisational fields (i.e. groups of organisations offering similar products or services) initially exhibit diversity but, over time, tend toward homogeneity. DiMaggio and Powell (1991) identified three drivers of this convergence: coercive isomorphism (due to political or regulatory pressures), mimetic isomorphism (as a response to uncertainty) and normative isomorphism (linked to professional standards and education). Building on the theoretical foundations of institutional theory and isomorphism, this study aims to

- identify the institutional pressures that limit the adoption of precast structural systems
- identify the key stakeholders capable of driving institutional change
- propose a framework to support broader adoption and inform policy in developing countries.

To achieve these objectives, this study adopts a qualitative approach, using semi-structured interviews with construction professionals and structural engineers from Peru's AEC industry. A combined deductive and inductive analytical strategy was applied, enabling both theory-driven categorisation and data-driven insights. This approach supports the identification of barriers through the lens of institutional isomorphism that shapes the adoption of precast structures.

This chapter contributes to the ongoing debate on how to scale up the use of modern construction solutions in developing countries. By emphasising the role of the institutional environment and the actors who shape it, it proposes a necessary shift in perspective to unlock the adoption of construction innovation and advance towards more modern, safe and sustainable buildings and infrastructure.

15.2. Perspectives and concepts

15.2.1 Institutional theory

Institutional theory provides a powerful lens for explaining why organisations adopt or resist technologies, practices or organisational models, even when these clearly offer improvements in efficiency or productivity (Chandler and Hwang, 2015). From this perspective organisational decisions are not guided solely by technical or economic factors but are also deeply embedded in the institutional context. This context comprises regulatory frameworks, social norms, cultural expectations and shared beliefs that collectively shape what is considered legitimate or appropriate (Loosemore and Richard, 2015). Organisations are thus seen as actors within structured organisational fields, where formal and informal rules determine acceptable conduct and influence decision-making beyond mere performance optimisation.

DiMaggio and Powell (1991) identified three primary sources of institutional pressure that act upon organisations: coercive, mimetic and normative.

- *Coercive pressures* emerge from formal mechanisms such as legislation, technical regulations, public policies and contractual requirements (Masocha and Fatoki, 2018). These forces compel organisations to adopt certain practices to comply with legal mandates or to secure access to resources managed by regulatory entities (Tam *et al.*, 2007).
- *Mimetic pressures* arise under conditions of uncertainty, when the pathway to achieving organisational objectives is unclear (Horak *et al.*, 2018). In such contexts organisations tend to emulate models perceived as successful or legitimate within their field, not

necessarily because these models are more efficient, but because imitation offers a way to manage risk and align with prevailing norms (Arashpour *et al.*, 2015; Hosseini *et al.*, 2018).

- *Normative pressures* are rooted in shared professional values and standards. These are transmitted through education, professional accreditation, industry networks and the influence of technical associations. Such pressures shape how professionals define best practices, justify decisions and evaluate innovations, promoting the adoption of solutions that conform to what is considered appropriate within the professional community (Nadim and Goulding, 2011; Pan *et al.*, 2007).

Together, these institutional pressures may explain the persistence of conventional practices in the construction sector, as well as the barriers to adopting more industrialised and innovative solutions, even when these are demonstrably superior on technical or economic grounds.

15.2.2 Rationale for using institutional theory

Institutional theory enables a structurally informed analysis of technological adoption by recognising that organisational decisions are embedded within social, regulatory and cultural contexts (Blismas *et al.*, 2006). In sectors such as construction, where change tends to be slow and institutional inertia is strong, this perspective is particularly relevant (Goodier and Gibb, 2007). Specifically, applying institutional theory to the adoption of precast structural systems facilitates a deeper understanding of how the absence of adequate technical standards, the limited visibility of prior applications and deficiencies in professional training may function as structural barriers that cannot be overcome solely through technical evidence or economic arguments (Jaillon and Poon, 2009; Luo *et al.*, 2015).

Understanding adoption, or lack of, from an institutional perspective may provide a robust analytical foundation for both diagnosis and the design of effective interventions to support technological innovation within the construction sector (Wuni *et al.*, 2020).

15.2.3 Institutional factors affecting the adoption of precast structures

Existing scholarly work on the adoption of precast concrete structural systems reveals a series of persistent institutional barriers that constrain their large-scale implementation across various international contexts – for instance, coercive pressures are primarily manifested through the absence of authoritative technical standards, the limited presence of government-led promotion initiatives and the weak institutional capacity of certification agencies. A summary is presented in Table 15.1.

Jayawardana *et al.* (2023) observed that the lack of integration of prefabricated systems into national construction frameworks and building codes creates ambiguity and discourages adoption. This finding is echoed by Akinradewo *et al.* (2023), who identified the absence of clear standards and robust regulatory mechanisms as a significant source of uncertainty for designers and contractors.

Similarly, Saad *et al.* (2025) highlighted that the lack of financial incentives and institutional support diminishes the economic viability of these systems, particularly in markets dominated by traditional methods. Furthermore, Ali *et al.* (2025) stressed that institutional weaknesses in oversight and certification processes undermine stakeholder trust. These structural limitations perpetuate limited adoption governed by outdated regulatory frameworks.

Mimetic pressures also emerge as a recurring pattern. Several studies have shown that, in the absence of reference experiences or sufficient empirical evidence, developers and contractors tend to replicate traditional models – for instance, Akinradewo *et al.* (2023) argued that this trend intensifies in environments where successful applications of prefabricated construction are scarcely visible, reinforcing perceptions of risk and inhibiting innovation. Saad *et al.* (2025) likewise claimed that the preference for conventional methods is not necessarily driven by technical or economic considerations, but rather by the imitation of what is perceived as the *standard* or *safest* practice within the field.

This mimetic behaviour is further reinforced by the lack of technical dissemination campaigns and low client demand, which often stems from unawareness of the benefits of precast systems or persistent misconceptions about their performance. Altogether, these factors create an environment where uncertainty is addressed not through technical evidence but through the replication of dominant practices, thereby perpetuating institutional inertia.

Some studies on normative pressures point to a significant disconnect between technological advancement and education and professional standards. Jayawardana *et al.* (2023) and Akinradewo *et al.* (2023) emphasised that insufficient technical training in precast or modular construction and the limited involvement of professional associations in disseminating specialised knowledge represent notable obstacles. Chourasia *et al.* (2023) reported that in many emerging contexts, the lack of sustained investment in research and development (R&D) prevents the adaptation of internationally developed solutions, leaving professionals without locally validated tools.

This situation is further exacerbated by limited collaboration between government, academia and industry, which, according to León *et al.* (2024), weakens the feedback mechanisms needed to build technical capabilities for the design, assembly and certification of these systems. As a result, a vicious cycle emerges, characterised by a lack of standardisation and technical resistance.

Table 15.1 presents a summary of the coercive, mimetic and normative forces found in the literature on precast structural adoption.

15.3. Methodological approach

To explore the institutional factors influencing the adoption of precast concrete structures in developing countries, this research adopted a qualitative approach, combining a review of relevant literature with empirical data from semi-structured interviews. A qualitative strategy was selected because it enables in-depth exploration of institutional complexity, actor

Table 15.1 Coercive, mimetic and normative pressures (continued on next page)

Institutional pressures		Akinradewo *et al.* (2023)	Tatari (2023)	Jayawardana *et al.* (2023)	Saad *et al.* (2025)	Guribie and Alkanam (2021)	Kosbar *et al.* (2023)	Bendi *et al.* (2021)	Daget and Zhang (2019)	Ali *et al.* (2025)	Suliman and Rankin (2021)	Ortega *et al.* (2023)	Ali *et al.* (2023)	León *et al.* (2024)	Chourasia *et al.* (2023)	Oti-Sarpong *et al.* (2022)	Trigunarsyah and Santoso (2020)
Coercive	COE1. Lack of authoritative technical standards within the market	✓		✓	✓	✓	✓	✓	✓	✓	✓	✓	✓	✓	✓	✓	✓
	COE2. Weak governmental role and absence of public policies	✓		✓	✓	✓	✓	✓	✓	✓		✓	✓	✓	✓	✓	✓
	COE3. Weak institutional capacity of regulatory and certification bodies				✓				✓					✓			
	COE4. Lack of financial or regulatory incentives	✓		✓	✓	✓	✓	✓	✓	✓		✓	✓	✓	✓	✓	✓

Mimetic	MIM1. Low client demand	✓		✓	✓	✓	✓	✓	✓	✓		✓	✓	✓	✓	✓	✓
	MIM2. Uncertainty and resistance to change	✓	✓	✓	✓	✓	✓	✓	✓	✓	✓	✓	✓	✓	✓	✓	✓
	MIM3. Limited visibility of successful case studies within the market			✓	✓		✓		✓	✓	✓	✓	✓	✓	✓	✓	✓
Normative	NOR1. Lack of investment in research and development	✓		✓	✓		✓	✓	✓	✓	✓	✓	✓	✓	✓	✓	✓
	NOR2. Weak collaboration among government, industry and academia			✓	✓		✓	✓	✓			✓		✓		✓	✓
	NOR3. Lack of engagement by professional bodies in regulation, dissemination and training				✓									✓			

Source: Authors' creation

perceptions and context-specific barriers, which are often challenging to capture quantitatively (Denzin and Lincoln, 2018).

Firstly, a literature review on institutional factors influencing industrialised construction and precast systems was conducted. Using the institutional perspective as a theoretical lens, the factors in Table 15.1 were derived.

Data were collected through forty semi-structured interviews with professionals from Peru's construction sector, organised into two groups. The first group consisted of professionals working in construction organisations (N = 11) and precast systems suppliers (N = 7). In contrast, the second group included senior structural engineers (N = 22) with experience in building structural design.

These participants were deliberately selected to capture perspectives from key actors on both the supply and demand sides of precast systems, as well as from experts responsible for ensuring structural safety in high-seismic contexts. All interviewees are active professionals within the AEC industry and their profiles ensure coverage of the main organisational and technical perspectives relevant to the adoption of precast structural systems.

In the first group, interviews explored topics such as market supply, technical capacity, regulatory challenges, and organisational and environmental barriers. Each interview began with an open-ended question inviting the participant to share a direct experience of implementing precast components, followed by eight open-ended questions that examined related themes. Each interview concluded with a reflective question on what changes would be necessary to enhance the use of prefabrication in their professional environment.

Interviews with the second group also followed a semi-structured design, this time focused on capturing the perspective of structural design professionals regarding the use of precast concrete structures. These interviews were organised into three sections: control questions to gather background information (such as years of experience, current role and prior involvement with precast), open-ended questions addressing the theoretical dimensions, and a final, open question on their vision of construction industrialisation.

The interviews were conducted individually, recorded with prior informed consent and subsequently transcribed. As a result, 289 pages of empirical data were collected. Deductive thematic analysis was conducted using the factors outlined in Table 15.1. Inductive analysis was also performed to identify potential factors not found in the literature. Based on findings from both the literature and empirical data and informed by professional experience, an industry-level responsibility matrix was developed to support the broader adoption of structural precast systems in developing countries. This framework aligns institutional actions with key institutional actors, thus serving as a strategic tool to guide implementation efforts and foster systemic change.

15.4. Results and discussion

The qualitative data supported the institutional factors identified in Table 15.1. In addition, two factors inductively emerged for the data analysis and interpretation. Firstly, the limitations imposed by construction permitting authorities (COE5) and, secondly, the lack of technical knowledge in structural engineering applied to precast systems due to traditional university curricula (NOR4).

Furthermore, the data revealed differences in the degree to which certain factors were validated by the different actor groups interviewed. Table 15.2 presents the level of empirical support for the institutional forces, as perceived by the different stakeholder groups.

Table 15.2 Institutional blockers to precast concrete structures adoption

	Institutional Pressures	Construction Professionals (N = 11)	Precast Systems Suppliers (N = 7)	Senior Structural Engineers (N = 22)
Coercive	COE 1. Lack of authoritative technical standards within the market	✓	✓	✓
	COE 2. Weak governmental role and absence of public policies	✓	✓	⚠
	COE 3. Weak institutional capacity of regulatory and certification bodies	⚠	✓	⚠
	COE 4. Lack of financial or regulatory incentives	✓	✓	⚠
	COE 5. Limitations imposed by construction permitting authorities*	✓	⚠	✓
Mimetic	MIM 1. Low client demand	✓	✓	⚠
	MIM 2. Uncertainty and resistance to change	✓	⚠	✓
	MIM 3. Limited visibility of successful case studies within the market	✓	✓	✓
Normative	NOR 1. Lack of investment in research and development	✓	✓	✓
	NOR 2. Weak collaboration among government, industry and academia	✓	✓	✓
	NOR 3. Lack of engagement by professional bodies in regulation, dissemination and training	⚠	✓	✓
	NOR 4. Knowledge gap in precast structures linked to traditional university curricula*	⚠	⚠	✓

* Additional factors found in the empirical data
✓ Strongly supported by the group
⚠ Partially or weakly supported by the group
Source: Authors' creation

15.4.1 Coercive pressures

Construction professionals highlighted the lack of authoritative technical standards and regulatory incentives as persistent barriers to greater precast adoption. They described frequent delays caused by outdated regulatory frameworks and the lack of government-led initiatives supporting industrialised construction methods. Many of them perceived that public bodies remain passive or indifferent to the modernisation of the sector.

From the perspective of precast systems suppliers, coercive pressures were closely linked to the limited institutional capacity for certification and inspection. They noted the absence of structured government support to facilitate market entry and scale-up. This regulatory gap forces firms to navigate uncertain, often arbitrary approval procedures. In addition, they lamented the lack of financial incentives or tax benefits for companies willing to invest in innovative construction technologies.

Senior structural engineers also acknowledged the lack of authoritative technical standards as a relevant coercive challenge. Their concern centred on the technical ambiguity encountered when integrating precast elements into traditional structural design workflows. Several interviewees noted that, in the absence of local standards, they rely on international codes, which, in turn, lead to subjective interpretation by construction permitting authorities, increasing project risk and legal exposure. This concern is particularly acute in a high-seismicity country like Peru that lacks a national seismic standard for precast connections. Without clear, locally validated design codes, engineers cannot ensure the safety and performance of connections under seismic loads, further reinforcing professional hesitation.

Participants from all three groups reported that bureaucratic obstacles delay project approvals and discourage the use of non-conventional systems such as precast structures. The absence of uniform technical criteria across municipalities, combined with public officials' limited technical expertise, exacerbates project uncertainty and undermines developers' confidence. This coercive barrier reveals the institutional misalignment between innovation and regulation.

15.4.2 Mimetic pressures

The data confirmed that mimetic pressures remain a persistent barrier to the diffusion of precast systems. One of the most frequently cited challenges among the three actor groups was the low level of client demand. This lack of market force reinforces conventional practices and inhibits innovation. Respondents indicated that clients, especially in private developments, often perceive precast systems as unfamiliar or high-risk due to their limited visibility.

Construction professionals emphasised that client reluctance is exacerbated by the absence of reference projects showcasing successful precast applications. In the absence of local case studies, stakeholders tend to default to traditional construction methods. Moreover, they suggested that uncertainty regarding long-term performance further reinforces resistance to adoption. According to this group, the industry needs flagship projects and demonstrable case studies to reshape market perceptions.

Precast system suppliers shared similar concerns, highlighting that developers and project owners are often unaware of innovative precast solutions. Suppliers also pointed to professional inertia and the lack of pilot projects in the public sector as limiting factors. They proposed public-private partnerships to build demonstrative infrastructure that could catalyse broader acceptance.

Senior structural engineers acknowledged that mimetic behaviour is also present in their profession. Some interviewees noted that design practices remain conservative, with a preference for traditional in-situ concrete construction, due to the lack of client demand. Even when precast offers evident advantages, engineers tend to replicate standard approaches unless institutional mandates or client demands prompt them to innovate. The limited dissemination of technical knowledge and the lack of continuing education on precast methods further reinforce this pattern.

15.4.3 Normative pressures

From the perspective of construction professionals, normative pressures (such as expectations shaped by industry norms and professional practices) are reflected in the scarcity of applied research and the limited collaboration between industry and academia. They observed that construction-related research is underfunded and fragmented, hindering the development of new technical practices. Some proposals are being established for strategic partnerships between universities and organisations to foster innovation systems capable of addressing real-world challenges in precast construction.

Precast systems suppliers highlighted the lack of structured training programmes and professional development pathways for these technologies. They noted that even when companies invest in innovation, the industry lacks the educational infrastructure needed to sustain knowledge transfer. For suppliers, this normative weakness translates into a persistent skills gap that undermines the integration of precast structures into planning, design and construction.

Senior structural engineers expressed concerns over their peers' limited familiarity with precast design, particularly in Peru's seismic context. Some interviewees also reported difficulties mentoring younger engineers who had never been exposed to such systems during their university education. This factor reflects a systemic disconnect between higher education and the evolving demands of the industry. Interviewees emphasised that current academic programmes devote little or no attention to industrialised construction techniques, leaving graduates ill-equipped to apply them in practice. This finding highlights a long-standing institutional gap with significant implications for national innovation capacity.

Finally, structural engineers also reported deep-rooted resistance within the profession to embracing change in the absence of formal endorsement from academic and professional institutions.

15.5. Implications

The findings of this study reveal that the main barriers to the adoption of precast structural systems in Peru include a lack of authoritative technical standards, limited investment in research and development, low client demand, and widespread uncertainty and resistance to change, reflecting a low level of market maturity. These institutional weaknesses manifest across all three isomorphic pressures.

The absence of clear regulatory guidelines (coercive), the limited dissemination of successful case studies (mimetic) and the insufficient provision of technical training (normative) hinder the establishment of the necessary conditions for systemic change. In contrast to developed countries such as New Zealand, Japan or the United States, where these aspects have been progressively strengthened through public policies, technical standards and strategic investments (Kosbar *et al.*, 2023; León *et al.*, 2024; Ortega *et al.*, 2023), Peru's construction sector remains fragmented and structurally unprepared to facilitate the widespread adoption of precast systems.

Moreover, while developing regions such as Ethiopia face comparable institutional barriers, seismic risks there are less prominent, allowing adoption debates to focus more on cost, capacity and standardisation (Akinradewo *et al.*, 2023). By contrast, other developing countries with significant seismic exposure, such as Chile or Colombia, have made more progress in updating codes and promoting innovation through pilot projects (León *et al.*, 2024; Ortega *et al.*, 2023).

In Peru, however, the combination of weak institutions with high seismic exposure intensifies the uncertainty around precast systems, underscoring the urgency of strengthening institutional foundations before any meaningful technological transition can occur.

The institutional barriers identified in Peru are consistent with those reported in other developing countries such as Colombia, Chile, Ethiopia, India, Ghana and Saudi Arabia. Several studies confirm that the absence of government-led incentives, the weakness of regulatory enforcement and the limited coordination among sector stakeholders are recurrent problems in these contexts (Akinradewo *et al.*, 2023; Ali *et al.*, 2025; Ortega *et al.*, 2023; Suliman and Rankin, 2021).

Similarly, persistent gaps in professional training and a lack of collaboration between academia and industry remain significant limitations in Ghana and Ethiopia (Akinradewo *et al.*, 2023), while in India and Saudi Arabia, public misconceptions regarding the cost and performance of precast systems reinforce resistance to change (Luo *et al.*, 2024; Suliman and Rankin, 2021).

However, the Peruvian case presents additional challenges. In particular, the country's high seismic risk imposes stringent technical requirements for structural integrity. However, there is currently no national design and construction code addressing this critical issue. Unlike countries such as New Zealand, Japan, Italy and the United States, which have developed robust seismic standards for precast systems, the Peruvian context lacks specific guidelines to ensure the structural safety of these systems under seismic conditions. In New Zealand, for example, the NZS 3101:2006 *Concrete Structures Standard* includes specific provisions for precast elements in seismic zones, reflecting a regulatory environment aligned with the country's seismic characteristics. Similarly, Japan, one of the most seismically active country in the world, has developed detailed guidelines through the Architectural Institute of Japan (AIJ), such as *Behaviours and Design Method on Vertical Joint of Wall Panel Precast Concrete Structures* (1989) and *AIJ Guidelines for Seismic Design of Reinforced Concrete Foundation Members* (2017), which specifically address the seismic behaviour of precast elements and their connections, providing unified and well-defined design approaches.

In Europe, Italy incorporates such systems into its *Norme Tecniche per le Costruzioni*, which establishes minimum performance requirements for precast structural elements in seismic regions, thereby supporting their successful application in high-risk areas. Meanwhile, California, the US state with the most advanced seismic code, sets out detailed requirements for 'ordinary precast structural walls' in its *2022 Building Code*, complemented by local seismic ordinances that specifically regulate these systems.

The existence of such authoritative standards in highly seismic countries generates trust among designers, builders and authorities, facilitating the large-scale adoption of precast systems by providing clear regulatory frameworks that technically validate their use in seismic areas. In contrast, Peru's lack of specific technical standards for precast systems in seismic zones creates regulatory uncertainty, which limits their acceptance among structural professionals, who tend to prioritise traditional solutions already validated by the National Building Code (RNE).

The findings of this study demonstrate that institutional pressures do not operate in isolation but rather intersect and reinforce one another, thereby perpetuating the limited adoption of precast structural systems. Figure 15.1 illustrates this interaction among coercive,

Figure 15.1 Overlap between institutional blockers (see codes in Table 15.2)

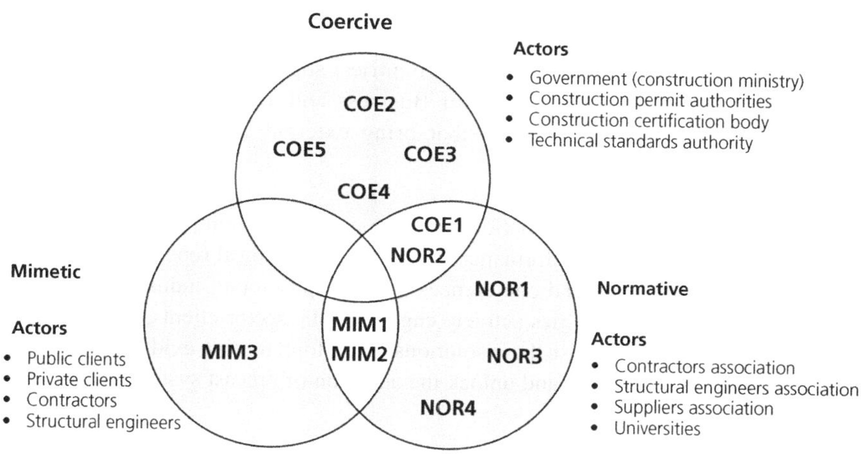

Source: Authors' creation

mimetic and normative pressures, emphasising the distinct barriers associated with each type and the overlapping dynamics between them. This convergence suggests that unlocking systemic change requires coordinated action across multiple fronts, particularly in areas where challenges intersect.

A clear example of this convergence is found in the interaction between coercive and normative pressures, represented by COE1 (lack of authoritative technical standards) and NOR2 (weak collaboration among government, industry and academia). The absence of technical standards reflects the lack of institutional coordination required to generate, validate and disseminate new knowledge.

This situation points to the need to establish a multi-stakeholder working group, led by the Ministry of Construction and supported by representatives from the Peruvian Chamber of Construction (CAPECO), the Institute of Peruvian Engineers (CIP), the national association of precast suppliers, construction permitting authorities and universities with Civil Engineering departments. The purpose of this group would be to identify knowledge gaps, align R&D efforts with national priorities and formulate a shared roadmap for standard development. Without such a collaborative platform, efforts to establish technical norms risk remaining fragmented and disconnected from sectoral realities. In this regard, addressing coercive gaps requires normative leadership and the institutionalisation of cross-sectoral dialogue. The creation of this working group would constitute the foundational step towards regulatory innovation and industrial alignment.

Another critical interaction arises between mimetic and normative pressures, particularly with MIM1 (low client demand) and MIM2 (uncertainty and resistance to change). These barriers are deeply rooted in the lack of technical evidence and visible success stories, which foster a culture of scepticism and risk-aversion. Overcoming this challenge requires building trust through evidence-based research and dissemination. This involves generating empirical data on the structural performance, productivity and sustainability of precast systems through well-documented pilot projects. In much the same way that BIM gained traction in several developing countries, supported by foreign government agencies such as the UK's Department for Business and Trade, precast systems could benefit from international partnerships that bring external credibility and specialised technical expertise.

Such collaborations could result in national or regional demonstrator projects that serve as live laboratories for validating performance, supported by technical reports, academic publications, professional seminars and conferences aimed at promoting industrialised construction. It is crucial that such activities actively engage public sector clients, who can serve as first adopters and help to legitimise these solutions. Therefore, turning evidence into influence is essential to reduce uncertainty and unlock the adoption of precast systems in markets that remain resistant to change.

To unlock the potential of precast structural systems and overcome the institutional barriers limiting their adoption, it is imperative to implement coordinated actions across regulatory, professional, academic and market groups. The following is a proposed list of key actions that can be promoted at the industry level.

(a) *Establish a government-led taskforce on industrialised construction* coordinated by the Ministry of Construction, to guide public policy, regulation, training and technological innovation agendas in the construction sector.
(b) *Set up a working group to coordinate research initiatives* to identify gaps in knowledge and professional practice in the areas of structural design, on-site execution and quality control for precast systems.
(c) *Set up a technical committee on structural integrity* focused on developing specific seismic standards for precast elements, including connection detailing and structural validation methods.
(d) *Set up a working group to standardise technical criteria for building permits* to harmonise the requirements imposed by building authorities and streamlining the approval processes for research-validated non-conventional construction technologies.
(e) *Design incentive schemes to promote the adoption of industrialised solutions* such as tax benefits, fast-track permitting mechanisms or strategies for maximising land use efficiency.
(f) *Promote applied research programmes*, developed through partnerships between government, academia, suppliers and the private sector, focused on addressing specific challenges related to the adoption of precast solutions.
(g) *Design and implement training programmes in structural precast construction* for industry professionals, integrating dedicated modules into university curricula, technical education and training activities.
(h) *Develop and document national pilot projects*, gathering data on structural performance, productivity, cost-efficiency and sustainability.
(i) *Create a national digital knowledge platform on industrialised construction* to centralise and disseminate technical standards, case studies, best practice guidelines, specialised literature and results from applied research.
(j) *Design a dissemination and internationalisation strategy*, incorporating the organisation of seminars, workshops and conferences, the publication of technical reports and the engagement with global cooperation networks from leading countries such as Japan, New Zealand, Italy or the United Kingdom.

Figure 15.2 presents a responsibility matrix linking the ten proposed institutional actions to the key stakeholders involved in adopting precast structural systems. Stakeholders are categorised according to their expected level of involvement: A (leading), B (supporting) and C (participating), as suggested by Kassem and Succar (2017). This visual representation functions as a strategic institutional framework, enabling a more precise identification of leadership roles and collaboration needs. The matrix highlights the Ministry of Construction's central role as the coordinating body, while also recognising the critical contributions of universities, professional associations, construction permitting authorities and industry groups. By making the distribution of responsibilities visible, this framework supports the design of concrete institutional interventions to enable the first steps towards industrialised construction.

Figure 15.2 Responsibility matrix for institutional adoption of precast structures in developing countries

Actions

Actions	Establish a national taskforce on industrialised construction	Set up a working group to coordinate research initiatives	Set up a technical committee on structural integrity	Set up a working group to standardise technical criteria for building permits	Design incentive schemes to promote adoption	Promote applied research programmes	Design and implement training programmes	Develop and document national pilot projects	Create a national digital knowledge platform	Design a dissemination and internationalisation strategy
Government (Ministry of construction)	A	A	B	A	A	A	A	A	B	A
Construction permit authorities	B	B	B	B	–	B	C	–	–	C
Construction certification body	B	B	B	B	–	B	C	–	–	C
Technical standards authority	B	B	A	B	–	B	C	–	B	C
Public clients	C	C	–	C	B	C	–	C	–	–
Private clients	C	C	–	C	B	C	–	C	–	–
Contractors association	C	B	B	C	C	B	B	B	C	B
Structural engineers association	C	B	B	C	C	B	B	B	C	B
Suppliers association	C	C	C	C	C	C	C	C	C	B
Universities	C	B	B	B	–	A	A	A	A	C

A → Leading; B → Supporting; C → Participating

Source: Authors' creation

15.6. Conclusions

This chapter examined the limited adoption of structural precast structures in a developing market through the lens of institutional theory, focusing on the coercive, mimetic and normative pressures that influence organisational behaviour. Drawing on 40 semi-structured interviews, the study identified 12 key institutional barriers: ten consistent with previous research and two specific to the Peruvian context. These barriers interact with one another, reinforcing an institutional environment that is generally unfavourable to technological innovation and the shift towards industrialised construction.

The findings reveal that the absence of locally validated technical standards, limited collaboration between academia and industry, and low client demand act as mutually reinforcing constraints. In the Peruvian context, characterised by high seismic risk, the lack of specific standards for the design and connection of precast components exacerbates regulatory uncertainty and professional resistance. This contributes to a technical culture defined by risk-aversion and a continued reliance on traditional in-situ concrete solutions. In response to these conditions, an institutional responsibility matrix was developed to provide a structured visual representation of the expected roles of key actors across the construction sector. This strategic framework clarifies which stakeholders are best positioned to lead, support or participate in efforts to overcome the identified institutional barriers. By aligning institutional responsibilities with specific challenges, the matrix provides a practical basis for designing coordinated interventions and strengthening inter-institutional collaboration.

The proposed framework contributes to the growing body of literature on the institutional conditions required to scale up industrialised construction in developing contexts. Beyond the Peruvian case, its insights are practically transferable to other countries facing high seismic risks or institutionally fragmented environments, where similar barriers may hinder the adoption of precast structural systems. Future research could also complement this framework with quantitative methods to assess the relative influence of institutional pressures or the impact of specific interventions led by public and private actors.

The main limitation of this study lies in its exclusive focus on the institutional environment. The technological and organisational dimensions were not purposefully addressed to focus on the institutional factors. Future research should therefore broaden the analytical scope to include these additional dimensions, as well as consider longitudinal studies to assess how institutional conditions evolve and influence the sustained implementation of industrialised solutions.

Acknowledgement

The authors wish to thank Claudia Torres and Edgar Vilcamich for their support with data collection.

REFERENCES

Akinradewo O, Aigbavboa C, Aghimien D, Oke A and Ogunbayo B (2023) Modular method of construction in developing countries: the underlying challenges. *International Journal of Construction Management* **23(8)**: 1344–1354, 10.1080/15623599.2021.1970300.

Ali AH, Kineber AF, Elshaboury N, Arashpour M and Daoud AO (2025) Analysing multifaceted barriers to modular construction in sustainable building projects: a comprehensive evaluation using multi-criteria decision making. *International Journal of Construction Management* **25(1)**: 1–17, 10.1080/15623599.2023.2299557.

Ali AH, Kineber AF, Elyamany A, Ibrahim AH and Daoud AO (2023) Exploring stationary and major modular construction challenges in developing countries: a case study of Egypt. *Journal of Engineering, Design and Technology* **23(2)**: 569–598, 10.1108/JEDT-03-2023-0099.

Arashpour M, Wakefield R, Blismas N and Minas J (2015) Optimisation of process integration and multi-skilled resource utilisation in off-site construction. *Automation in Construction* **50**: 72–80, 10.1016/j.autcon.2014.12.002.

Bendi D, Rana M and Arif M (2021) Assessing off-site readiness in construction organisations: cases from India. *Construction Innovation* **21(4)**: 708–733, 10.1108/CI-01-2021-0005.

Bhatta J, Dhakal RP and Sullivan TJ (2022) Seismic performance of a rocking precast concrete cladding panel system under lateral cyclic displacement demands. *Journal of Earthquake Engineering* **27(1)**: 1–30, 10.1080/13632469.2022.2033359.

Blismas N, Pasquire C and Gibb A (2006) Benefit evaluation for off-site production in construction. *Construction Management and Economics* **24(2)**: 121–130, 10.1080/01446190500184444.

Chandler D and Hwang H (2015) Learning from learning theory. *Journal of Management* **41**: 1446–1476, 10.1177/0149206315572698.

Chourasia A, Singhal S and Manivannan M (2023) Prefabricated volumetric modular construction: A review on current systems, challenges and future prospects. *Prefabricated and Precast Structures in Developing Countries* **12(1)**: 55–66, 10.1061/PPSCFX.SCENG-1185.

Daget YT and Zhang H (2019) Decision-making model for the evaluation of industrialised housing systems in Ethiopia. *Engineering, Construction and Architectural Management* **26(6)**: 1134–1154, 10.1108/ECAM-05-2018-0212.

Denzin N and Lincoln YS (2018), *The SAGE Handbook of Qualitative Research*, 5th ed. SAGE Publications, Thousand Oaks, CA, USA.

DiMaggio PJ and Powell WW (1991) The iron cage revisited: institutional isomorphism and collective rationality in organisational fields. In *The New Institutionalism in Organizational Analysis* (Powell WW and DiMaggio PJ (eds)). University of Chicago Press, Chicago, IL, USA, pp. 63–82.

Goodier C and Gibb A (2007) Future opportunities for off-site in the UK. *Construction Management and Economics* **25(6)**: 585–595, 10.1080/01446190601071821.

Guribie FL and Alkanam N (2021) Structural equation modelling of the critical driving forces of off-site construction in Ghana. *Built Environment Project and Asset Management* **11(5)**: 945–964, 10.1108/BEPAM-03-2021-0051.

Hong J, Li Z, Zhang Y and Zhao D (2023) Simulating cost risks for prefabricated construction in developing countries using Bayesian networks. *Journal of Construction Engineering and Management* **149(2)**: 04023015, 10.1061/JCEMD4.COENG-13259.

Horak S, Arya B and Ismail K (2018) Organisational sustainability determinants in different cultural settings: a conceptual framework. *Business Strategy and the Environment* **27**: 528–546, 10.1002/BSE.2018.

Hosseini MR, Martek I, Zavadskas EK *et al.* (2018) Critical evaluation of off-site construction research: a scientometric analysis. *Automation in Construction* **87**: 235–247, 10.1016/j.autcon.2017.12.002.

Jaillon L and Poon CS (2009) The evolution of prefabricated residential building systems in Hong Kong: a review of the public and the private sector. *Automation in Construction* **18(3)**: 239–248, 10.1016/j.autcon.2008.09.002.

Jayawardana J, Kulatunga AK, Jayasinghe J and Sandanayake M (2023) Environmental sustainability of off-site construction in developed and developing regions: a systematic review. *Journal of Architectural Engineering* **29(3)**: 04023027, 10.1061/JAEIED.AEENG-1420.

Kassem M and Succar B (2017) Macro BIM adoption: comparative market analysis. *Automation in Construction* **81**: 286–299, 10.1016/j.autcon.2017.04.005.

Kosbar MM, Elbeltagi E, Mahdi I, Kassem M and Ehab A (2023) Off-site manufacturing: determining decision-making factors. *Buildings* **13(11)**: 2856, 10.3390/buildings13112856.

León W, Guevara J and García-López N (2024) Exploring the limitations and opportunities of industrialised construction in Colombia from a Lean perspective. In *Proceedings of the 32nd Annual Conference of the International Group for Lean Construction (IGLC32)* (Costa DB, Drevland F & Florez-Perez L (eds)), pp. 401–415, 10.24928/2024/0174.

Loosemore M and Richard J (2015) Valuing innovation in construction and infrastructure: getting clients past a lowest price mentality. *Engineering, Construction and Architectural Management* **22(1)**: 38–53, 10.1108/ECAM-03-2014-0035.

Luo L, Ding Z, Niu J, Zhang L and Liao L (2024) A digital project management framework for transnational prefabricated housing projects. *Buildings* **14(9)**: 2915, 10.3390/buildings14092915.

Luo L, Mao C, Shen L and Li Z (2015) Risk factors affecting practitioners' attitudes toward the implementation of an industrialised building system: a case study from China. *Engineering, Construction and Architectural Management* **22(6)**: 622–643, 10.1108/ECAM-04-2014-0048.

Maqbool R, Rashid Y, Altuwaim A, Shafiq MT and Oldfield L (2024) Coping with skill shortage within the UK construction industry: scaling up training and development systems. *Ain Shams Engineering Journal* **15**: 102396, 10.1016/j.asej.2023.102396.

Masocha R and Fatoki O (2018) The impact of coercive pressures on sustainability practices of small businesses in South Africa. *Sustainability* **10(9)**: 3032, 10.3390/SU10093032.

Nadim W and Goulding JS (2011) Off-site production: a model for building down barriers. *Engineering, Construction and Architectural Management* **18(1)**: 82–101, 10.1108/09699981111098702.

Ortega J, Mesa HA and Alarcón LF (2023) The interrelationship between barriers impeding the adoption of off-site construction in developing countries: the case of Chile. *Journal of Building Engineering* **73**: 106824, 10.1016/j.jobe.2023.106824.

Oti-Sarpong K, Shojaei RS, Dakhli Z and Burgess G (2022) How countries achieve greater use of off-site manufacturing to build new housing: identifying typologies through institutional theory. *Sustainable Cities and Society* **84**: 104015, 10.1016/j.scs.2021.103403.

Pan W, Gibb AG and Dainty AR (2007) Perspectives of UK housebuilders on the use of off-site modern methods of construction. *Construction Management and Economics* **25(2)**: 183–194, 10.1080/01446190600827058.

Saad S, Rasheed K, Ammad S *et al.* (2025) Off-site modular construction adoption in developing countries: partial least square approach for sustainable future. *Ain Shams Engineering Journal* **16(1)**: 101975, 10.1016/j.asej.2024.103228.

Scott WR (2004) Institutional theory: contributing to a theoretical research program. In *Great Minds in Management: The Process of Theory Development* (Smith K and Hitt M (eds)). Oxford University Press, Oxford, UK.

Suliman A and Rankin J (2021) Maturity-based mapping of technology and method innovation in off-site construction: conceptual frameworks. *Journal of Information Technology in Construction (ITcon)* **26**: 381–399, 10.36680/j.itcon.2021.021.

Tam VW, Tam CM, Zeng SX and Ng WC (2007) Towards adoption of prefabrication in construction. *Building and Environment* **42(10)**: 3642–3654, 10.1016/j.buildenv.2006.10.003.

Tatari A (2023) Simulating cost risks for prefabricated construction in developing countries using Bayesian networks. *Journal of Construction Engineering and Management* **149(6)**: 04023042, 10.1061/JCEMD4.COENG-13259.

Tornatzky LG and Fleischer M (1990) *The Processes of Technological Innovation*. Lexington Books, Lexington, MA, USA.

Trigunarsyah B and Santoso TP (2020) Adopting off-site construction into the Saudi Arabian construction industry. *Smart and Sustainable Built Environment* **9(4)**: 431–447, 10.1680/jsmic.21.00004.

Wuni IY, Shen GQ and Hwang BG (2020) Risks of modular integrated construction: a review and future research directions. *Frontiers of Engineering Management* **7(1)**: 63–80, 10.1007/s42524-019-0059-7.

Zhao H and Takahashi N (2025) Loss estimation of precast prestressed concrete frame structure with cost-efficient mortise tenon connection under earthquake. *Frontiers in Built Environment* **11**: 1612096, 10.3389/fbuil.2025.1612096.

Zhou X, Shao X, Yan W, Wang B and Li X (2024) Seismic behaviour of precast prestressed concrete beam–column joints using high-performance materials. *Scientific Reports* **14(1)**: 10529, 10.1038/s41598-024-63083-y.

Emmanuel I Daniel and Fidelis A Emuze
ISBN 978-1-83708-437-1
https://doi.org/10.1108/978-1-83708-436-420261019

Chapter 16
Lean and design for manufacturing and assembly (DfMA) in construction: perspectives from Australia

Rachel Couper[1], Laura Gutierrez-Bucheli[1], Ivana Kuzmanovska[2], Gao Shang[3] and Duncan Maxwell[1]

[1]*Future Building Initiative, Department of Architecture, Monash University, Caulfield East, Australia* 0000-0002-3762-7996; 0000-0003-3764-7855; 0000-0002-9039-1441
[2]*Future Building Initiative, Department of Architecture, Monash University, ARC Centre for Next-Gen Architectural Manufacturing, UNSW* 0009-0000-6718-8587
[3]*Faculty of Architecture, Building and Planning, The University of Melbourne, Parkville, Australia* 0000-0002-4161-5592

Abstract

Lean is a production management philosophy focused on minimising waste and maximising customer value through a holistic and flexible view of the supply chain. In contrast, design for manufacture and assembly (DfMA) is a design-led approach aimed at enhancing the speed, simplicity and efficiency of manufacturing and assembly processes. Both methodologies, originating in the mid-twentieth century, have profoundly influenced modern manufacturing. Since the 1990s, lean and DfMA have attracted attention as potential solutions to low productivity in the construction industry. However, their adoption, particularly in Australia, remains limited despite evidence of their benefits. This study examines potential synergies between lean and DfMA within Australia's construction sector. Focusing on a recent mass engineered timber project, it investigates how lean and DfMA principles were applied and assesses their combined potential to generate greater value. The findings reveal that several lean tools and DfMA principles were implemented, with collaboration identified as crucial for successful integration. The research also highlights project-specific and industry-wide barriers to broader application. It concludes that viewing lean and DfMA as interconnected rather than separate approaches can unlock their full potential, driving industry-wide transformation and supporting their long-term adoption across the entire construction value chain.

Keywords: lean construction; DfMA; modern methods of construction; mass engineered timber

16.1. Introduction

Modern methods of construction (MMC) is an approach that is often framed as being a promising solution for improving productivity in the Australian construction industry. As the sector continues to struggle with persistent underperformance in cost, time, waste and quality management compared to other industries, MMC provides a framework for adopting more efficient, standardised and process-driven practices (Aitchison *et al.*, 2018).

Two different MMC approaches that are gaining in popularity are lean construction (LC) and DfMA. When applied to the construction industry, DfMA can optimise design for off-site manufacture and streamlined on-site assembly (Kuzmanovska, 2020), while LC can enhance production efficiency through waste minimisation and continuous improvement (Koskela *et al.*, 2002). While both strategies address inefficiencies and offer pathways toward more productive and sustainable construction practices, the synergies between the two are often overlooked.

This research considers how a synthesised application of the two could unlock significant productivity gains across the construction sector through the analysis of a multistorey mass engineered timber (MET) project in Australia. Using cross laminated timber (CLT) and glue laminated timber (glulam), MET (or mass timber) projects are considered a form of MMC because they involve prefabricated, standardised components that are typically manufactured and processed in off-site factories, resulting in less waste, increased speed of on-site assembly and improved safety (Aitchison *et al.*, 2018). It is through the factory based, manufacturing and off-site production processes of MET construction projects that a sense of alignment with both lean and DfMA principles is generated.

Both lean and DfMA originated in manufacturing and several scholars have viewed DfMA as either a complementary design methodology to LC or as a lean tool itself, highlighting persistent conceptualisation differences of both strategies. (Chick, 2013; Gao *et al.*, 2019; Tan *et al.*, 2020; Vaz-Serra *et al.*, 2021). However, despite a substantial body of literature on LC and a growing interest in DfMA, neither approach has seen widespread adoption in Australia's construction industry.

Implementation remains limited, largely due to cultural and organisational barriers (Kuzmanovska, 2020). Moreover, confusion persists because DfMA is often conflated with prefabrication and lean is inconsistently applied without sustained organisational cultural support. Given this, and the lack of analysis of local projects in the literature, this chapter aims to address the gap by examining the challenges of delivering a multi-storey MET building, with the aim of identifying where both DfMA and lean principles were adopted and how effectively.

The aim of the chapter is to offer a structured reflection on lessons learned and highlight opportunities where better integration of lean and DfMA could enhance project outcomes. The structure of this chapter is set as follows: Section 16.2 reviews lean production (LP) and DfMA in construction, as well as their synergies; Section 16.3 outlines the methodological approach and research context; Section 16.4 presents the results and discussion; Section 16.5 provides implications for practice and research; and Section 16.6 presents the conclusions of the study.

16.2. Perspectives and concepts

16.2.1 LP to LC

LP originated in the automotive industry post-World War II when Toyota refined Ford's existing systems to develop the Toyota Production System (TPS). Central to TPS is the philosophy of achieving total waste elimination in pursuit of maximal efficiency (Toyota, n.d.). This production system gained widespread recognition with the 1990 publication of *The Machine That Changed the World* (Womack *et al.*, 1990). This five-year research documented TPS's advantages over traditional mass production and established the foundational principles of LP, outlining their applicability across diverse industries.

The principles of lean were subsequently introduced to the construction sector through the seminal work of Koskela (1992), who redefined construction as a production system to be managed through the lenses of transformation, flow and value generation (TFV). However, a review of academic and industry publications reveals a persistent lack of consensus regarding a definitive definition of LC (e.g. Mossman, 2018). For the purposes of this chapter, the authors propose the following working definition of LC as 'a production management approach that enables minimising waste while aiming towards total waste elimination and generating value by being open to change – that is, flexible and by improving the supply chain as a whole'.

Observing the parallels between manufacturing and construction challenges, Glenn Ballard and Greg Howell established the lean Construction Institute (LCI) in 1997 to advance and disseminate lean thinking within the built environment. More recently, LC practices have gained significant attention because of their proven advantages in achieving production efficiency (Ghosh and Burghart, 2021; Yu *et al.*, 2013). While some view it as a direct application of LP principles, others see it as an approach adapted to the dynamic and project-based nature of the construction sector.

Despite the lack of a formal definition of LC, existing studies have proposed various frameworks, guidelines and roadmaps for LC implementation such as the lean Construction Wheel (Diekmann *et al.*, 2004), socio-technical models (Paez *et al.*, 2005) and comprehensive model like the Toyota Way (Gao and Low, 2014). These frameworks suggest that successful LC implementation requires both technical tools and cultural organisational transformation. In other words, effective deployment goes beyond applying tools like the Last Planner System, 5S, Kanban and value stream mapping; it also demands organisational change such as collaboration, transparency, leadership and continuous improvement.

Furthermore, there is a large body of LC literature that explores the barriers to adoption and identifies critical success factors for effective LC implementation (e.g. Albalkhy and Sweis, 2021).

Key barriers were identified, including the fragmented nature of construction, lack of leadership support, poor supplier integration, limited lean knowledge, misaligned implementation strategies, cultural resistance and rigid traditional mindsets. These barriers suggest that successful implementation requires not just tools, but a supportive organisational environment and aligned project delivery systems.

Critical factors for LC implementation include implementation strategy, management support, stakeholder relationship, buy-ins from workers, organisational culture and contract allocation and procurement selections (Chick, 2013). Importantly, lean cannot be delivered top-down or in isolation. Moreover, recent literature has also explored the synergy between lean and digital technologies. Tools such as building information modelling (BIM), Internet of Things (IoT) and real-time data platforms are increasingly seen as enablers of lean implementation, supporting improved planning, defect prevention and decision-making (McHugh *et al.*, 2022). This convergence, often referred to as 'Digital lean', presents a promising pathway to overcome traditional implementation barriers.

16.2.2 Design for manufacture and assembly – emergence in construction

Design for Manufacture and Assembly (DfMA) is a data-informed approach to product design that integrates production considerations to dramatically improve cost, speed and quality (Bogue, 2012). Originating in the manufacturing industry in the 1970s and coined by Geoffrey Boothroyd, DfMA is a combination of design for manufacturing (DfM) and design for assembly (DfA) techniques, emphasising assembly efficiency as a key factor in reducing overall production costs (Boothroyd, 1987; Bralla,1999).

While highly successful in reducing assembly times and production costs as well as improving quality and reliability in the manufacturing industries, it was not until the turn of the millennium that the first signs of interest appeared in applying DfMA to construction, with the approach only recently attracting more widespread attention (Fox *et al.*, 2001). Consequently, the construction industry lacks a unified and shared understanding of DfMA when it is applied to the design and construction of buildings (Gao *et al.*, 2019).

Compounding the issue, literature on DfMA in construction also includes a wide range of interpretations. Some view DfMA as a systematic design approach similar to value engineering, others as a design evaluation method (Gao *et al.*, 2019). Further, DfMA is often conflated with prefabrication, because many practitioners primarily associate it with off-site manufacturing (Trinder, 2018). This is misguided because prefabrication refers specifically to the off-site assembly of components, whereas DfMA is a broader methodology that can apply to both on-site and off-site production, whether standardised or bespoke. The distinction is important because prefabrication can occur without applying full DfMA principles and DfMA can be implemented without prefabrication (Kuzmanovska, 2020).

DfMA is seen as a key enabler of MMC through the promotion of early-stage design collaboration to optimise building systems for both off-site manufacture and on-site assembly (Lu *et al.*, 2020). The benefits of DfMA, such as improved speed, quality, efficiency, reduced waste and safer construction, are central goals of MMC initiatives (Laing O'Rourke, 2013; Lu *et al.*, 2020). Industry organisations such as the Royal Institute of British Architects (RIBA) and Singapore's Building and Construction Authority (BCA) are making efforts to raise awareness of DfMA by proposing roadmaps and guides to integrate DfMA workflows (BCA, 2018; RIBA, 2013; 2021).

Popular DfMA principles that have been incorporated into industrialised construction terminology include standardisation, modularisation, part minimisation, easy handling, clear physical and visual access, simple interfaces, error-proof design, assembly from above, part weight reduction, proper clearances, and the use of jigs and templates.

However, current DfMA practices in construction often directly apply manufacturing guidelines without considering the distinct nature of construction projects (scale of product, complexity, unstable site conditions), including differences in supply chains, production processes, product types and site assembly requirements (Kuzmanovska, 2020; Lu *et al.* 2021; Tan *et al.*, 2021). Additional challenges include the lack of best practice knowledge about what constitutes a DfMA design method and the lack of construction-specific data to support the data-driven design evaluation processes required to emulate manufacturing-derived DfMA methods (Kuzmanovska, 2020).

Industry-wide changes and support are needed to facilitate DfMA implementation in construction. These include raising awareness, building capabilities and demand through training, especially cross-disciplinary training for design teams (Langston and Zhang, 2021; Staub-French *et al.*, 2018). Also, government incentives, as seen in the UK, Singapore and Hong Kong, can be a significant driver (Langston and Zhang, 2021). At the project level, success hinges on factors such as allowing sufficient design time, novel collaborative contracting models, early stakeholder involvement, appropriate fee structures and robust data management practices (Kuzmanovska 2020). This also might include utilising past project data and prototyping to minimise risks (Langston and Zhang, 2021).

Where DfMA has been adopted as a more integrated and optimised process, such as in the Leadenhall Building and the Dagenham Park Church of England School by Laing O'Rourke, the successful outcomes have demonstrated its advantages, including early-stage design collaboration to optimise building systems for off-site manufacture and on-site assembly, resulting in reduced risk and improved efficiency (Laing O'Rourke, 2013).

16.2.3 Synergies between lean and DfMA

This research has identified that there is limited existing empirical knowledge regarding the integrated application of lean and DfMA in construction, particularly within the context of MMC in Australia. Despite this, it is possible to identity key commonalities in their objectives: namely the reduction of waste, maximising value and improving productivity. Lean targets efficiency in workflows and processes (Koskela *et al.*, 2002), while DfMA simplifies and refines design for manufacture and assembly (Kuzmanovska, 2020). This alignment creates strong synergies across the project life cycle.

Both approaches contribute to waste reduction in complementary ways (Ng and Hall, 2019). Lean eliminates inefficiencies throughout the value chain and DfMA reduces complexity and material use through more efficient design (Ng and Hall, 2019). Together, they address waste caused by fragmented teams and poorly coordinated workflows, ensuring value is considered from the outset.

The synergies between lean and DfMA also support collaboration and knowledge transfer (Ng and Hall, 2019). Lean promotes cooperation across disciplines (Meng, 2019), while DfMA encourages early coordination between design and production (Gerth *et al.*, 2013). These shared principles enable smoother communication, clearer value expectations and more cohesive project delivery (Langston and Zhang, 2021). In the context of MMC, common objectives of lean and DfMA have the potential to enhance standardisation, support continuous improvement and align digital design with practical execution (Aitchison *et al.*, 2018).

The following analysis of a multistorey mass engineered timber (MET) project in Australia illustrates the benefits of integrating lean and DfMA across an MMC project, showing how lean's emphasis on collaboration and waste reduction complements DfMA's focus on design and production efficiency. Early stakeholder involvement and strong alignment between key partners enabled effective implementation and despite some challenges, the project highlights how their combined use, especially in MET projects, can improve outcomes through better teamwork and streamlined processes.

16.3. Methodological approach

16.3.1 MET construction project analysis

An applied research approach was selected to thoroughly examine the practical application and integration of lean and DfMA within a real construction project. The scope of the research was designed to address a need articulated by two stakeholder companies who, having recognised the individual benefits of lean and DfMA, sought the research team's expertise to understand how these methodologies could be more effectively synergised in a practical setting.

16.3.2 Project structure

The MET project analysed in this study was a multistorey tower that formed part of a significant urban renewal initiative. A research project funded by the Australian Government's Cooperative Research Centre tracked the design, manufacture and construction of the project. Completed between 2018 and 2019, the ten-story office building, reaching 52 m in height, stands as one of Australia's tallest timber structures. The scale of the project and the production processes used to develop it renders it an exemplary case for examining the practical application of lean and DfMA on a live project from the initial conceptual phase through to completion.

A major Australian construction organisation played a multifaceted role in the project, uniquely operating as both the developer and lead contractor through its diverse business arms. This direct organisational link facilitated the early involvement of the project manager and construction manager during the initial conceptual phase, working closely with an internal design team specialising in mass timber. Within this same overarching organisation, a specialised timber manufacturing department functioned as the timber contractor. This early cross-departmental engagement fostered a deeply collaborative environment to produce the project's conceptual design package. The internal design team subsequently transitioned to become the design arm for the timber contractor, ensuring design continuity from early conceptualisation through to the final delivery of the timber structure.

In contrast, the services contractor, an independent company, maintained a more distant relationship with the main project group, with their involvement commencing significantly later in the delivery phase following a successful tender bid. While a professional working relationship between the lead contractor and the services subcontractor facilitated the early sharing of relevant project information (e.g. architectural renders), the delayed engagement of the services contractor limited their capacity to influence critical design decisions. This temporal disconnect ultimately contributed to design rework and inefficient production processes later in the project life cycle.

16.3.3 Applied research methodology

Applied research was adopted to analyse how lean and DfMA were integrated and manifested in an MET construction project (Hedrick *et al.*, 1993). The primary data source for this analysis involved semi-structured interviews conducted with key participants from the project team. Participants were strategically drawn from the lead contractor, the timber contractor and the services contractor to ensure a comprehensive representation of stakeholder perspectives. These interviews were structured around three core themes to ensure systematic coverage while allowing for emergent insights: *(a)* project overview (including participants' roles in the project, stakeholder communications, challenges and successes); *(b)* understanding of lean and its implementation in the project; and *(c)* understanding of DfMA and its implementation in the project.

These themes provided a flexible framework, allowing for in-depth discussions while ensuring that core aspects of lean and DfMA, both generally and project-specifically, were addressed. Purposive sampling was employed to select participants, ensuring a diverse range of stakeholder perspectives, reflecting varied roles, responsibilities and interests within the project, were captured (e.g. Patton, 2015).

16.4. Results and discussion

16.4.1 How lean construction was utilised in the MET construction project

Lean construction principles were primarily applied in the construction project as a process support for the contract, aiming to reduce waste, increase value-adding activities to diverse customers (project stakeholders) and optimise overall project performance through a focus on streamlined flow and coordination.

Interviewees primarily defined lean through the lens of waste elimination. The integrated prefabrication leader from the services subcontractor stated that 'It's all about reducing waste and streamlining our processes'. A recurring theme in discussions about waste was inventory, particularly concerning product and materials. Examples provided included inefficient panel utilisation, ordering excess spare panels, discarding offcuts and managing surplus CLT packaging. Other forms of waste mentioned were rework (defects), inefficient transportation (e.g. sending empty trucks or excessive material movement) and underutilised talent (e.g. searching for knowledge that already exists within the company). Another interviewee conceptualised waste more broadly as the environmental impact of global timber transportation (carbon miles).

When considering how lean specifically manifested in the MET construction project, those most confident in discussing lean principles were individuals working in the off-site prefabrication and mass timber processing factories. Their insights revealed four key lean principles in action: value identification and generation, waste elimination, flow and process optimisation, and continuous improvement.

Interviewees discussed the *value* added in the project in terms of benefits to the customer, themselves and the advantages of MET. Customer value was the most frequently discussed, which was articulated in terms of cost in particular, and also time and knowledge or experience. As the project manager from the timber subcontractor noted, 'Their primary value would be cost and programme. So, cost for me. That's the DNA, I guess, of the industry and this probably leads back into KPIs and contracts.' Similarly, the construction manager from the lead contractor stated, 'When I say value, I'm talking about cost, I'm not talking about overall value'.

The value attributed to MET included a quieter worksite, higher productivity, an improved reputation among the wider industry, and the appeal of an added-value product that could attract higher rental yields. For the project team, value was associated with awards, recognition and strengthened relationships with other stakeholders. However, a critical observation was the absence of a unified value proposition or a common understanding of lean's vision across the project team, which appeared to contribute to a lack of a disciplined and rigorous implementation plan.

The interviewees consistently pointed to a significant focus on *waste reduction* through the identification and removal of non-value-added activities. This was primarily evident in the off-site mass timber processing factory, where a redesign process successfully eliminated the need for panel flipping. On-site, scaffolding was entirely removed and the material delivery sequence achieved a 90% match with the assembly sequence, streamlining logistics. However, the interviews also revealed that the unpredictability of the site program often led to inaccuracies in material delivery timing, indicating a persistent challenge in optimising the supply chain.

Participants also highlighted a focus on *overall flow and process optimisation*, understood as an attempt to eliminate 'surprises'. This was primarily achieved through meticulous sequencing and coordination, facilitated using 3D models in collaborative discussions with the site team. The timber subcontractor exemplified *continuous improvement* by consistently responding to on-site issues with an efficient two-week turnaround. Nevertheless, interviewees also acknowledged the challenges associated with applying continuous improvement to processes due to the dynamic nature of changing projects and teams.

These principles were further supported by tracking and collaborative discussions conducted within an Obeya room on the construction site. Derived from the Japanese word for 'large room', an Obeya room works as a collaborative workspace designed to foster enhanced communication and problem-solving among teams. While interviewees recognised the importance of just-in-time (JIT) strategies for embracing lean construction, they also noted that full JIT implementation was not always feasible due to the necessity of on-site material storage.

This suggests a potential disconnect where production may not be entirely driven by immediate construction demand, possibly leading to instances of overproduction.

16.4.2 How DfMA was utilised in the MET construction project

DfMA principles were strategically applied in the MET construction project as a key design driver, with the explicit objective of minimising the use of material, while reducing material waste and optimising transportation logistics (shipping and trucking). The success of DfMA implementation was notably influenced by the collaboration and early procurement strategies between the timber design consultancy and the lead contractor. This early engagement facilitated proactive design discussions and the resolution of assembly sequences, thereby optimising constructability from the conceptual phase.

There were inconsistent levels of understanding of DfMA among the interviewees. Generally, those who had a closer connection with the factory environment tended to have a better knowledge of DfMA and could speak to specific instances of when DfMA principles had been applied in the construction project: 'DfMA is about mindset, it's about teaching that thinking, seeing and touching things and understanding that it includes the whole process from end-to-end' (factory manager, timber subcontractor). There was also the perception that DfMA is inherent in the delivery methodology of mass timber projects: 'From my experience, it was inherent in how we were operating' (project manager, timber subcontractor). Others mispronounced the acronym, confusing it with DFMEA (design failure mode and effect analysis). Some referenced the interview questions sheet before pronouncing the abbreviation, suggesting a lack of familiarity and uncertainty with the concept.

DfMA implementation in the project was discussed mainly by the lead contractor and timber subcontractor teams. The services subcontractor team instead talked about DfMA in construction more broadly. This could reflect the limited ability of the services subcontractor to provide DfMA input early on in the design process but could also be affected by the nature of the interviews themselves and the fact that not all services subcontractor interviewees were directly involved in the project.

Among the DfMA principles applied in the project were part count reduction as a design driver, provision for large tolerances (such as on panel corners and at panel connections) for improved machining and easier site assembly, material reduction through optimisation of reinforced laminated veneer lumber (LVL), standardisation of CLT mother panel types and through a kit-of-part approach, and the use of jigs for easier assembly of lift cores. Interviewees noted that this was achieved through collaboration facilitated by design workshops between architects and the timber delivery team. Additionally, the site manager's willingness to learn and implement mass timber methods of construction, early procurement strategies and continuous improvement to resolve assembly sequence, manufacturing and coordination issues early on contributed to the successful DfMA outcomes reported.

Interviewees emphasised the importance of early decisions at the project management level to adopt DfMA. This required promoting an integrated delivery process where designers and developers were involved in design discussions from the outset. Early procurement of the site team was also seen as beneficial, facilitating more design-related collaboration. However, interviewees cautioned that these design-related discussions relied heavily on digital workflows, 3D models and BIM. While these digital tools enabled late-phase design changes, interviewees highlighted a potential downside: a lack of consideration for the practical consequences of modifications.

Interviewees expressed concern that changes made to a digital model ('changing a line in a drawing') in the later phases of the project without a proper understanding of the flow-on effects could have serious unintended implications for the delivery of the project.

16.4.3 Discussion – synthesising lean and DfMA

The above findings show that while both lean construction and DfMA principles were applied by project teams during the construction project, their implementation varied in depth across different stakeholder groups, presenting both significant successes and persistent challenges.

Lean principles were predominantly applied as a process support focused on waste reduction, value enhancement and flow optimisation, particularly evident within the prefabrication facilities. Key successes included the elimination of panel flipping in the factory, the removal of scaffolding on-site and a highly synchronised material delivery sequence. Value was largely interpreted through a cost lens by many stakeholders, although the benefits of mass timber (e.g. quiet worksite, productivity) were also recognised.

However, the absence of a unified lean vision across the entire project team proved problematic, contributing to a fragmented and unaligned implementation. Similarly, while DfMA served as a crucial design driver for material minimisation (e.g. reducing costly LVL) and optimising logistics, its understanding was inconsistent, with deeper knowledge concentrated among factory-connected personnel.

Successes included specific DfMA applications such as part count reduction, provision of adequate tolerances, and attempts at standardisation and kit-of-parts. Both lean and DfMA efforts were significantly aided by early collaboration, procurement strategies and the use of digital tools such as 3D models and Obeya rooms. However, the late involvement of the services contractor limited their DfMA input, leading to rework. Furthermore, despite recognising the importance of early decision-making and integrated processes, the reliance on digital workflows for late-stage design changes sometimes overlooks practical implications.

Ultimately, the interviews highlight a collective recognition among parties that integrating lean and DfMA concepts more formally and comprehensively, particularly through fostering a unified understanding and ensuring early, cross-disciplinary engagement, is essential for unlocking further efficiencies and avoiding recurring pitfalls in future complex timber construction projects. This implementation relies on two forms of commitment: a defined theoretical framework to guide activities and clear recommendations for industry practitioners to effectively implement LC and DfMA in unison.

16.4.3.1 Supporting strategy through a theoretical framework

The findings of the analysis reveal a nuanced and often inconsistent application of LC and DfMA within the context of MMC in Australia. While certain stakeholders regarded these concepts as intrinsic to MET construction, the project exposed significant variation in understanding and application across different teams. Rather than representing a comprehensive integration, the efforts to implement LC and DfMA largely emerged as bottom-up initiatives, fragmented by disciplinary silos and constrained by procurement structures.

The requirement for integration of LC and DfMA in theory is underpinned by a shared focus on value maximisation and waste reduction. In the MET construction project, DfMA played a strategic role during design, particularly through simplification of part counts, incorporation of tolerances to ease site assembly and standardisation efforts intended to facilitate off-site production. In parallel, lean was deployed to support flow and process efficiency, most evident in the prefabrication facility and through sequencing efforts on-site. However, the limited early involvement of key stakeholders, most notably the services subcontractor, highlights the enduring challenge of aligning design and delivery processes in construction projects.

A key factor influencing the depth of DfMA integration was the close business relationship between the developer, lead contractor and timber design consultant, who subsequently expanded to become the timber supplier, all operating under a single umbrella company. This synergistic alignment afforded a significantly greater capacity for DfMA input in the design and execution of the MET structure compared to the building services.

This disparity in DfMA application likely contributed to variations in financial outcomes and perceptions of project success on site – for instance, the regional manager of the services subcontractor voiced critical concerns, stating: ‘If they’d have engaged us six months earlier… and worked with us at that stage, I think we could have made their penetration process a lot less painful’. This comment further confirms the fact that challenges arose because the services subcontractor’s involvement was largely confined to the delivery phase, excluding them from the crucial conceptualisation and early planning stages. This highlights how traditional procurement approaches often inhibit genuine collaboration and innovation, leading to a temporary organisational structure for the project team and stakeholders that is not conducive to long-term developmental trajectories or holistic value creation.

While there is a prevalent perception that DfMA and lean principles are inherent to MET project delivery, the findings suggest this may not be entirely accurate and the analysis demonstrates that actual delivery still depends on deliberate and systematic integration across all disciplines. While MET projects inherently achieve certain DfMA objectives (such as early design resolution) and lean objectives (such as fewer on-site processes), the interviews revealed challenges – for instance, projects embracing MET could not universally follow standardisation principles and in some cases, this did not align with the company’s core business model. This divergence could lead to several issues, including lengthy evaluation processes for alternative design solutions, additional costs associated with manufacturing and holding yards, the need for stronger cranes to lift larger components and site limitations concerning storage and transport. Furthermore, a shortage of familiarity and skill in the general workforce posed an additional barrier to realising the full benefits of these methodologies.

In a broader sense, this research supports the premise that lean can function as a powerful production management-based methodology for project delivery, capable of generating value for all stakeholders and significantly reducing waste. This waste often arises from the silos created between various project stakeholders: architects, designers, engineers, general contractors, trade contractors and suppliers. However, its successful implementation is contingent upon a clear definition of value by each stakeholder and a shared understanding of the value expected to be delivered and received within the project value chain. Crucially, lean thrives when contractual methods foster collaboration, as evidenced in this project, by aligning the owner, design team and construction team, underpinned by an organisational culture that promotes mutual respect among all individuals involved.

16.5. Implications

In theoretical terms, the findings of this research affirms that LC and DfMA are complementary, with DfMA acting as a strategic design tool that supports lean principles of flow, value delivery and waste minimisation. LC provides an overarching management and delivery philosophy that guides project execution and operational decision-making. Together, they support a more industrialised, design-led construction paradigm, a necessary evolution to achieve the productivity and quality improvements sought through MMC.

Based on these insights and to support strategic implementation, a theoretical framework is proposed, consisting of the following core elements.

- *Articulated and shared value definitions,* co-developed and maintained by all stakeholders throughout the project life cycle.
- *DfMA-led design integration* enabling consideration of manufacturing and assembly constraints from the earliest design stages and working towards a successful design freeze.
- *Lean as an overarching delivery philosophy* guiding both high-level project organisation and detailed operational decisions.
- *Cross-disciplinary early engagement* facilitating integrated procurement models and collaborative contracting approaches.
- *Digital and data-enabled feedback systems* supporting continuous improvement and enabling design-intent alignment with construction realities.

This framework seeks to move beyond isolated technical improvements to foster a more holistic, systems-oriented approach that balances efficiency with adaptability and broader value outcomes. It positions LC and DfMA not as separate or static tools but as interdependent strategies that, when effectively integrated, can support a transformation towards more resilient and sustainable construction practices.

16.5.1 Supporting operations through recommendations for practice

From a practical perspective, this applied research illustrates both the opportunities and the challenges of integrating LC and DfMA within construction projects, particularly in the context of MMC. These outcomes are framed through five clear recommendations for construction industry practitioners to support the operationalisation of LC and DfMA within their organisations and with their collaborators to drive improved performance.

Practice Recommendation 1: Integrated engagement and procurement

Projects must be approached with a view towards integration and a focus on project delivery models or alliance contracting that enable and embed collaboration structurally rather than relying on ad hoc goodwill.

Practice Recommendation 2: Establishing a shared value case

Project teams must commit to a collective understanding of value, revisited throughout delivery, which is therefore fundamental to achieving coherent outcomes and avoiding localised optimisations that undermine project-wide benefits.

Practice Recommendation 3: Supporting new skills and knowledge

Organisations should focus on supporting the workforce through educational and professional development initiatives that prioritise MMC competencies and capabilities, supported by industry bodies and policy instruments where appropriate.

Practice Recommendation 4: Integrating digital tools with practice-based knowledge

Taking an integrative view of digitalisation, means that practitioners should approach digital workflows critically, ensuring technology enhances, rather than replaces, the relational and experiential aspects of design and delivery, thereby benefiting from extant practice-based knowledge.

Practice Recommendation 5: Cultural commitment

Lean and DfMA must be embedded as organisational philosophies rather than confined to isolated projects and approached as technical tools. This applied research indicates that where these cultural underpinnings were stronger, integration was more successful and more enduring, showing promise for future industry practice.

For future research, there is scope to further develop and empirically validate integrated frameworks that operationalise LC and DfMA synergies across different typologies and project scales. The contractual and organisational conditions necessary for successful integration warrant deeper exploration, as do the long-term impacts on cost, sustainability and end-user experience.

Additionally, emerging technologies, such as real-time feedback systems that leverage digital twins and artificial intelligence/machine learning support, offer potential to support continuous improvement and closer alignment of design and delivery phases, although their practical limitations require careful examination.

Ultimately, the integration of LC and DfMA represents an evolution in construction practice rather than a mere technical enhancement. By fostering early collaboration, aligning value definitions and embedding continuous learning, these approaches have the potential to support a more adaptive, resilient and responsive construction sector. Such a shift is increasingly necessary as the industry seeks to address complex challenges, including environmental imperatives and changing societal expectations. Through deliberate and integrated adoption, LC and DfMA offer a pathway towards a construction industry that is not only more efficient but also more capable of delivering long-term value to clients, communities and the environment.

16.6. Conclusions

This applied research demonstrates that lean and DfMA are highly complementary approaches with significant potential to improve productivity, collaboration and waste reduction in construction, particularly in projects involving mass engineered timber and MMC. The analysis of the MET construction project demonstrates how lean's production management principles, focused on value creation and waste elimination across the project value chain, can be effectively married with DfMA's strategic design and production optimisation to deliver substantial efficiencies and stakeholder value.

Successful integration of lean and DfMA depends on clear, shared definitions of value among all stakeholders, early and meaningful involvement of designers, contractors and subcontractors, and contractual models that foster genuine collaboration rather than siloed operations. The research highlights the importance of early procurement and digital workflows such as BIM to facilitate this integration, while also recognising limitations where digital changes do not always translate seamlessly into construction realities.

The findings reveal that while bottom-up lean and DfMA initiatives occur within organisations, a unified vision and top-down support are essential to sustain benefits throughout the full project life cycle and, importantly, between projects to leverage continual improvement. The close alignment between key project players under one organisational umbrella significantly enhanced DfMA adoption in MET design and execution, but also revealed disparities when other disciplines, such as building services, were excluded early on.

This study contributes to closing the gap in both academic and practical knowledge by showcasing the integrated application of lean and DfMA within a real project. Traditionally, these methodologies are applied in isolation and few studies have explored their combined potential. The research illustrates that these methodologies could work as interconnected strategies for unlocking MMC's full potential. However, industry-wide changes, including cultural shifts, cross-disciplinary training, early engagement, appropriate incentives and improved data management practices are necessary to support their wider adoption.

Future research should focus on developing integrated frameworks, exploring contractual and cultural enablers and assessing long-term impacts on project success and sustainability. Ultimately, harnessing the synergy between lean and DfMA offers a promising pathway toward a more efficient, collaborative and sustainable construction industry, one that is fit for purpose in today's modern age.

Acknowledgements

This research was funded by the Building 4.0 Cooperative Research Centre. The authors would like to thank all the participants who generously shared their time and insights as interviewees for this research.

REFERENCES

Aitchison M, Bashirzadeh Tabrizi T, Beim A *et al.* (2018) *Prefab Housing and the Future of Building: Product to Process. Illustrated edn*. Routledge, London, UK.

Albalkhy W and Sweis R (2021) Barriers to adopting lean construction in the construction industry: a literature review. *International Journal of lean Six Sigma* **12(2)**: 210–236.

Bogue R (2012) Design for manufacture and assembly: background, capabilities and applications. *Assembly Automation* **32(2)**: 112–118.

Boothroyd G (1987) Design for assembly – the key to design for manufacture. *The International Journal of Advanced Manufacturing Technology* **2(3)**: 3–11.

Bralla JG (ed.) (1999) *Design for Manufacturability Handbook*, 2nd ed. McGraw-Hill, New York, NY, USA.

BCA (Building Construction Authority) (2018) *Design for Manufacture and Assembly: Prefabricated Mechanical, Electrical and Plumbing (MEP) Systems*. BCA, Singapore.

Chick G (2013) *Implementing lean in Construction: A lean Guide for Client Organisations*. CIRIA, London, UK.

Diekmann J E., Krewedl M, Balonick J, Stewart T and Won S (2004) *Application of lean Manufacturing Principles to Construction*. Construction Industry Institute, Boulder, CO, USA.

Fox S, Marsh L and Cockerham G (2001) Design for manufacture: a strategy for successful application to buildings. *Construction Management and Economics* **19(5)**: 493–502.

Gao S and Low SP (2014). *Lean Construction Management*. Springer, Singapore.

Gao S, Jin R and Lu WS (2019) Design for manufacture and assembly in construction: a review. *Building Research & Information* **48(5)**: 538–550.

Gerth R, Boqvist A, Bjelkemyr M and Lindberg B (2013) Design for construction: utilizing production experiences in development. *Construction Management and Economics* **31(2)**: 135–150.

Ghosh S and Burghart J (2021) lean construction: experience of US contractors. *International Journal of Construction Education and Research* **17(2)**: 133–153.

Hedrick TE, Bickman L and Rog DJ (1993) *Applied Research Design: A Practical Guide*. SAGE Publications, Newbury Park, CA, USA.

Koskela L (1992), Application of the New Production Philosophy to Construction. *CIFE Technical Report No. 72*. Stanford University, Stanford, CA, USA.

Koskela L, Howell G, Ballard G and Tommelein I (2002) The foundations of lean construction. In *Design and Construction: Building in Value* (Best R and de Valence G (eds)). Butterworth-Heinemann (now Routledge), Oxford, UK, pp. 211–226.

Kuzmanovska I (2020), Building Design for Manufacture and Assembly (BDfMA): Reframing the Design-Production Relationship in Industrialised Construction. PhD Thesis, Monash University.

Laing O'Rourke. (2013) The future of DfMA is the future of construction' (Industry Publication No. 1; *Engineering Excellence Journal*, p. 76). Laing O'Rourke.

Langston C and Zhang W (2021) DfMA: towards an integrated strategy for a more productive and sustainable construction industry in Australia. *Sustainability* **13(16)**.

Lu W, Tan T, Xu J *et al.* (2020) Design for manufacture and assembly (DfMA) in construction: the old and the new. *Architectural Engineering and Design Management*, **17(3–4)**, pp. 158–176. https://doi.org/10.1080/17452007.2020.1768505.

Lu W, Tan T, Xu J *et al.* (2021) Design for manufacture and assembly (DfMA) in construction: the old and the new. *Architectural Engineering and Design Management* **17(1–2)**: 77–91.

McHugh K, Dave B and Koskela L (2022) On the role of lean in digital construction. In *Industry 4.0 for the Built Environment (Structural Integrity*, vol 20) (Bolpagni M, Gavina R and Ribeiro D (eds)). Springer, Cham, Switzerland.

Meng X (2019) lean management in the context of construction supply chains. *International Journal of Production Research* **57(11)**: 3784–3798.

Mossman A (2018) What Is lean Construction: Another Look. In *Proceedings of the 26th Annual Conference of the International Group for lean Construction* (IGLC 26).

Ng MS and Hall DM (2019) Toward lean management for digital fabrication: a review of the shared practices of lean, DfMA and DFAB. In *Proc. 27th Annual Conference of the International Group for lean Construction (IGLC)* (Pasquire C and Hamzeh FR (eds)). Dublin, Ireland, pp. 725–736.

Paez O, Salem S, Solomon J and Genaidy A (2005) Moving from lean manufacturing to lean construction: toward a common sociotechnological framework. *Human Factors and Ergonomics in Manufacturing & Service Industries* **15(2)**: 233–245.

Patton MQ (2015) *Qualitative Research & Evaluation Methods*, 4th ed. Sage Publications, Thousand Oaks, CA, USA.

RIBA (Royal Institute of British Architects) (2013) *RIBA Plan of Work 2013: Designing for Manufacture and Assembly*. Royal Institute of British Architects (RIBA), London, UK.

RIB (2021) *DfMA Overlay to the RIBA Plan of Work—Mainstreaming Design for Manufacture in Construction*. Royal Institute of British Architects (RIBA), London, UK.

Staub-French S, Zadeh P, Calderon F *et al.* (2018) *Building Information Modelling (BIM) and Design for Manufacturing and Assembly (DfMA) for Mass Timber Construction*. BIM Topics Research Lab, Vancouver, Canada.

Tan T, Lu W, Tan G *et al.* (2020) Construction-Oriented Design for Manufacture and Assembly (DfMA) Guidelines. *Journal of Construction Engineering and Management* **146(8)**: 04020085.

Toyota (no date) Production System, Toyota Global. https://global.toyota/en/company/vision-and-philosophy/production-system/ (accessed 07/07/2025).

Trinder L (2018) Design for manufacture and assembly: its benefits and risks in the UK water industry. *Proceedings of the Institution of Civil Engineers - Management, Procurement and Law* **171(4)**: 152–163.

Vaz-Serra P, Wasim M and Egglestone S (2021) Design for manufacture and assembly: a case study for a prefabricated bathroom wet wall panel. *Journal of Building Engineering* **44**: 102849.

Womack JP, Jones DT and Roos D (1990) *The Machine That Changed the World: The Story of lean Production—Toyota's Secret Weapon in the Global Car Wars That Is Now Revolutionizing World Industry*. Rawson Associates, New York, NY, USA.

Yu HT, Al-Hussein M, Al-Jibouri S and Telyas A (2013) lean transformation in a modular building company: a case for implementation. *Journal of Management in Engineering* 29 (1): 103–111.

Emmanuel I Daniel and Fidelis A Emuze
ISBN 978-1-83708-437-1
https://doi.org/10.1108/978-1-83708-436-420261020

Chapter 17
Modern methods of construction in Africa: towards effective adoption

Wakisa Simukonda[1], Fidelis A. Emuze[2] and Emmanuel I. Daniel[3]

[1]*School of Built Environment, Malawi University of Business and Applied Sciences, Malawi*
0000-0003-1940-0840
[2]*Department of Engineering Technology, Texas State University, Texas, USA*
0000-0001-7714-4457
[3]*School of Architecture and Built Environment, University of Wolverhampton, UK*
0000-0002-5675-1845

Abstract

Africa faces persistent housing and infrastructure deficits driven by rapid urbanisation, population growth and the limitations of traditional construction methods (TCM). Modern methods of construction (MMC), including modular, prefabricated and industrialised systems, provide an off-site, industrialised alternative that enhances productivity, quality and sustainability. However, adoption across the continent remains limited. This chapter draws on a desktop-based review of secondary data from six MMC case studies in Nigeria, Ghana, Namibia, Egypt and Mozambique, employing an epistemological approach combining interpretivism and critical realism to capture both mediated stakeholders' perspectives and the structural mechanisms influencing MMC adoption. Findings highlight key challenges, including financing constraints, regulatory gaps, skills shortages and cultural barriers, alongside success factors such as government leadership, supportive policy frameworks, capacity building, local manufacturing and public awareness. International collaborations and public–private partnerships (PPPs) emerge as effective strategies for technology transfer, risk mitigation and scalability. Based on these insights, the chapter proposes a framework for MMC adoption anchored on five interlinked pillars: policy and regulation; finance and investment; capacity and knowledge development; local manufacturing and supply chain integration; and cultural and public acceptance. The framework offers a strategic roadmap to accelerate affordable housing delivery, stimulate industrial growth and enhance urban resilience across Africa.

Keywords: adoption; Africa; construction; financing; housing; modular

17.1. Introduction

The construction industry in Africa plays a pivotal role in socio-economic development, contributing between 3% and 10% of national gross domestic product (GDP) and providing extensive employment opportunities, particularly for low-skilled labour (AfDB, 2019; UNECA, 2022). It underpins infrastructure development, industrialisation and urbanisation, which are key pillars for achieving the African Union's Agenda 2063: The Africa We Want (African Union, 2015). Despite its significance, the sector remains fragmented, predominantly informal and heavily reliant on traditional construction methods (TCM) (Gier *et al.*, 2025). These methods, often labour-intensive, slow and resource-inefficient, are increasingly inadequate for meeting the continent's escalating housing demands (Ofori, 2015).

Rapid population growth and accelerated urbanisation are intensifying housing pressures. Currently, over 700 million Africans reside in urban areas, a number projected to rise substantially by 2050 (Africa Centre for Strategic Studies, 2025). Globally, the United Nations estimates that nearly 3 billion people, around 40% of the population, will require adequate housing by 2030 (UN-Habitat, 2024). In Africa, persistent housing deficits exemplify the scale of the challenge. Nigeria faces an estimated shortfall of 17 million units (Okonjo-Iweala, 2014), while Egypt and South Africa continue to contend with sprawling informal settlements (CAPMAS, 2024; Marutlulle, 2021). Traditional construction approaches have struggled to bridge these gaps due to slow delivery, high costs and inefficiencies (Ibimilua and Ibitoye, 2015).

MMC have emerged as a strategic response to these challenges. MMC encompasses prefabricated, modular and industrialised construction systems that enable faster, higher-quality and more sustainable housing delivery (Goodier and Gibb, 2007). By integrating off-site manufacturing, digital design tools and standardised processes, MMC offers opportunities to reduce material waste, optimise labour and improve overall project efficiency (Li and Paik, 2024). MMC adoption aligns with Sustainable Development Goal (SDG) 11, which promotes inclusive, safe and resilient urban settlements, as well as the International Covenant on Economic, Social and Cultural Rights (ICESCR).

This chapter critically examines the adoption and implementation of MMC in Africa, drawing on secondary case studies from Nigeria, Ghana, Namibia, Egypt and Mozambique. The analysis focuses on the challenges and success factors influencing MMC uptake, highlighting the roles of government leadership, regulatory frameworks, innovative financing, capacity development, local manufacturing and cultural acceptance. The chapter seeks to answer the overarching question: *What insights do secondary case studies of MMC projects provide regarding the challenges and success factors influencing MMC adoption in Africa?*

The insights from the findings inform the development of a framework for the effective adoption and implementation of MMC, aimed at addressing the continent's persistent housing deficit. By situating MMC within Africa's housing and infrastructure landscape, this chapter provides a foundation for understanding how industrialised construction methods can accelerate affordable housing delivery, stimulate industrial development and enhance urban resilience across the continent.

17.2. Perspectives on modern methods of construction

17.2.1 Traditional construction methods and their limitations

TCM, also known as conventional construction, has long dominated Africa's built environment. These methods depend on on-site manual labour and sequential assembly of materials such as brick, block, timber and concrete (Gibb, 1999). They provide flexibility for design modification but are often associated with inefficiencies in cost, time and quality (Ball, 1999). Material wastage rates under TCM range between 10% and 30% (Poon *et al.*, 2001), while the lack of standardisation contributes to inconsistent workmanship and reduced structural quality (Agyekum *et al.*, 2020).

Additionally, reliance on site-based activities exposes projects to delays from weather conditions and occupational safety risks (Gibb and Isack, 2003). These limitations have motivated the search for alternative construction systems capable of achieving higher productivity, better quality control and greater sustainability, leading to the emergence of MMC.

17.2.2 Historical evolution of MMC

Although the term 'modern methods of construction' suggests a recent innovation, its origins trace back centuries. In medieval Europe, timber components were precut off-site and assembled rapidly on location (Green, 2022). The Industrial Revolution in the eighteenth and nineteenth centuries marked a turning point, as prefabricated iron and steel elements were produced for structures such as bridges and railway stations (Satya, 2008). The Crystal Palace Exhibition Hall (1851) in London remains a landmark example of early modular construction (Green, 2022). Following the two world wars, housing shortages in Europe accelerated the use of prefabrication. The UK's post-World War II 'Prefab' and 'Airey' houses, designed for mass production and rapid deployment, laid the groundwork for modern industrialised housing.

17.2.3 Policy evolution and conceptual frameworks

The formal recognition of MMC within policy and research discourse was catalysed by key UK government reports, notably Latham's Constructing the Team (1994) and Egan's Rethinking Construction (1998). Both emphasised collaboration, efficiency, sustainability and customer focus, which are principles central to MMC. These reports inspired similar reforms worldwide, promoting off-site manufacturing, lean construction and integrated project delivery (Blismas *et al.*, 2006; Payne and Serin, 2023).

MMC now encompasses a broad spectrum of techniques, including modular construction, panelised systems, volumetric units, 3D printing and hybrid systems, all designed to enhance productivity and sustainability (Abanda *et al.*, 2017; Pan and Sidwell, 2011). Globally, the method is increasingly recognised as a response to housing shortages, labour constraints and environmental pressures (Abanda *et al.*, 2017; Pan and Sidwell, 2011).

17.2.4 Global adoption and the housing imperative

In the twenty-first century, MMC has emerged as a central strategy to address the escalating global housing crisis. UN-Habitat (2024) estimates that nearly 3 billion people, approximately

40 per cent of the world's population, will require adequate housing by 2030, necessitating the construction of 96 000 affordable homes daily. Developed countries such as the United Kingdom, the United States, Germany and Sweden have institutionalised MMC within national housing policies to enable faster, cost-effective and higher-quality housing delivery (Payne and Serin, 2023; White *et al.*, 2020) – for example, the UK Government's 'Accelerated Construction' programme and the 'Home Building Fund' aim to stimulate MMC growth (Lord's Library, 2025). Similarly, in Ireland, MMC is promoted to improve construction efficiency and address housing shortages (NESC, 2024). These initiatives reflect a global consensus on the potential of MMC to meet urgent housing demands through scalable, efficient and sustainable construction approaches.

17.2.5 MMC in the African context

In Africa, the adoption of MMC is primarily driven by urgent demand for affordable housing amid rapid population growth and urban expansion. Countries such as Nigeria, Egypt and South Africa recognise MMC's potential to reduce housing deficits, although financial, technical and institutional challenges persist (Bakhaty *et al.*, 2024; Makinde, 2014). National housing programs are increasingly integrating prefabricated and modular construction to improve scalability, efficiency and cost-effectiveness.

These initiatives align with global frameworks, including UN-Habitat, the International Covenant on Economic, Social and Cultural Rights (ICESCR) and Sustainable Development Goal 11, which advocate inclusive, resilient and sustainable urban settlements.

Realising MMC's full benefits requires an enabling policy environment, targeted capacity development, innovative financing mechanisms and multi-stakeholder collaboration (Makalima and Sokhetye, 2025). This chapter examines the challenges and success factors influencing MMC adoption in Africa and proposes a framework to guide countries in addressing housing shortages through a scalable, MMC-driven approach.

17.3. Methodological approach

The primary aim of this book chapter is to explore the key challenges and success factors shaping the adoption and implementation of MMC across Africa, to formulate a framework and practical recommendations to promote effective MMC uptake and alleviate the continent's persistent housing shortages. The chapter adopts a desktop research design (Karimi *et al.*, 2024; Scolobig *et al.*, 2023) and employs a dual epistemological approach integrating interpretivism and critical realism (Bogna *et al.*, 2020). This combination enables a comprehensive understanding of observable events and underlying causal mechanisms, including structural linkages (Peters *et al.*, 2013).

Through an interpretivist lens, the research explores stakeholders' interpretations of MMC adoption challenges and success factors, drawing exclusively on secondary published case materials, including academic articles, policy reports and institutional frameworks. This approach captures the nuances of MMC adoption and implementation in Africa without requiring primary data collection. The critical realist perspective situates these interpretations within broader institutional, organisational and policy contexts. As Sayer (1992) notes, this facilitates an arguably pragmatic account of causality, enabling insight into the underlying structures and causal mechanisms (Easton, 2010; Lawan, 2021).

By combining these epistemological perspectives, the chapter moves beyond description to provide explanatory insights into how and why specific factors enable or hinder MMC implementation in Africa, offering a hermeneutic link between observed realities and underlying causal mechanisms.

The analysis draws on secondary case studies chosen to highlight best practices and contextual variations in MMC adoption. Case studies were initially identified through online searches and subsequently screened for credibility and relevance. Only cases appearing in peer-reviewed journal articles, published by reputable news outlets (e.g. national newspapers), republished on official government websites or documented by recognised MMC contractors and consultants, were included in the analysis. Notable examples include the following.

(a) *Peridot Parkland Estate, Lagos, Nigeria* is a PPP delivering 251 affordable housing units through MMC, highlighting the role of government facilitation, financing and regulatory oversight (Edge, 2020).
(b) *FastHouse Modular Housing, Ghana* demonstrates international collaboration in modular construction, emphasising knowledge transfer, cost efficiency and rapid assembly (Wuni and Shen, 2019).
(c) *Namibia's Smart Construction Pilot in Windhoek* delivered 100 units in 6 months, addressing cultural and financing challenges through smart technology, PPPs, community sensitisation and affordable financing models (WEF, 2021).
(d) *Adom City Estate in Accra, Ghana*, delivered 2000 modular homes in 12 months, overcoming awareness and policy gaps through high-quality construction, efficient assembly, community education and advocacy for MMC-supportive policies (African land, 2020). Another MMC project in Ghana, namely, *Ashaiman Modular Housing Project* (ongoing), focuses on affordable housing, overcoming high setup costs and local resistance through international partnerships, builder training programs and government incentives for modular construction adoption (African land, 2020).
(e) *Mozambique's low-cost prefabricated housing* delivered 96 units in 2019, featuring energy-efficient technologies. Challenges included high setup costs and local resistance, addressed through community sensitisation and affordable financing models (House-Matic, 2019).
(f) *Egypt's CSCEC Apartment Project in Alaman* delivered 237 modular units. Challenges included high setup costs and local resistance, which were mitigated through smart technologies, PPPs, community sensitisation and affordable financing models (GS Housing, 2022).

These case studies of MMC projects are synthesised to identify recurring challenges and success factors in MMC adoption in Africa. Drawing on global and African experiences, the chapter provides practical recommendations, informing a framework and actionable strategies to promote sustainable, scalable and contextually relevant adoption across the continent.

17.4 Results and discussion

17.4.1 Challenges of MMC adoption in Africa

The documented case studies of MMC projects provide insights into the challenges of adopting MMC in Africa.

17.4.1.1 High initial investment

High initial investments present a major obstacle to MMC adoption across Africa. Projects like the Peridot Parkland Estate in Nigeria and the CSCEC modular apartment project in Egypt demonstrate that setting up modular production facilities demands significant capital for equipment, logistics and specialised labour (Edge, 2020; GS Housing, 2022). Likewise, the Ashaiman Modular Housing Project in Ghana and Namibia's Smart Construction Pilot involved large upfront costs for plant setup, technology acquisition and workforce training, often surpassing traditional construction budgets (African Land, 2020; WEF, 2021).

These financial requirements pose particular challenges for small-medium enterprises (SMEs), which dominate Africa's construction industry and typically lack access to affordable long-term financing. Without targeted incentives, concessional loans or PPPs, high start-up costs hinder scalability, confine adoption to pilot projects and limit private sector involvement, preventing MMC from becoming a widespread construction method across the continent.

17.4.1.2 Limited access to finance

Access to finance is a crucial factor in the adoption of MMC in Africa. Projects such as Mozambique's low-cost prefabricated housing and Egypt's CSCEC modular development show that high initial costs for factory setup, equipment and imported components often surpass the capacity of local developers (GS Housing, 2022; House-Matic, 2019).

In Namibia and Ghana, financial barriers also reduce end-user affordability, as limited mortgage options and high interest rates restrict demand for modular homes (WEF, 2021; Wuni and Shen, 2019). Perceptions of MMC as high risk, along with weak regulatory and certification systems, further limit credit access from public and private lenders. While PPP models, such as Nigeria's Peridot Parkland Estate, offer some relief, funding remains inconsistent and reliant on donors.

Addressing these barriers requires government-supported credit schemes, concessional financing and policies that mitigate investment risks, supporting the growth of sustainable, MMC-based affordable housing across Africa.

17.4.1.3 Limited market demand

Low market demand remains a major barrier to MMC adoption across Africa. Case studies like Adom City Estate and FastHouse Ghana demonstrate that although modular projects can provide affordable, high-quality homes quickly, their adoption is limited due to low awareness, cultural preferences for traditional brick-and-mortar construction and limited consumer financing (African Land, 2020; Wuni and Shen, 2019).

In Mozambique and Namibia, slow sales were worsened by limited mortgage access and skepticism toward prefabricated materials (House-Matic, 2019; WEF, 2021). Fragmented housing markets further hinder economies of scale, increase production costs and discourage private investment. On the other hand, government-led initiatives, such as Nigeria's Peridot Parkland Estate PPP, achieved higher adoption through guaranteed demand and supportive regulations.

Expanding MMC requires coordinated policy efforts, large-scale public housing programs and community engagement to build trust, boost demand and integrate modular construction into mainstream African housing markets.

17.4.1.4 Limited research and development

A major obstacle to sustainable MMC growth in Africa is limited investment in research and development (R&D) and digital innovation. Case studies from Ghana, Namibia and Egypt show that most MMC projects rely heavily on imported technologies, emphasising the lack of local innovation systems (GS Housing, 2022; WEF, 2021; Wuni and Shen, 2019).

The FastHouse Ghana project demonstrates the advantages of international collaboration for knowledge transfer, but technology adaptation and capacity building still fall short. Low adoption of building information modelling (BIM), computer-aided design (CAD) and automated manufacturing continues due to insufficient training, poor digital literacy and high software costs (Osei-Tutu *et al.*, 2021). Unlike China's government-funded modular research centres, African countries lack sustained public–private R&D investment, which limits local material innovation and cost reduction.

Strengthening partnerships between universities, contractors and policymakers can support indigenous MMC research, leading to affordable, locally suitable and technologically advanced construction solutions.

17.4.1.5 Lack of government support

The lack of strong government support significantly hampers MMC adoption in Africa. In Ghana, MMC is not fully recognised in policy, which limits its implementation (African Land, 2020). Globally, MMC prospers where governments serve as anchor clients, creating stable demand through public housing and infrastructure projects (Goodier and Gibb, 2007).

In many African countries, limited policy frameworks, incentives and technical standards discourage private investment in factories and supply chains (Wuni and Shen, 2020). Outdated building codes for conventional methods further slow approval processes for prefabricated systems (Bello *et al.*, 2023). Without tax incentives, concessional financing or subsidies, high upfront costs remain a barrier, especially for SMEs (Hong *et al.*, 2018). Gaps in design standards and certification schemes in Nigeria and South Africa weaken quality assurance and reinforce social inertia (Ortega *et al.*, 2023; Wuni and Shen, 2020). Experiences from Malaysia and Singapore demonstrate that proactive regulation and government engagement can promote large-scale MMC adoption, a model Africa could follow.

17.4.1.6 Knowledge and skills gap

The successful adoption of MMC in Africa depends on specialised expertise in design, manufacturing, logistics and installation. Many countries, such as Nigeria, face significant shortages of trained professionals due to limited MMC-focused education, little research and few hands-on learning opportunities (Edge, 2020). This skills gap results in a lack of qualified designers, contractors, manufacturers and workers, leading to higher costs, lower quality and inadequate inspection systems.

MMC also requires precise control of tolerances, complex module interfaces and error correction, skills often missing locally. As a result, reliance on expatriate expertise raises costs and hampers sustainability (Bello *et al.*, 2023; Ortega *et al.*, 2023). Closing this gap calls for focused investment in education and training, the creation of research and innovation hubs, partnerships between industry and academia,, and pilot projects to build local expertise and expand MMC adoption across Africa.

17.4.1.7 Technical and cultural considerations

Both technical and cultural factors limit MMC adoption in Africa. Technically, MMC needs advanced manufacturing, precise tolerances and efficient logistics, but poor infrastructure, unreliable power and insufficient transport disrupt supply chains (Wuni and Shen, 2020). The absence of local standards and testing facilities also weakens developer and regulator confidence (Bello *et al.*, 2023).

Culturally, homeownership is strongly connected to permanence and traditional masonry construction, making MMC seem experimental, costly or unsuitable (Ortega *et al.*, 2023; Zhang *et al.*, 2014). Resistance from local builders, as seen in Egypt, reinforces these views (GS Housing, 2022). Such attitudes influence client and policymaker preferences (African Land, 2020). When combined with the construction sector's fragmented and conservative nature, these barriers sustain cultural inertia, limiting MMC adoption despite its potential to address housing and infrastructure gaps.

17.4.2 Key success factors for MMC adoption in Africa

The obstacles hindering the widespread adoption of MMC in Africa call for coordinated action from government, industry and academia to redirect its trajectory. A desk-based review of MMC projects in Africa reveals that MMC success is predicated on several key factors outlined below.

17.4.2.1 International collaborations and capacity building

International collaborations between global construction firms and African governments are crucial for advancing MMC across the continent. These partnerships enable technology and skills transfer, introducing local industries to modular systems, automation and digital design tools (Modular Building Institute, 2023). They help overcome capacity constraints, enabling faster, higher-quality housing and infrastructure delivery, while improving financial credibility and attracting private investment (Quarrying Africa, 2025).

Such collaborations also support localisation of MMC technologies, adapting them to Africa's climatic, cultural and material contexts (African Review, 2024). Examples include EchoStone Nigeria partnering with a Swedish firm, Ghana collaborating with Limavady from Northern Ireland, and a UK-based Etopia pilot project in Namibia (Edge, 2020; WEF, 2021; Wuni and Shen, 2019). Evidence from China and Malaysia shows structured training enhances efficiency (Hong *et al.*, 2018), indicating Africa can benefit from vocational and academic upskilling initiatives (Bello *et al.*, 2023).

These partnerships offer a pathway for cost-effective, rapid and resilient construction to meet urbanisation and housing goals.

17.4.2.2 Strong government support and policy frameworks

Evidence from case studies highlights the pivotal role of government in driving large-scale MMC adoption in Africa. Governments influence uptake through policy, regulatory frameworks, public procurement and investment incentives. Projects in Ghana, Nigeria and Egypt demonstrate that streamlined approvals, funding facilitation and inclusion of MMC in national housing strategies accelerate adoption (Edge, 2020; WEF, 2021; Wuni and Shen, 2019).

Tanzania's proposed MMC housing project, for example, calls for 'government approval and support' for a 5000-unit PPVC complex in Dar es Salaam (Nair, 2025). International experience reinforces this. In Singapore and Hong Kong, government-led programmes created predictable demand, set MMC standards, mandated modular use in public projects and offered fiscal incentives (Lu *et al.*, 2018). In Africa, public project allocation and incentives such as tax breaks, concessional loans and land allocation can similarly enhance market confidence and private sector participation (Ortega *et al.*, 2023).

17.4.2.3 Investment in local manufacturing and supply chains

Developing local prefabrication and modular manufacturing facilities is critical for reducing reliance on imported components, lowering project costs and enhancing long-term sustainability. Experiences from Malaysia demonstrate that localised MMC factories can improve housing affordability, generate employment and build skilled labour capacity (Musa *et al.*, 2014). For Africa, regional MMC factories can overcome logistical challenges, create economies of scale and foster industrial systems (Wuni and Shen, 2020).

The evidence from documented case studies shows a significant gap in local manufacturing capacity, constraining MMC adoption and increasing dependency on imports. A proposed MMC mass housing project in Dar es Salaam, Tanzania, advocates for a state-of-the-art prefabrication facility to establish a self-sustaining supply chain, reduce import reliance, stimulate local economic activity and provide long-term employment.

Structured training programs for the workforce will enhance technical capacity and enable sustainable industrial growth in the construction sector (Nair, 2025).

17.4.2.4 Access to finance and innovative funding models

The high capital outlay required for MMC facilities and logistics remains a major barrier to broader adoption globally. Countries with mature MMC industries have addressed this through innovative financing models. Hong *et al.* (2018) note that China, Singapore, Brazil and India have leveraged PPPs, concessional loans and tax incentives to attract private investment, a model applicable to African housing and infrastructure (Bakhaty *et al.*, 2024). In Tanzania, a modular housing proposal using prefabricated prefinished volumetric construction

(PPVC) outlines a two-tier financing model combining international and local financial institutions, private investment, PPPs, and government grants or subsidies (Nair, 2025). Case studies in Nigeria, Ghana and Egypt show PPPs as a cornerstone of MMC implementation. Such mechanisms can de-risk private investment, enabling strategic investment in local MMC manufacturing. This approach is essential for scaling modern construction in Africa, stimulating economic growth and strengthening local technical capacity.

17.4.2.5 Public awareness and cultural acceptance

Perceptions that MMC are temporary or low-quality have hindered adoption in Africa (Zhang *et al.*, 2014). In Namibia, Ghana and Mozambique, cultural norms and public opinion often label MMC buildings as inferior, creating resistance (African Land, 2020; House-Matic, 2019; WEF, 2021). Addressing these perceptions requires combining technical demonstration with community engagement.

Public awareness campaigns, pilot projects and stakeholder consultations have been shown to shift attitudes effectively. MMC pilot projects in India and Mauritius demonstrated durability, efficiency and cost-effectiveness, building user confidence (Abanda *et al.*, 2017). Similarly, case studies in Namibia, Ghana and Nigeria highlight community sensitisation initiatives that educate societies on MMC benefits and foster cultural acceptance (House-Matic, 2019; WEF, 2021; Wuni and Shen, 2019).

Building social acceptance alongside technical adoption is essential for mainstreaming MMC in Africa, ensuring that modern construction methods are socially embraced, technically sustainable and culturally accepted (Ortega *et al.*, 2023).

17.4.3 Strategies for MMC expansion in Africa: government, policy and financing

Government involvement, supportive policy frameworks and innovative financing are pivotal for scaling MMC across Africa (See Table 17.1). Governments, as major public-sector clients, can create stable demand for modular housing, stimulating market confidence. The Peridot Parkland Estate in Nigeria illustrates how government-facilitated PPPs enabled delivery of 251 modular units through regulatory oversight, financial support and streamlined approvals (Edge, 2020). Similarly, Egypt's CSCEC Apartment Project benefited from active government engagement, ensuring compliance and access to financing (GS Housing, 2022). International experience from Singapore, Malaysia and China shows that government mandates, fiscal incentives and integration of MMC in housing strategies reduce investor risk and attract private participation (Gao *et al.*, 2018; Lu *et al.*, 2018). Using public projects as anchor clients provides predictable demand, encouraging large-scale adoption and industrial investment.

Policy and governance mechanisms are equally critical. The success of FastHouse Ghana and Namibia's Smart Construction Pilot underscores the importance of training, technical standards and community engagement in ensuring quality and cultural acceptance (WEF, 2021; Wuni and Shen, 2019). Africa, however, faces shortages of skilled professionals in modular design, digital fabrication and assembly.

Table 17.1 Key African MMC projects and enabling strategies

Case Study	Country	Units / Timeline	Government / Policy Role	Financing / PPP Models
Peridot Parkland Estate	Nigeria	251 units	Government facilitation, regulatory oversight and approvals	PPP; public-private risk sharing
Smart Construction Pilot	Namibia	100 units / 6 months	Policy support, community sensitisation	PPP with affordable end-user financing
Adom City Estate	Ghana	2000 units / 12 months	Advocacy for MMC-supportive policies, community engagement	Mixed financing with international collaboration

Source: Authors' creation

Coordinated initiatives such as MMC-focused curricula, vocational programs, accreditation schemes and BIM adoption can bridge this gap, strengthen local competence and reduce reliance on expatriate expertise (Marinelli *et al.*, 2022; Wuni and Shen, 2020). Governance also streamlines approvals and enforces quality assurance, mitigating perceived risks and supporting industrialisation.

Financing is a fundamental enabler. Case studies from Nigeria, Namibia and Egypt show that blended public-private funding, particularly by way of PPPs, supports risk-sharing and private-sector efficiency (Edge, 2020; GS Housing, 2022; WEF, 2021). Concessional loans, tax incentives, credit guarantees and dedicated public funds can further de-risk investment, support factory establishment, supply chains and workforce training (World Bank Group, 2025).

Aligning financing with industrial and housing policies ensures scalable, sustainable and locally integrated MMC, delivering cost-effective, high-quality and resilient housing while stimulating economic growth and technical capacity.

Government intervention is critical for the expansion of MMC in Africa. By driving policy and governance reforms and supporting innovative financing mechanisms, governments can create an enabling environment that fosters the widespread adoption of MMC. Evidence from case studies indicates that sustained collaboration among government, industry and academia is essential for raising awareness, addressing structural challenges and ensuring the successful implementation of MMC across the continent and beyond (See Figure 17.1).

17.4.4 Towards a framework for MMC adoption in Africa

Drawing from the reviewed case studies, a coherent framework for adopting and implementing MMC in Africa emerges (See Figure 17.2). The framework is designed to address the continent's acute housing and infrastructure gaps by providing an integrated, context-sensitive model for scaling up MMC adoption. The cases demonstrate that success depends on multi-level collaboration among government, industry, academia and communities, supported by clear policies, innovative financing and public awareness.

Figure 17.1 MMC challenges and adoption strategies in Africa

Source: Authors' creation

Figure 17.2 A framework for effective MMC adoption in Africa

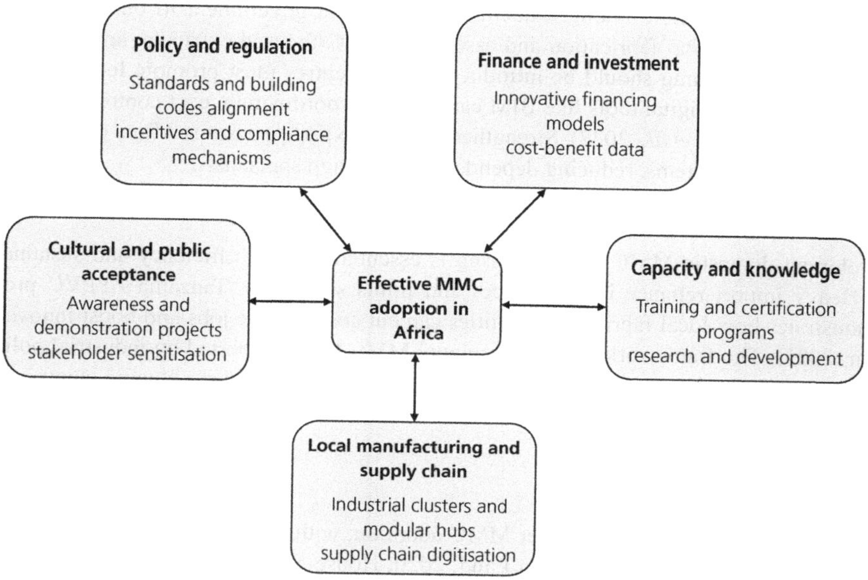

Source: Authors' creation

17.4.4.1 Policy and institutional foundations

Government leadership is vital for MMC adoption. Projects like Nigeria's Peridot Parkland Estate and Namibia's Smart Construction Pilot show that policy direction, financing and regulatory oversight drive success (Edge, 2020; WEF, 2021). Governments should integrate MMC into housing strategies, update building codes and establish certification standards. Lessons from Singapore's DfMA and the UK's Construction 2025 highlight how regulatory clarity spurs innovation (BCA, 2020).

In Africa, developing standardisation and approval systems will build confidence, ensure quality and align MMC with national sustainable construction goals.

17.4.4.2 Financial and economic enablers

High upfront costs limit MMC adoption, particularly for SMEs (African Land, 2020; GS Housing, 2022). The framework prioritises innovative financing, especially PPPs that combine government incentives with private efficiency. Successful examples include Nigeria's Peridot Parkland Estate and Egypt's CSCEC Modular Project (Edge, 2020; GS Housing, 2022). Governments should introduce grants, concessional loans and tax incentives to attract investment, promote factory establishment and fund R&D. Such mechanisms reduce financial risks, improve affordability and stimulate sustainable MMC market growth across Africa.

17.4.4.3 Capacity building and knowledge development

Limited technical skills constrain MMC diffusion (Wuni and Shen, 2019). The framework emphasises collaboration among academia, industry and government to build capacity in modular design, digital fabrication and assembly. MMC-focused curricula, apprenticeships and continuous training should be introduced. R&D centres must promote locally adapted innovations, while digital tools like BIM can enhance coordination, cost control and quality assurance (Marinelli *et al.*, 2022). Strengthening local expertise ensures Africa develops self-sustaining MMC systems, reducing dependency on foreign specialists.

17.4.4.4 Local manufacturing and supply chains

Developing domestic MMC manufacturing is essential for cost efficiency and sustainability. Heavy import reliance inflates prices and limits scalability. Tanzania's PPVC project demonstrates how local fabrication facilities can cut costs, create jobs and boost innovation (Nair, 2025). The framework proposes regional MMC hubs supported by industrial policies to achieve economies of scale. Such hubs would supply prefabricated components across borders, enhance technology transfer and build resilient supply chains essential for long-term MMC expansion in Africa.

17.4.4.5 Public awareness and cultural acceptance

Cultural misconceptions often hinder MMC adoption, with many perceiving modular housing as inferior or temporary (African Land, 2020; House-Matic, 2019). To address this, the framework promotes awareness campaigns, pilot projects and community engagement. Demonstration housing, as seen in Namibia's Smart Construction Pilot and Ghana's Adom City Estate, builds trust and showcases MMC's quality and durability (WEF, 2021; Wuni and Shen, 2019).

Public education and stakeholder sensitisation are vital to shifting perceptions and promoting mainstream MMC acceptance across Africa.

17.4.4.6 Integrated governance and collaboration

The proposed framework advocates a multi-stakeholder governance model integrating the roles of government, private sector, academia and communities. Governments should coordinate through central MMC agencies or task forces that align policy, finance, research and implementation. Industry should lead in technology development and standardisation, while academia provides evidence-based innovation and workforce training. Public–private–academic–community partnerships (PPACP) can serve as the operational platform for coordination, knowledge exchange and monitoring of MMC adoption outcomes.

17.4.4.7 Framework summary

The proposed framework for MMC adoption in Africa rests on five interlinked pillars: policy and regulation to establish enabling legislation and standards; finance and investment to leverage PPPs, incentives and concessional funding; capacity and knowledge to strengthen skills, research and digital tools; local manufacturing to develop regional factories and supply chains for cost efficiency; and cultural and public acceptance to promote awareness and stakeholder engagement.

Systematic implementation of this framework provides a strategic roadmap for Africa, integrating social, economic and institutional dimensions to address housing deficits, stimulate industrial growth and position MMC as a transformative tool for sustainable development across the continent.

As shown in Figure 17.2 and explained in Table 17.2, the successful implementation of MMC in Africa relies on a comprehensive framework that addresses both systemic enablers and local challenges. The framework highlights five interconnected components, policy and regulation, finance and investment, cultural and public acceptance, local manufacturing and supply chain and capacity and knowledge, that serve as key pillars for promoting MMC adoption. Each component identifies important points for action, from supportive governance and

Table 17.2 Framework for advancing MMC in Africa

Component	Description /Focus	Key Elements (Examples)
Policy and Regulation	Establishing supportive governance structures and standards that enable MMC integration.	Standards and building codes alignment to accommodate modular and prefabricated systems. Incentives and compliance mechanisms (e.g. tax rebates, quality assurance frameworks).
Finance and Investment	Strengthening financial systems and incentives to enhance MMC adoption.	Innovative financing models such as PPPs, green bonds and concessional loans. Lifecycle cost–benefit frameworks to attract institutional investors.
Cultural and Public Acceptance	Promoting trust, awareness and acceptance of MMC among stakeholders and communities.	Demonstration projects showcasing MMC performance and design flexibility. Stakeholder sensitisation and public education campaigns on MMC benefits.
Local Manufacturing and Supply Chain	Building localised production capacity and efficient logistics for MMC components.	Development of modular industrial clusters and regional production hubs. Supply chain digitisation using BIM and IoT for real-time logistics coordination.
Capacity and Knowledge	Enhancing human capital, education and research to sustain MMC innovation.	Training and certification programs for professionals and tradespeople. Collaborative research and innovation hubs focused on MMC adaptation to African contexts.

Source: Authors' creation

financing strategies to workforce development and societal involvement. Collectively, these elements offer a clear roadmap for policymakers, industry practitioners and researchers to encourage innovation, sustainability and growth in Africa's construction sector.

17.5. Implications

The MMC case studies highlight the transformative potential of MMC in addressing Africa's housing and infrastructure deficits. The evolution of MMC from early timber-framed prefabrication in Europe to modern modular and industrialised systems demonstrates its adaptability, scalability and efficiency. In the African context, MMC presents a practical alternative to TCM, capable of delivering affordable, high-quality and sustainable housing to meet the demands of rapid urbanisation.

A key implication is the central role of government in enabling MMC adoption. Evidence shows that government action through policy formulation, regulation and financial incentives is essential to stimulate private participation. Acting as anchor clients and promoting PPPs, governments can de-risk investment and support large-scale, cost-efficient delivery. Equally important are regulatory and institutional reforms. The absence of standardised codes, certification systems and quality assurance frameworks constrains MMC's credibility. Lessons from the UK, USA, Singapore and Malaysia reveal that codified standards build investor confidence and ensure quality.

Alongside this, Africa's shortage of MMC-skilled professionals underscores the need for targeted capacity-building through partnerships among government, academia and industry to strengthen local expertise. Cultural perceptions also influence uptake, as MMC is sometimes viewed as foreign or inferior. Awareness campaigns, pilot projects and community participation are crucial to shifting attitudes and demonstrating MMC's safety, durability and cost benefits.

Overall, mainstreaming MMC in Africa requires coordinated action across five fronts: government leadership, innovative finance, regulatory reform, capacity building and cultural adaptation. When aligned, these elements can elevate MMC from a niche innovation to a scalable, standardised construction solution, advancing housing delivery, infrastructure development and socio-economic transformation across the continent.

17.6. Conclusions and future research directions

MMC provide an effective solution to Africa's housing and infrastructure deficits. Evidence from Nigeria, Ghana, Namibia and Egypt shows that MMC can deliver affordable, high-quality and sustainable housing more efficiently than traditional methods. Government leadership is vital, with supportive policies, regulations and PPPs driving investment and reducing risks. However, weak building codes, lack of certification systems and limited technical skills constrain adoption. Collaboration among governments, academia and industry is needed to build capacity and local expertise. Public awareness, pilot projects and community engagement are equally crucial for improving trust, acceptance and market demand.

Future research should focus on evaluating MMC's cost-effectiveness, sustainability and socio-economic impacts in African contexts. Comparative studies of financing models, policy frameworks and localised adaptation of MMC technologies can inform scalable strategies. Longitudinal studies tracking workforce development, industrial transformation and housing outcomes are also recommended.

In conclusion, integrating government leadership, innovative financing, technical capacity development, regulatory reform and cultural adaptation provides a strategic pathway for mainstreaming MMC. When implemented holistically, MMC can accelerate affordable housing delivery, enhance urban infrastructure and contribute to sustainable socio-economic development across Africa.

REFERENCES

Abanda FH, Tah JHM and Cheung FKT (2017) BIM in off-site manufacturing for buildings. *Journal of Building Engineering* **14**: 89–102.

African Land (2020) The Rise of Modular Housing in Ghana: A Simple Self-Assembly Solution. https://african.land/blog/article/the-rise-of-modular-housing-in-ghana-a-simple-self-assembly-solution-b1065?utm_source=chatgpt.com (accessed 29/09/2025).

African Review (2024) Smart Building Solutions Transform Eastern Africa. https://africanreview.com/construction/smart-building-solutions-transform-eastern-africa (accessed 22/10/2025).

African Union (2015) *Agenda 2063: The Africa We Want*. African Union Commission, Addis Ababa, Ethiopia. https://au.int/en/agenda2063/overview (accessed 28 April 2026).

Ball M (1999) Chasing a snail: innovation and housebuilding firms' strategies. *Housing Studies* **14(1)**: 9–22.

Bello W, Akinlolu M and Oyedele L (2023) Barriers to offsite construction adoption in Africa: a systematic review. *Journal of Engineering, Design and Technology* **21(2)**: 479–498.

Blismas NG, Pasquire CL and Gibb AGF (2006) Benefit evaluation for off-site production in construction. *Construction Management and Economics* **24**: 121–130.

Bogna D, Koskela L and Rooke J (2020) *Research Methods in Built Environment: Exploring the Interpretivist Paradigm*. Routledge, London, UK.

Building and Construction Authority (BCA) (2020) Buildability and Constructability Framework. https://www1.bca.gov.sg/buildsg/productivity/buildability-buildable-design-and-constructability (accessed 10/06/2025).

CAPMAS (2024) Economic Census Egypt 2018. Central Agency for Public Mobilization and Statistics, Cairo, Egypt. https://censusinfo.capmas.gov.eg/Metadata-en-v4.2/index.php/catalog/405/related_materials (accessed 26/05/2025).

Easton G (2010) Critical realism in case study research. *Industrial Marketing Management* **39**: 118–128.

Edge (2020) Peridot Parkland Estate. https://edgebuildings.com/project-studies/peridot-parkland-estate/ (Accessed 12/10/2025).

Egan J (1998) Rethinking Construction. https://constructingexcellence.org.uk/wp-content/uploads/2014/10/rethinking_construction_report.pdf (accessed 05/06/2025).

Gao S, Low SP and Nair K (2018) Design for manufacturing and assembly (DfMA): a preliminary study of factors influencing its adoption in Singapore. *Architecture, Engineering and Design Management* **14**: 440–456.

Gibb AGF (1999) *Off-site fabrication: Prefabrication, pre-assembly and modularisation*. Whittles Publishing, Caithness, UK.

Gibb AGF and Isack F (2003) Re-engineering through pre-assembly: client expectations and drivers. *Building Research and Information* **31(2)**: 146–160.

Gier A, Gottlieb S and Frederiksen N (2025) Construction clients as drivers of change? How communities filter institutional pressures for sustainability. *Construction Management and Economics* **43(8)**: 599–616.

Goodier CI and Gibb AGF (2007) Future opportunities for offsite in the UK. *Construction Management and Economics* **25(6)**: 585–595.

Green SD (2022) Modern methods of construction: reflections on the current research agenda. *Buildings and Cities* **3(1)**: 653–662.

GS Housing (2022) Prefab house – CSCEC Egypt apartment project. https://www.gshousing-group.com/projects/prefab-house-cscec-egypt-apartment-project/?utm_source=chatgpt.com (accessed 30/09/2025).

Hong J, Shen GQ, Li Z, Zhang B and Zhang W (2018) Barriers to promoting prefabricated construction in China: a cost–benefit analysis. *Journal of Cleaner Production* **172**: 649–660.

House-Matic (2019) Low-Cost Prefabricated Houses in Mozambique. https://www.house-matic.com/projects/low-cost-housing-mozambique.html (accessed 20/10/2025).

Ibimilua AF and Ibitoye OA (2015) Housing policy in Nigeria: an overview. *American International Journal of Contemporary Research* **5(2)**: 53–59.

Karimi F, Zhang G and Skitmore M (2024) Qualitative synthesis in construction management: integrating desktop research and interpretivism. *International Journal of Project Management* **42(3)**: 234–246.

Latham M (1994) *Constructing the Team – The Final Report of the Government/Industry Review*. HMSO, London, UK. https://constructingexcellence.org.uk/wp-content/uploads/2014/10/Constructing-the-team-The-Latham-Report.pdf (accessed 02/06/2025).

Lawan A (2021) Critical realism: what you should know and how to apply it. *Qualitative Research Journal* **21(3)**: 320–333.

Li Z and Paik SM (2024) Resource cost optimization strategies in off-site manufacturing for modern construction methods: a lean theory perspective case study in Ireland. *International Journal of Construction Management*, 1–9, 10.1080/15623599.2025.2490169

Lu W, Chen K, Xue F and Pan W (2018) Searching for an optimal level of prefabrication in construction: an analytical framework. *Journal of Cleaner Production* **201**: 236–245.

Makalima M and Sokhetye AM (2025) Social housing in South Africa's urban landscape: addressing land access and sustainability challenges in Johannesburg, Cape Town and Durban. *Regional Science and Environmental Science* **2(2)**: 1–13 .

Makinde OO (2014) Housing delivery system, need and demand. *Environment, Development and Sustainability* **16(1)**: 49–69.

Marinelli M, Konanahalli A, Dwarapudi R and Janardhanan M (2022) Assessment of barriers and strategies for the enhancement of off-site construction in India: an ISM approach. *Sustainability* **14(11)**.

Marutlulle NK (2021) A critical analysis of housing inadequacy in South Africa and its ramifications. *Africa's Public Service Delivery and Performance Review* **9(1)**: 1–26.

Modular Building Institute (2023) Modular Builders in Africa. https://www.modular.org/modular-construction-in-africa/?utm_source=chatgpt.com (accessed 22/10/2025).

Musa H, Ismail F and Ali A (2014) Industrialised Building System (IBS) in Malaysia: opportunities and challenges. *Journal of Construction in Developing Countries* **19(1)**: 23–41.

Nair P (2025) *Tanzania Modular Housing Project: MMC prefabrication facility proposal*. Ministry of Infrastructure Development, Dar es Salaam, Tanzania.

NESC – National Economic and Social Council (2024) *Modern Methods of Construction in Ireland: Addressing Housing Supply*. NESC, Dublin, Republic of Ireland. https://www.nesc.ie/app/uploads/2024/09/166_modern_methods_of_construction.pdf (accessed 22/10/2025).

Ofori G (2015) Nature of the construction industry, its needs and its development: a review of four decades of research. *Journal of Construction in Developing Countries* **20(2)**: 115–135.

Okonjo-Iweala N (2014) *Unleashing the housing sector in Nigeria and Africa*. Keynote Address at the *6th Global Housing Finance Conference*, World Bank, Washington, DC, USA, 28 May.

Ortega J, Mesa HA and Alarcón LF (2023) The interrelationship between barriers impeding the adoption of off-site construction in developing countries: the case of Chile. *Journal of Building Engineering* **73**.

Osei-Tutu E, Badu E and Owusu-Manu D (2021) Barriers to Building Information Modelling (BIM) adoption in developing countries: a Ghanaian construction industry perspective. *Journal of Engineering, Design and Technology* **19(6)**: 1444–1463.

Pan W and Sidwell R (2011) Demystifying the cost barriers to offsite construction in the UK. *Construction Management and Economics* **29(11)**: 1081–1099.

Payne S and Serin B (2023) The Potential Role of Modern Methods of Construction in Addressing Systemic Supply Issues. https://housingevidence.ac.uk/wp-content/uploads/2024/01/20230310_MMC_PayneSerin_V4.pdf (accessed 17/05/2025).

Peters LD, Pressey AD, Vanharanta M and Johnston WJ. (2013) Constructivism and critical realism as alternative approaches to the study of business networks: convergences and divergences in theory and in research practice. *Industrial Marketing Management* **42(3)**: 336–346.

Poon CS, Yu ATW and Ng LH (2001) *A Guide for Managing and Minimizing Building and Demolition Waste*. Hong Kong Polytechnic University, Hong Kong.

Quarrying Africa (2025) How Modular Solutions Are Transforming the Construction Industry in SA. https://quarryingafrica.com/how-modular-solutions-are-transforming-the-construction-industry-in-sa/?utm_source=chatgpt.com (accessed 22/10/2025).

Satya L (2008) British Imperial Railways in Nineteenth Century South Asia. *Economic and Political Weekly* **43(47)**: 69–77.

Sayer A (1992) *Method in Social Science: A Realist Approach*, 2nd ed. Routledge, London, UK.

Scolobig A, Linnerooth-Bayer J, Martin J *et al.* (2023) Transformative adaptation through nature-based solutions: a comparative case study analysis in China, Italy and Germany. *Regional Environmental Change* **23**: 69.

UN-Habitat (2024) *The Housing Gap Is Widening: 2024 Annual Report*. United Nations Human Settlements Programme, Nairobi, Kenya. https://unhabitat.org/news/02-jun-2025/2024-annual-report-the-housing-gap-is-widening (accessed 22/10/2025).

United Nations Economic Commission for Africa (UNECA) (2022) Expanding Municipal Finance in Africa. https://www.uneca.org/expanding-municipal-finance-in-africa (accessed 10/10/2025).

White JT, Kenny T, Samuel F and *et al.* (2020) Delivering design value: The housing design quality conundrum. *UK Collaborative Centre for Housing Evidence*, Glasgow, Scotland.

World Bank Group (2025) Government Support in PPP Projects. https://ppp.worldbank.org/public-private-partnership/government-support (accessed 13/06/2025).

World Economic Forum (WEF) (2021) How Innovative Construction Could Ease Namibia's Housing Crisis. https://www.weforum.org/stories/2021/10/how-innovative-construction-ease-namibias-housing-crisis/ (accessed 28 April 2026).

Wuni IY and Shen GQ (2019) *Making a case for modular integrated construction in West Africa: Rethinking housing supply in Africa*. West Africa Built Environment Research (WABER) Conference, 10th anniversary, 5–7 August 2019, Accra, Ghana.

Wuni IY and Shen GQ (2020) Barriers to adopting modular integrated construction: systematic review and meta-analysis, integrated conceptual framework and strategies. *Journal of Cleaner Production* **249**: 119347.

Zhang X, Skitmore M and Peng Y (2014) Exploring the challenges to industrialized residential building in China. *Habitat International* **41**: 176–184.

Modern Methods of Construction: Foundations, Processes and Applications

Emmanuel I Daniel and Fidelis A Emuze
ISBN 978-1-83708-437-1
https://doi.org/10.1108/978-1-83708-436-420261021

Chapter 18
Advancing modern methods of construction in Africa: an investigation of prefabrication practices in Nigeria

Nenpin Dimka[1,2], Louis Gyoh[2], Emmanuel I. Daniel[2], Obas John Ebohon[3] and Mariam Akinlolu[4,5]

[1]*London South Bank University, UK*
[2]*University of Wolverhampton, UK* 0000-0002-3200-4458; 0000-0002-8257-9380; 0000-0002-5675-1845
[3]*Canterbury Christ Church University, UK* 0000-0003-3282-6002
[4]*London Metropolitan University, UK*
[5]*Nelson Mandela University, South Africa* 0000-0002-8357-096X

Abstract

Modern methods of construction (MMC) have the potential to revolutionise the construction industry in Africa by addressing challenges of inefficiency, low productivity and quality inconsistencies. This chapter focuses on adopting prefabrication as a key component of MMC, using an African contextual lens to explore trends and challenges. The chapter investigates prefabrication technologies' knowledge and implementation levels among small and medium-sized enterprises (SMEs) that dominate sub-Saharan Africa's housing sector. Using a quantitative research approach, descriptive and inferential statistics were used to examine the differences in the knowledge and application of prefabrication technologies among various professional groups. Findings reveal critical knowledge gaps, barriers to implementation and opportunities for scaling prefabrication practices, which are essential for increased MMC adoption. Evidence-based insights into optimising prefabrication to enhance construction processes in emerging economies are presented. Focusing on Nigeria, the chapter offers pragmatic and strategic directions for advancing MMC adoption across Africa, promoting economic growth, sustainability and innovation in the region's construction sector.

Keywords: modern methods of construction; prefabrication; emerging economies; SMEs; offsite manufacturing; Africa

18.1. Introduction

The construction industry across sub-Saharan Africa faces unprecedented challenges in meeting the housing demands of rapidly growing urban populations. With over 1.6 billion people worldwide lacking access to adequate housing, representing one-fifth of the global population (UN-Habitat, 2005, 2016), the crisis is particularly acute in emerging economies where small and medium-sized enterprises (SMEs) dominate the construction landscape.

Nigeria, exemplifying this demographic trend with a population projected to reach 400 million by 2050 (UN-Habitat, 2017), confronts one of the world's most significant housing deficits, estimated at approximately 24 million units (Makinde, 2014). The traditional construction industry's challenges extend beyond mere capacity constraints. The sector consumes nearly half of the world's material resources, generates the most significant waste stream, consumes 40% of global energy and produces 33% of global carbon emissions (Chukwu *et al.*, 2019; Wang *et al.*, 2021). Furthermore, productivity has remained stagnant, with the conventional labour-intensive model hindered by a fragmented supply chain, skills shortages, an ageing workforce and high casualty rates (Farmer, 2016; McKinsey and Company, 2017).

Off-site manufacturing, particularly prefabrication, has emerged as a potential solution to address these multifaceted challenges (Chakravarty, 2022). Prefabrication offers controlled production environments that can mitigate project delays, enhance production efficiency, reduce costs, improve safety and elevate the quality of building components while minimising waste (Mostafa *et al.*, 2020). As a resource-optimised building solution, prefabrication has the potential to optimise material usage, facilitate recycling and reuse and provide energy-efficient dwellings at a rapid pace, thus addressing current deficits and future demands (Oti-Sarpong *et al.*, 2022).

The historical precedent for prefabrication's effectiveness is well-established. During the post-World War II era, prefabricated housing solutions provided rapid and cost-effective responses to housing displaced populations and returning soldiers. Notable examples include the Levitt brothers' mass production of prefabricated homes in Levittown developments and the widespread use of prefabricated concrete panel systems in European multi-storey residential buildings (Frampton and Cava, 1992; Gans, 1967).

Despite extensive research on prefabrication in developed nations, significant knowledge gaps persist regarding its application and implementation pathways in emerging economies, particularly in sub-Saharan Africa (Dimka *et al.*, 2024). A systematic literature review and meta-analysis of 1998 articles from Scopus and Web of Science databases (2011–2021) revealed critical research gaps in the context of emerging economies. The analysis identified that fewer topics were researched for emerging economies compared to developed nations, with a notable absence of studies examining the specific challenges and opportunities faced by SMEs in these contexts.

Furthermore, the analysis revealed methodological gaps, with mixed-methods approaches being less represented (only 28% of reviewed papers) and limited use of comprehensive data sources that could capture the complexity of prefabrication adoption in resource-constrained environments. Most significantly, 75% of studies in emerging economies neglected to specify the level of analysis, creating ambiguity about which prefabrication categories are most suitable for different contexts.

SMEs are vital to the construction industry in both developed and developing economies, constituting a large percentage of firms in the sector and driving innovation adoption (Adedeji *et al.*, 2019; Rhodes, 2018). In Nigeria and similar emerging economies, SMEs dominate the construction landscape, making their engagement critical to propagating innovative techniques such as prefabrication. These enterprises often pioneer innovations and serve as important drivers of change, underscoring their significance in the industry's technological transformation (Vuran and Vuran, 2012). However, the unique challenges faced by construction SMEs in emerging economies, including limited access to capital, constrained technical expertise and fragmented supply chains, create distinct adoption patterns that differ significantly from those observed in developed markets (Revell, 2007). Understanding these patterns is essential for developing effective strategies to promote prefabrication adoption and enhance the sector's capacity to address housing shortages sustainably.

Given the massive housing needs, energy demands and SME predominance in populous emerging economies such as Nigeria, prefabrication merits urgent exploration. The intersection of current housing deficits, increasing demand for housing, impacts of climate change and the need for sustainable construction techniques creates a compelling case for investigating how prefabrication technologies can be effectively implemented in these contexts.

This study addresses these critical knowledge gaps by investigating the knowledge and implementation levels of prefabrication technologies among SMEs in Nigeria's housing sector. The findings contribute to a deeper understanding of how off-site manufacturing can be harnessed to enhance the efficiency and sustainability of housing construction in emerging economies, thereby advancing economic growth and social development in these regions while providing practical insights for policymakers, industry practitioners and development organisations seeking to promote innovative construction methods in sub-Saharan Africa.

Based on the identified knowledge gaps, this study pursues three specific objectives.

(*a*) To assess the knowledge levels of prefabrication technologies across different categories (component-based, non-volumetric, volumetric, hybrid and complete building systems) among construction professionals in Nigerian SMEs.
(*b*) To evaluate the implementation levels of prefabrication practices within Nigerian construction SMEs.
(*c*) To examine whether significant differences exist in knowledge and implementation levels of prefabrication technologies among different professional groups within the SME sector.

18.2. Methodological approach

18.2.1 Research approach and design

The research employed a quantitative approach with deductive reasoning as the primary methodological framework for the confirmation or disconfirmation of hypotheses (Collis and Hussey, 2019). The research design incorporated an investigation of knowledge levels and implementation patterns across different professional groups within Nigeria's construction SMEs. This design facilitated rigorous examination of existing theory while grounding research findings in empirical evidence through statistical analysis and hypothesis testing (Blaikie, 2010).

To achieve these objectives, the following research hypotheses were formulated.

- H1: there are significant differences in the knowledge levels of prefabrication technologies among different professional groups in Nigerian construction SMEs.
- H2: there are significant differences in the implementation levels of prefabrication practices among different professional groups in Nigerian construction SMEs.

The study population consisted of construction professionals working within SMEs in Nigeria's housing delivery system. The unit of analysis was individual professionals, with data analysed both at the individual level and aggregated by professional group to examine inter-professional differences in knowledge and implementation patterns. The sample frame was purposively created from built environment practitioners based in Nigeria, particularly those working in the sector's major urban areas, to optimise data collection while representing the target population.

Furthermore, the study adopted purposive sampling, a non-probability sampling technique involving the collection of data from respondents who best fulfil the purpose of the research (Hair *et al.*, 2010; Sekaran, 2010). This approach was selected because it involves selecting samples based on criteria other than random selection, offering time and cost-saving advantages while ensuring participants possessed relevant expertise in prefabrication and construction practices.

The selection criteria for participants included

- professional qualification in construction-related disciplines
- experience in housing construction projects
- knowledge of construction technologies and methods
- active involvement in small and medium-sized construction enterprises.

18.2.2 Sample size and response rate

A total of 153 questionnaires were distributed electronically to construction professionals across Nigeria. The study achieved 101 responses, representing a response rate of 66%, which exceeds the acceptable response rate of approximately 60% recommended by numerous studies (Babbie, 1990; Baruch, 1999) and meets Bryman's (2012) criteria for validity and reliability in construction research.

18.2.3 Prefabrication categories framework

The study employed Gibb's (2001) classification system to categorise prefabrication technologies, namely

- components: individual prefabricated elements such as doors, windows and structural elements

- non-volumetric assembly: pre-assembled units that do not require usable space, such as wall panels and structural frames
- volumetric assembly: enclosed units of usable space, including bathroom pods and kitchen modules
- hybrid systems: combinations of different prefabrication approaches
- complete building systems: fully prefabricated structures requiring minimal on-site works.

To analyse the data collected, the study adopted univariate and bivariate analyses. All statistical analyses were conducted using SPSS version 27. Univariate analysis was employed to describe and summarise individual variables, providing details regarding distribution, mean, central tendency and variability (Field, 2018). To test for normality, the Kolmogorov-Smirnov and Shapiro-Wilk tests were conducted to assess the normality of data distribution across all variables. These tests were essential for determining the appropriate statistical procedures for subsequent inferential analyses.

Based on normality test results revealing non-normal distribution across variables, non-parametric statistical tests were employed for inferential analysis. The Kruskal-Wallis test was adopted to examine significant differences among three or more independent groups (professional categories) by comparing scores on continuous variables (Field, 2013). An alpha level of 0.05 or less was used to indicate significant differences between groups. For variables showing significant differences in the Kruskal-Wallis test, Dunn-Bonferroni post-hoc tests were conducted to identify specific pairwise differences between professional groups while controlling for multiple comparisons.

18.3. Results and discussion

18.3.1 Demographic information

The study achieved a robust sample of 101 construction industry professionals, representing a 66% response rate that exceeds recommended thresholds for construction research validity (Babbie, 1990; Baruch, 1999). The demographic analysis revealed a diverse professional composition, with architects constituting the largest group (40.6%, n=41), followed by engineers (17.8%, n=18), builders (14.9%, n=15) and quantity surveyors (9.9%, n=10). This distribution aligns with previous studies on professional composition in construction innovation research (Blismas and Wakefield, 2009; Mostafa *et al.*, 2020), although the high proportion of architects should be considered when interpreting findings.

The sample demonstrated substantial industry experience, with 33.6% of respondents having 10–15 years of experience, providing a robust foundation for evaluating prefabrication implementation factors. Most respondents (47.5%, n=48) worked in organisations with 10–50 employees, confirming the study's focus on SMEs that dominate Nigeria's construction landscape.

18.3.2 Descriptive statistics and normality tests

18.3.2.1 Level of knowledge of prefabrication and off-site technology implementation

Table 18.1 presents the normality test and descriptive results for the knowledge of prefabrication implementation among construction SMEs. The assessment of knowledge of

Table 18.1 Prefabrication categories variables

Prefabrication Category	Mean	Median	Std. Dev.	Kolmogorov–Smirnov Statistic	p-value	Shapiro–Wilk Statistic	p-value
Components	3.50	4	1.79	0.25	< .001	0.78	< .001
Non-volumetric assembly	2.50	3	1.81	0.14	0.029	0.89	< .001
Volumetric assembly	2.18	2	1.91	0.20	< .001	0.85	< .001
Hybrid	2.02	2	1.91	0.22	< .001	0.83	< .001
Complete building	2.09	2	2.00	0.24	< .001	0.82	< .001

Source: Authors' creation

implementing prefabrication and off-site technologies across various categories revealed that component-based prefabrication ($M=3.5$, $SD=1.79$) was the most known. Volumetric assembly had the lowest mean score (2.02), indicating it is the least known prefabrication category. Results from the Kolmogorov-Smirnov and Shapiro-Wilk tests indicated a non-normal distribution ($p < .001$) for all the variables.

18.3.2.2 Level of implementation of prefabrication and off-site technologies

Table 18.2 presents the indicator variables measuring the level of prefabrication implementation. Respondents were required to indicate the extent to which prefabrication was implemented on their projects on a 5-point Likert scale ranging from 1 (very low) to 5 (very high). The mean scores for the seven variables ranged from 2.64 to 2.32. The highest-ranked variables were continuous improvement culture ($M=2.64$), on-site assembly expertise ($M=2.62$) and level of design for prefabrication ($M=2.59$). The relatively low mean scores across all categories (below 3.00) indicate low implementation of prefabrication practices. Results from the Kolmogorov-Smirnov and Shapiro-Wilk tests reveal a non-normal distribution of data ($p < .05$) for all the variables presented in Table 18.2.

18.3.2.3 Professional group differences in the knowledge of prefabrication/off-site technologies

Table 18.3 presents the results from the Kruskal-Wallis test assessing the mean scores for the prefabrication and off-site technology categories and their rank orders among the professional groups (architect, quantity surveyor, contractor, engineer, builder, project manager and fabricator). The chi-square, degree of freedom (df) and sig. value obtained revealed significant differences in the knowledge of prefabrication and off-site technologies. Significant differences were found (Chi-square$=15.65$, $p=0.016$) as the sig. value was below 0.05.

Table 18.2 Prefabrication levels indicators

Indicator	Mean	Rank	Median	Std. Dev.	Kolmogorov–Smirnov Statistic	p-value	Shapiro–Wilk Statistic	p-value
Level of design for prefabrication	2.59	3	2	1.19	0.23	< .001	0.89	< .001
In-house prefabrication capabilities	2.39	5	2	0.98	0.22	< .001	0.89	< .001
Supply chain integration	2.32	7	2	1.07	0.20	< .001	0.86	< .001
On-site assembly expertise	2.62	2	3	1.13	0.20	< .001	0.90	< .001
Technology adoption	2.39	5	2	1.05	0.20	< .001	0.88	< .001
Extent of standardisation and modularity	2.53	4	3	1.09	0.18	< .002	0.90	< .001
Continuous improvement culture	2.64	1	3	1.22	0.17	< .005	0.90	< .001

Source: Authors' creation

The assessment of the mean scores for the prefabrication categories revealed that volumetric (mean score=49.71) and hybrid (mean score=43.37) were the most familiar for the architect group; non-volumetric (mean score=73.05) and components (mean score=69.05) were the most familiar for the quantity surveyor group. For the contractor group, components (mean score=64.50) and non-volumetric (mean score=54.83) were the most common. The builder and fabricator group ranked the complete building category highest, while the engineer and project manager group ranked the volumetric category (mean score=53.21; 42.17) highest.

To identify where the significant differences lay between the profession groups, the Dunn-Bonferroni post-hoc test was conducted for the component constructs. The test performs pairwise comparisons by adjusting and multiplying each p-value by the total number of tests conducted. Although the Kruskal-Wallis test revealed statistically significant differences among groups, the Dunn-Bonferroni correction revealed no significant pairwise differences, as shown in Table 18.4. This suggests that although there were differences in the knowledge of prefabrication across professions, these differences are not substantial enough to pinpoint specific inter-professional disparities, therefore, H1 is rejected.

Table 18.3 Test statistics for profession and knowledge of prefabrication

Category	Chi-Square	df	p-value	Architect	Quantity Surveyor	Contractor	Engineer	Builder	Project Manager	Fabricator
Components	15.65	6	0.016	46.28 (4)	69.05 (1)	64.50 (2)	43.97 (5)	62.90 (3)	30.83 (7)	31.00 (6)
Non-volumetric	7.62	6	0.268	46.82 (6)	73.05 (1)	54.83 (2)	48.69 (5)	50.40 (4)	37.83 (7)	52.30 (3)
Volumetric	3.16	6	0.789	49.71 (5)	54.65 (2)	37.33 (7)	53.21 (3)	55.77 (1)	42.17 (6)	52.40 (4)
Hybrid	2.73	6	0.842	47.37 (6)	49.85 (4)	53.94 (3)	49.71 (5)	58.47 (1)	38.33 (7)	57.40 (2)
Complete building	13.85	6	0.031	43.99 (5)	43.40 (6)	54.28 (3)	48.71 (4)	64.73 (2)	38.50 (7)	81.90 (1)

Note: Mean scores are shown with rankings in parentheses.

Source: Authors' creation

Table 18.4 Analysis of Dunn–Bonferroni test

Comparison	Components	
	Test Static	Adj. Sig
architect – quantity surveyor	−22.77	0.419
architect – contractor	−18.22	1.000
architect – engineer	2.31	1.000
architect – builder	−16.62	0.990
architect – project manager	15.45	1.000
architect – fabricator	15.28	1.000
quantity surveyor – contractor	4.55	1.000
quantity surveyor – engineer	25.08	0.460
quantity surveyor – builder	6.15	1.000
quantity surveyor – project manager	38.22	0.764
quantity surveyor – fabricator	38.05	0.258
contractor – engineer	20.53	1.000
contractor – builder	1.60	1.000
contractor – project manager	33.67	1.000
contractor – fabricator	33.50	0.638
engineer – builder	−18.93	1.000
engineer – project manager	13.14	1.000
engineer – fabricator	12.97	1.000
builder – project manager	32.07	1.000
builder – fabricator	31.90	0.545
project manager – fabricator	−0.17	1.000

Source: Authors' creation

18.3.2.4 Professional group differences in the level of implementation of prefabrication/off-site technologies

Table 18.5 presents the mean scores and rank orders for the implementation levels of prefabrication and the Z-value and sig. value. Results of the test for significant differences in the level of implementation of prefabrication among the professional groups revealed no statistically significant differences in respondents' perceptions since the sig. value for all factors presented fell above the alpha value of 0.05, therefore, H2 is rejected.

The assessment of the prefabrication levels revealed the extent and standardisation of modularity (was ranked highest for the architect, mean score = 51.94) the builder group (mean score=58.50); on-site assembly expertise for the quantity surveyor (mean score=61.10) and contractor group (mean score=50.89); technology adoption for the engineer group (mean score=56.83) and supply chain integration (mean score=72.40) for the fabricator group.

Table 18.5 Test statistics for profession and level of implementation of prefabrication

Category	Chi-Square	df	p-value	Architect	Quantity Surveyor	Contractor	Engineer	Builder	Project Manager	Fabricator
Level of design for prefabrication	5.15	6	0.525	49.67 (4)	44.50 (7)	47.50 (5)	56.69 (3)	45.03 (6)	69.50 (1)	67.50 (2)
In-house prefabrication capabilities	5.20	6	0.518	51.70 (5)	34.30 (7)	46.00 (6)	53.11 (4)	56.40 (3)	57.33 (2)	60.10 (1)
Supply chain integration	4.49	6	0.610	49.12 (5)	43.30 (7)	47.56 (6)	53.94 (3)	50.93 (4)	59.67 (2)	72.40 (1)
On-site assembly expertise	4.26	6	0.642	47.59 (6)	61.10 (2)	50.89 (4)	47.75 (5)	56.73 (3)	36.83 (7)	62.00 (1)
Technology adoption	3.07	6	0.799	50.05 (5)	44.25 (6)	41.89 (7)	56.83 (3)	52.27 (4)	57.33 (2)	60.10 (1)
Extent of standardisation and modularity	6.33	6	0.387	51.94 (5)	31.60 (7)	47.78 (6)	52.64 (4)	58.50 (1)	56.17 (3)	56.40 (2)
Continuous improvement culture	9.99	6	0.125	46.30 (6)	40.85 (7)	46.61 (5)	56.75 (3)	50.33 (4)	88.33 (1)	65.70 (2)

Note: Mean scores are shown with rankings in parentheses.

Source: Authors' creation

18.3.3 Discussion

18.3.3.1 Knowledge levels of prefabrication technologies

The analysis revealed a clear hierarchy in technical comprehension across prefabrication categories, with component-based prefabrication systems being the most recognised (M=3.5, SD=1.79). This finding substantiates previous research by Pan *et al.* (2012) and extends Tschang's (2012) observations on technical learning patterns within the industry. Nearly three-quarters of respondents indicated high-level knowledge of component-based systems, suggesting this category serves as the primary entry point for prefabrication adoption among Nigerian SMEs.

The knowledge stratification followed a systematic pattern, with understanding levels decreasing as system complexity increased. Non-volumetric assembly occupied the second tier (M=2.5, SD=1.81), while volumetric assembly had the lowest mean score (M=2.02), indicating it as the least known prefabrication category. This hierarchical distribution framework supports Goulding *et al.*'s (2015) theory of incremental innovation adoption and extends Steinhardt *et al.*'s (2016) identification of how knowledge thresholds shape adoption decisions.

Significantly, the knowledge clustering patterns observed challenge traditional innovation diffusion models. The systematic decrease across complexity levels reflects constraints within emerging economies that prioritise organisational capacity over system sophistication. This finding extends Nadim and Goulding's (2011) theoretical framework by revealing how SMEs in emerging economies navigate technology adoption barriers differently from their larger counterparts in developed markets.

The study identified that informal knowledge networks serve as effective drivers for innovation among SMEs, contradicting assumptions about the necessity of formal knowledge transfer mechanisms (Wuni and Shen, 2020). Strong innovation activity among SMEs suggests that existing informal innovation networks can be more effective than formal capacity-building models when appropriate policy support acknowledges SMEs' dual nature as constrained innovators yet effective vehicles for adapting construction knowledge.

18.3.3.2 Implementation levels of prefabrication technologies

The analysis of implementation levels revealed distinctive patterns among Nigerian construction SMEs. The highest mean score was achieved for continuous improvement culture (M=2.64, SD=1.22), accompanied by strong on-site assembly expertise (M=2.62, SD=1.13). This finding challenges the path-dependency framework established by Mahapatra and Gustavsson (2008) by unveiling how SMEs in emerging economies establish innovative implementation shortcuts that circumvent conventional adoption stages.

The emphasis on operational adaptations reflects a unique approach where SMEs focus on embedding prefabrication within existing workflows rather than completely restructuring their operations. This pattern reinforces Steinhardt *et al.*'s (2016) observations about the industry's tendency towards incremental rather than transformational change, while revealing emerging economy-specific adaptation mechanisms.

Despite overall low implementation scores (all below 3.0 on a five-point scale), the data revealed opportunities for technological leapfrogging where emerging economy SMEs could bypass intermediate technology levels and adopt advanced prefabrication solutions. This phenomenon primarily occurred in digital design integration and automated assembly processes, where SMEs demonstrated surprising agility in adopting next generation BIM solutions that link directly with prefabrication processes.

The relatively low scores should not be interpreted solely as implementation failures but rather as strategic decisions to maximise impact relative to available resources. SMEs often create local variants of prefabrication technologies, developing context-specific solutions that achieve higher implementation success with lower technology investments.

18.3.3.3 Professional group differences in knowledge and implementation

The Kruskal-Wallis test revealed significant differences in knowledge across professional groups (Chi-square = 15.65, $p = 0.016$), although subsequent Dunn-Bonferroni post-hoc analysis showed no significant pairwise differences. This pattern suggests that, while differences exist in prefabrication knowledge across professions, these differences are not substantial enough to pinpoint specific inter-professional disparities. Therefore, we fail to reject the hypothesis that significant differences in the knowledge levels of prefabrication technologies exist among the different professional groups in Nigerian construction SMEs.

The analysis identified three unique knowledge distribution patterns specific to the emerging economy context. Firstly, the relatively balanced distribution across non-volumetric, volumetric and hybrid systems indicates resource constraints that incentivise general knowledge sharing over specialisation. Secondly, quantity surveyors demonstrated knowledge beyond traditional cost management skills, reflecting broader professional adaptation in emerging economies. Thirdly, builders ranked consistently across categories, highlighting integrated knowledge bases being established in resource-constrained environments.

The implementation analysis revealed no statistically significant differences between professional groups ($p > 0.05$), indicating an unexpected democratisation of prefabrication skills in emerging economies. This finding contrasts sharply with developed markets, where rigid professional role designations create clear capability boundaries. Instead, the data suggests professional role convergence driven by resource limitations rather than specialisation. Therefore, we reject the hypothesis, that significant differences in the implementation levels of prefabrication technologies exist among the different professional groups in Nigerian construction SMEs.

Fabricators demonstrated consistently high performance across categories, particularly in supply chain integration (mean rank difference = 38.92) and technology adoption (mean rank difference = 35.67), indicating their evolution toward integrated design-build capabilities that fill local market voids. Project managers showed notably higher performance in a continuous improvement culture (mean rank = 68.24), suggesting their role evolution toward innovation enablers within resource-constrained settings.

18.4. Policy implications and practical strategies

The findings from this study highlight critical areas where targeted policy interventions and practical implementation strategies can accelerate prefabrication adoption among SMEs in Nigeria and similar emerging economies. Government bodies may develop tiered support programmes that acknowledge the hierarchical knowledge pattern identified, where component-based systems are most familiar and volumetric systems least understood.

Financial interventions may address SME constraints through risk-sharing mechanisms including subsidised equipment leasing, tax incentives scaled to system complexity, and performance-based grants rewarding implementation success. Building codes and procurement regulations should accommodate prefabrication methods through streamlined approvals for standardised components and public procurement policies allocating minimum percentages of housing contracts to prefabrication-based SME bids.

Industry associations may leverage informal networks as effective innovation drivers by formalising peer-to-peer learning through SME roundtables, site visits and mentorship programmes. Given fragmented supply chains, associations should facilitate collaborative procurement consortia, develop supply chain directories connecting SMEs with reliable suppliers and create joint venture frameworks distributing risk and expertise.

Training institutions may embed prefabrication across core curricula, prioritising component-based and non-volumetric systems while developing contextualised emerging economy case studies. Practical skills development should include fabrication laboratories, SME partnership programmes and BIM training oriented toward prefabrication workflows to enable technological leapfrogging.

For practitioners, staged progression, beginning with component-based systems, minimises financial risk while building technical confidence. SMEs should adopt hybrid models that merge traditional methods with selective prefabrication integration, using standardised elements where they deliver maximum value. Leveraging innovation networks through collaborative arrangements enables cost-sharing for equipment, bulk material purchasing and pooled technical expertise. Supply chain integration requires establishing long-term partnerships with reliable suppliers that reduce procurement uncertainties and enable standardisation.

Cross-disciplinary learning within SME teams capitalises on the professional role convergence identified in this study. Coordinated action through national prefabrication bodies could monitor progress and track measurable indicators including SME progression through complexity levels and supply chain integration evolution.

18.5. Conclusions

This study investigated the knowledge and implementation levels of prefabrication technologies among small and medium-sized enterprises (SMEs) in Nigeria's housing sector, revealing significant details into how emerging economies navigate the adoption of modern construction methods. The research addressed critical knowledge gaps identified through a systematic literature review and provided empirical evidence of unique adoption patterns that challenge conventional innovation diffusion models.

The study findings revealed a noticeable decline in the knowledge levels across the refabrication categories as the prefabrication complexity increases, aligning with findings from previous studies that suggest the component-based prefabrication category as the entry point for many construction organisations delving into off-site methods.

Contrary to the rigid professional boundaries typically observed in developed markets, the study revealed a more dynamic and integrated approach among Nigerian construction professionals. The absence of significant differences between professional groups in implementation levels ($p>0.05$) indicates a democratisation of prefabrication skills driven by resource limitations. These findings challenge traditional theoretical frameworks that prioritise specialisation, instead highlighting how emerging economy contexts foster professional role convergence and collaborative knowledge development.

Findings from the study further highlight the proactive role of SMEs in the Nigerian construction industry as they engage with prefabrication technologies. Rather than being passive adopters, SMEs are innovative in tailoring these technologies to suit their specific circumstances. Although SMEs face resource limitations, they demonstrate great capacity to bypass the adoption of the integration of BIM and automated assembly. A strict culture of continuous improvement and solid expertise in on-site assembly allows SMEs to seamlessly embed prefabrication into existing workflows.

18.6. Recommendations

While the present study revealed that levels of adoption are moderate to low in areas like supply chain integration and dedicated technology adoption, it is essential to view these findings only in an emerging economy context. Although lower scores were achieved in other areas, this does not indicate failure but rather demonstrates carefully thought-out strategies targeted at optimising prefabrication within the resources available.

To maximise impact, while minimising delays and risks, SMEs are developing hybrid solutions that merge traditional methods and prefabrication. Furthermore, with the emergence of innovation networks, SMEs have demonstrated their ability to collaborate seamlessly, therefore challenging conventional beliefs regarding the correlation between the organisation's size and capacity to innovate. Through collaboration, SMEs can achieve high implementation levels, underscoring their ability to adapt and be resourceful in implementing prefabrication.

18.7. Limitations and future research

This study acknowledges several limitations that provide opportunities for future research. The reliance on self-reported data through questionnaires may introduce response bias, as participants' perceptions of their knowledge and implementation levels may differ from objective assessments. The purposive sampling approach, while appropriate for accessing relevant expertise, limits generalisability beyond the study context. The cross-sectional design captures prefabrication adoption at a single point, unable to track evolution over time or causality between knowledge and implementation.

Future research may adopt mixed methods approaches combining quantitative surveys with qualitative interviews and observational data to provide richer contextual understanding of

SME adoption processes. Longitudinal studies tracking SMEs over multiple years would illuminate how knowledge and implementation progress through prefabrication complexity levels. In-depth case studies examining successful and unsuccessful implementation attempts could reveal critical success factors and barriers not captured through survey methods. Comparative studies across multiple sub-Saharan African countries would identify region-specific as against universal adoption patterns, informing more targeted policy interventions.

REFERENCES

Adedeji YMD, Olotuah AO and Oloke OC (2019) Assessment of housing delivery in Nigeria: federal mortgage bank scenario. *Procedia Manufacturing* **35**: 1149–1154.

Babbie E (1990) *Survey Research Methods*, 2nd ed. Wadsworth Publishing, Belmont, CA, USA.

Baruch Y (1999) Response rate in academic studies: a comparative analysis. *Human Relations* **52(4)**: 421–438.

Blaikie N (2010) *Designing Social Research*, 2nd ed. Polity Press, Cambridge, UK.

Blismas NG and Wakefield R (2009) Drivers, constraints and the future of offsite manufacture in Australia. *Construction Innovation* **9(1)**: 72–83.

Bryman A (2012) *Social Research Methods*, 4th ed. Oxford University Press, Oxford, UK.

Chakravarty S (2022). Resource-constrained innovation in a technology-intensive sector: Frugal medical devices from manufacturing firms in South Africa. *Technovation* **112**: 102397.

Chukwu PK, Higginson ME and Khoshaba K (2019) A systematic review of the factors affecting the uptake of modular construction. *Journal of Building Engineering* **26**: Article 100929.

Collis J and Hussey R (2019) *Business Research: A Practical Guide for Undergraduate and Postgraduate Students*, 5th ed. Palgrave Macmillan, London, UK.

Dimka N, Gyoh L, Daniel E and Momoh J (2024) Offsite manufacturing for housing in emerging economies: an evaluation of current implementation levels. In *Transforming Construction with Off-site Methods and Technologies (TCOT)*, University of New Brunswick, 20-22 August 2024.

Farmer M (2016) *The Farmer Review of the UK Construction Labour Model: Modernise or Die*. Construction Leadership Council, London, UK.

Field A (2013) *Discovering Statistics Using IBM SPSS Statistics*, 4th ed. SAGE Publications, London, UK.

Field A (2018) *Discovering Statistics Using IBM SPSS Statistics*, 5th ed. SAGE Publications, London, UK.

Frampton K and Cava J (1992) *Studies in Tectonic Culture: The Poetics of Construction in Nineteenth and Twentieth Century Architecture*. MIT Press, Cambridge, MA, USA.

Gans H (1967) *The Levittowners: Ways of Life and Politics in a New Suburban Community*. Pantheon Books, New York, NY, USA.

Gibb AGF (2001) Standardization and pre-assembly: distinguishing myth from reality using case study research. *Construction Management and Economics* **19(3)**: 307–315.

Goulding JS, Pour Rahimian F, Arif M and Sharp MD (2015) New offsite production and business models in construction: priorities for the future research agenda. *Architectural Engineering and Design Management* **11(3)**: 163–184.

Hair JF, Black WC, Babin BJ and Anderson RE (2010) *Multivariate Data Analysis*, 7th ed. Prentice Hall, Upper Saddle River, NJ, USA.

Mahapatra K and Gustavsson L (2008) Multi-storey timber buildings: breaking industry path dependency. *Building Research and Information* **36(6)**: 638–648.

Makinde OO (2014) Housing delivery system, need and demand. *Environment, Development and Sustainability* **16(1)**: 49–69.

McKinsey and Company (2017) *Reinventing Construction: A Route to Higher Productivity*. McKinsey Global Institute, NY, USA.

Mostafa S, Chileshe N and Abdelhamid T (2020) Lean and agile integration within offsite construction using discrete event simulation: A systematic literature review. *Construction Innovation* **20(2)**: 213–235.

Nadim W and Goulding JS (2011) Offsite production: a model for building down barriers. A European construction industry perspective. *Engineering, Construction and Architectural Management* **18(1)**: 82–101.

Oti-Sarpong K, Shojaei RS *et al.* (2022) How countries achieve greater use of offsite manufacturing to build new housing: identifying typologies through institutional theory. *Sustainable Cities and Society* **76**, 103403.

Pan W, Gibb AGF and Dainty ARJ (2012) Strategies for integrating the use of off-site production technologies in house building. *Journal of Construction Engineering and Management* **138(11)**: 1331–1340.

Revell A (2007). The ecological modernisation of SMEs in the UK's construction industry. *Geoforum* **38(1)**: 114–126.

Rhodes C (2018) *Construction Industry: Statistics and Policy. House of Commons Library Briefing Paper Number 01432*. House of Commons Library, London, UK.

Sekaran U (2010) *Research Methods for Business: A Skill Building Approach*, 5th ed. John Wiley and Sons, Chichester, UK.

Steinhardt DA, Manley K and Miller W (2016) Profiling the nature and context of the Australian prefabricated housing industry. *Construction Economics and Building* **16(2)**: 37–52.

Tschang FT (2012) When does an innovation type become a standard? A stage-based model. *Technology Analysis and Strategic Management* **24(8)**: 833–848.

UN-Habitat (2005) *Financing Urban Shelter: Global Report on Human Settlements 2005*. Earthscan, London, UK.

UN-Habitat (2016) *World Cities Report 2016: Urbanization and Development - Emerging Futures*. United Nations Human Settlements Programme, Nairobi, Kenya.

UN-Habitat (2017) *The New Urban Agenda*. United Nations Human Settlements Programme, Nairobi, Kenya.

Vuran B and Vuran E (2012) An analysis on innovation and technology transfer in Turkish construction industry. *Procedia - Social and Behavioural Sciences* **41**: 167–178.

Wang M, Wang CC, Sepasgozar S and Zlatanova S (2021) A systematic review of digital technology adoption in off-site construction: current status and future directions towards industry 4.0. *Buildings* **10(11)**: Article 204.

Wuni IY and Shen GQ (2020) Critical success factors for management of the early stages of prefabricated prefinished volumetric construction project life cycle. *Engineering, Construction and Architectural Management* **27(9)**: 2315–2333.

Emmanuel I Daniel and Fidelis A Emuze
ISBN 978-1-83708-437-1
https://doi.org/10.1108/978-1-83708-436-420261022

Index

Note: Italics in page numbers indicates figures.

Printed and bound by CPI Group (UK) Ltd, Croydon, CR0 4YY
21/09/2026
14966034-0007